Design, Modeling, Optimization and Control of Flotation Process

Design, Modeling, Optimization and Control of Flotation Process

Editors

Fardis Nakhaei
Ahmad Hassanzadeh
Luis A. Cisternas

MDPI

Basel • Beijing • Wuhan • Barcelona • Belgrade • Novi Sad • Cluj • Manchester

Editors

Fardis Nakhaei
Department of Mining & Explosives Engineering
Missouri Univeristy
Rolla
United States

Ahmad Hassanzadeh
Department of Geoscience and Petroleum
Norwegian University of Science and Technology (NTNU)
Trondheim
Norway

Luis A. Cisternas
Department of Chemical Engineering and Mineral Process
Universidad of Antofagasta
Antofagasta
Chile

Editorial Office
MDPI
St. Alban-Anlage 66
4052 Basel, Switzerland

This is a reprint of articles from the Special Issue published online in the open access journal *Minerals* (ISSN 2075-163X) (available at: www.mdpi.com/journal/minerals/special_issues/DMOCFP).

For citation purposes, cite each article independently as indicated on the article page online and as indicated below:

Lastname, A.A.; Lastname, B.B. Article Title. *Journal Name* **Year**, *Volume Number*, Page Range.

ISBN 978-3-7258-1396-4 (Hbk)
ISBN 978-3-7258-1395-7 (PDF)
doi.org/10.3390/books978-3-7258-1395-7

Contents

About the Editors

Fardis Nakhaei

Dr. Fardis Nakhaei serves as a post-doctoral research fellow in the Department of Mining and Explosives Engineering at the Missouri University of Science and Technology, USA. His research interests encompass a wide array of topics, including flotation, critical minerals, coal processing, iron and steel manufacturing, comminution, dewatering, control systems, artificial intelligence, and image processing. With over 12 years of combined industrial and academic experience, Dr. Nakhaei possesses a strong foundation in mineral processing.

Dr. Nakhaei earned his Ph.D. from Amirkabir University of Technology, Iran, in 2018 and subsequently spent three years as a post-doctoral research fellow at North–West University, South Africa. Throughout his career, he has authored more than 50 papers in esteemed international journals. In addition to his research contributions, Dr. Nakhaei serves as a reviewer for renowned journals such as Minerals and Minerals Engineering, etc. Furthermore, Dr. Nakhaei serves on the editorial board for *Mineral Processing* and *Extractive Metallurgy Review*, and he also acts as a guest editor for *Minerals*.

Ahmad Hassanzadeh

Dr. Ahmad Hassanzadeh works as a post-doctoral research fellow at the Geoscience and Petroleum Department of the Norwegian University of Science and Technology (NTNU) and is actively involved in the application of hydrogen in mineral processing and the valorization of bauxite residue from the aluminum industry. Since 2021, he has served as the Chief Research Officer (CRO) at Maelgwyn Mineral Services Ltd., a United Kingdom-based company. Over the last couple of years, he has actively worked on the technological development of pneumatic flotation cells, the application of micro-nano-bubbles in flotation, recovering iron from bauxite residues, processing lithium-ion batteries, and modeling flotation kinetics. His areas of interest are grinding, classification, microwave and ultrasound pretreatment processes, flotation, gold and silver leaching, surface and colloidal chemistry-related concepts, and macroscopic and microscopic aspects of particle-bubble interactions. Since 2021, Dr. Ahmad Hassanzadeh has been listed as one of the top 2% of cited scientists in the world, according to Elsevier and Stanford University. He published 80 scientific articles in peer-reviewed journals and presented and published more than 50 conference papers at various symposiums. Since 2019, he has been the editor of the *Physicochemical Problems of Mineral Processing* (*PPMP*) journal and has served as a guest editor in the *Journal of Minerals*. Dr. Hassanzadeh was awarded the MEI's Young Mineral Processing Prize and Mineral's Travel Award in 2020.

Luis A. Cisternas

Luis A. Cisternas is a Mineral Process and Chemical Engineering Department professor at the University of Antofagasta, Chile. Professor Cisternas's principal research interest is the use of a systems approach to solving problems in mineral processing. In particular, his research combines the development of systematic (computer-aided) methods and tools with experimental works for solving problems in the mining industries, which can be classified in terms of the following topics: modeling and optimization; design and analysis of mineral processes; water resources; critical materials; and circular economy.

Professor Cisternas has held several positions, including Associate Dean of Research (1997–2001) and Graduate School Director (2001–2004) at the University of Antofagasta; Executive Director of CICITEM (2008–2011); and Program Leader Water & Energy, CSIRO Chile (2012–2018). He is a member of the editorial board of the following journals: *Minerals* (MPDI), *Mineral Processing and Extractive Metallurgy Review* (Taylor & Francis), and *International Journal of Mining Science and Technology* (Elsevier). He is also the associate editor of *Green and Smart Mining Engineering* (GSME).

Professor Cisternas has published more than 150 peer-reviewed journal articles, more than 150 conference papers, more than 30 book chapters, five books, and has made more than 15 oral presentations as an invited presenter. He has also given invited lectures and seminars at companies and academic institutions, and he has organized several workshops and symposiums on topics related to mineral process modeling and optimization. By December 2023, 20 MSc students and 15 Ph.D. students had submitted and successfully defended their theses under the supervision of Professor Cisternas.

minerals MDPI

Editorial

Editorial for Special Issue "Design, Modeling, Optimization and Control of Flotation Process"

Fardis Nakhaei [1,*], Ahmad Hassanzadeh [2] and Luis A. Cisternas [3]

1 Department of Mining & Explosives Engineering, Missouri University of Science and Technology, McNutt Hall, Rolla, MO 65409, USA
2 Department of Geoscience and Petroleum, Faculty of Engineering, Norwegian University of Science and Technology, Andersens veg 15a, 7031 Trondheim, Norway; ahmad.hassanzadeh@ntnu.no
3 Department of Chemical Engineering and Mineral Process, Universidad of Antofagasta, Antofagasta 1240000, Chile; luis.cisternas@uantof.cl
* Correspondence: fardis.nakhaei@mst.edu

Flotation is a significant and widely used processing technique, effectively separating valuable and gangue minerals. Mineral producers have long sought ways to consistently monitor and control the flotation process to ensure optimal conditions for effective mineral separation [1]. However, establishing such a sophisticated control system entails significant financial investments in equipment, resulting in substantial costs. Additionally, ongoing maintenance needs to be performed to uphold the system's high standards, which will incur additional expenses. To address these challenges, predictive models emerge as effective and economically viable solutions to handling the intricacies of the flotation process [2].

In the flotation process, numerous linear and nonlinear relationships exist between the operating parameters, chemical reagents, and minerals [3]. These connections can be evaluated using various experimental and numerical methods. Over recent decades, a variety of intelligent computing and statistical techniques, including machine learning, genetic algorithms (GAs), artificial neural networks (ANNs), fuzzy systems, and image processing, have been employed to predict the flotation process outcome and facilitate the process control [4]. It is worth noting that the crucial role of these techniques in achieving sustainable development across various industrial sectors was recognized more than a decade ago.

check for updates

Citation: Nakhaei, F.; Hassanzadeh, A.; Cisternas, L.A. Editorial for Special Issue "Design, Modeling, Optimization and Control of Flotation Process". *Minerals* **2024**, *14*, 391. https://doi.org/10.3390/min14040391

Received: 3 April 2024
Accepted: 8 April 2024
Published: 10 April 2024

The papers featured in this Special Issue of *Minerals*, titled 'Design, Modeling, Optimization, and Control of the Flotation Process', explore innovative approaches for modeling, optimizing, and controlling some flotation processes. These techniques aim to enhance efficiency by maximizing the recovery of valuable minerals, while minimizing energy and reagent consumption in relevant studied flotation processes. The manuscripts published in this issue can be grouped into three broad categories.

The initial category focused on hydrodynamic and kinetic models to examine the attachment of mineral particles to bubbles. The attachment process is intricate, involving multiple physical and chemical interactions, such as adsorption, desorption, and chemical reactions. These mechanisms were numerically represented through a variety of models, employing sets of differential equations that illustrate the concentration of minerals and chemical reagents within the flotation cell as time progresses [5–7].

The second group concentrated on modeling and optimizing the flotation process and developing strategies to enhance our understanding of the effects of gas dispersion and its regulation using dynamic models, multivariable linear models, and image analysis [8,9]. These studies explored the stability of the froth zone under varying flotation conditions and highlighted the significance of the relationship between particle and bubble sizes as critical factors impacting successful collection, froth transport processes, and the flotation rate and efficiency.

The third group of papers in this Special Issue delved into modeling and optimizing the flotation process performance utilizing advanced computational tools and algorithms such as the response surface methodology (RSM), GA, ANN, deep learning, and fuzzy systems [10–14]. Some of the studies published in this Special Issue aimed to address the challenges of recovering target elements more effectively using these methods. Generally, the efficiency of the flotation process is assessed by examining the characteristics of the concentrate, specifically its grade and recovery, which are crucial economic and technical indicators for process management and improvement. Determining these parameters typically involves time-consuming procedures. While the grade can be constantly monitored using an XRF analyzer, recovery is usually determined with mass balancing techniques [15]. However, the online measurement of these parameters using X-ray analyzers, although feasible, requires expensive and complex equipment, along with ongoing maintenance, which justifies the preference for models predicting the key performance indexes derived from secondary variables [16]. Given the nonlinear and intricate nature of the flotation process, coupled with the involvement of numerous variables and a limited understanding of its physicochemical principles, accurately forecasting metallurgical performance parameters presents a significant challenge.

The techniques outlined in the Special Issue have the potential to significantly improve the optimization and control of flotation operation, aiming to maximize the process efficiency. In the near future, with the increasing demand for metals and ongoing research advancements, these innovative computational techniques are expected to become effective solutions for monitoring and controlling the flotation process. We anticipate that this Special Issue will serve as a platform for future multidisciplinary research in modeling, designing, and optimizing the flotation process.

Acknowledgments: The Guest Editors express their gratitude to all the contributing authors for submitting their valuable work and for their cooperation in ensuring the timely publication of this issue. They also extend their appreciation to the reviewers, Assistant Editors, and the Editorial Board for their contributions and efforts in bringing this Special Issue to fruition.

Conflicts of Interest: The authors declare no conflicts of interest.

References

1. Nakhaei, F.; Ghodrati, S.; Hoseinzadeh, S.; Valipoori Goodarzi, H.; VandGhorbany, O. Combined effect of operating variables on performance and kinetic of lead flotation. *Sep. Sci. Technol.* **2023**, *58*, 75–85. [CrossRef]
2. Szmigiel, A.; Apel, D.B.; Skrzypkowski, K.; Wojtecki, L.; Pu, Y. Advancements in Machine Learning for Optimal Performance in Flotation Processes: A Review. *Minerals* **2024**, *14*, 331. [CrossRef]
3. Bergh, L.G.; Yianatos, J.B. The long way toward multivariate predictive control of flotation processes. *J. Process Control* **2011**, *21*, 226–234. [CrossRef]
4. Nakhaei, F.; Irannajad, M.; Mohammadnejad, S. A comprehensive review of froth surface monitoring as an aid for grade and recovery prediction of flotation process. Part A: Structural features. *Energy Sources Part A Recovery Util. Environ. Eff.* **2023**, *45*, 2587–2605. [CrossRef]
5. Evdokimov, S.I.; Golikov, N.S.; Zadkov, D.A.; Voitovich, E.V.; Kondratiev, V.V.; Petrovskiy, A.A.; Konyukhov, V.Y.; Gladkikh, V.A. Studying the Flotation of Gold-Bearing Ores Using Carrier Minerals. *Minerals* **2024**, *14*, 88. [CrossRef]
6. Evdokimov, S.I.; Golikov, N.S.; Pryalukhin, A.F.; Kondratiev, V.V.; Mishedchenko, A.; Kuzina, A.V.; Bryukhanova, N.N.; Karlina, A.I. Studying Flotation of Gold Microdispersions with Carrier Minerals and Pulp Aeration with a Steam—Air Mixture. *Minerals* **2024**, *14*, 108. [CrossRef]
7. Doubra, P.; Carelse, C.; Chetty, D.; Manuel, M. Experimental and Modelling Study of Pt, Pd, and 2E+Au Flotation Kinetics for Platreef Ore by Exploring the Influence of Reagent Dosage Variations. *Minerals* **2023**, *13*, 1350. [CrossRef]
8. Vinnett, L.; Cornejo, I.; Yianatos, J.; Acuña, C.; Urriola, B.; Guajardo, C.; Esteban, A. The Correlation between Macroscopic Image and Object Properties with Bubble Size in Flotation. *Minerals* **2022**, *12*, 1528. [CrossRef]
9. Betancourt, F.; Bürger, R.; Diehl, S.; Gutiérrez, L.; Martí, M.C.; Vásquez, Y. A Model of Froth Flotation with Drainage: Simulations and Comparison with Experiments. *Minerals* **2023**, *13*, 344. [CrossRef]
10. Jovanović, I.; Nakhaei, F.; Kržanović, D.; Conić, V.; Urošević, D. Comparison of Fuzzy and Neural Network Computing Techniques for Performance Prediction of an Industrial Copper Flotation Circuit. *Minerals* **2022**, *12*, 1493. [CrossRef]

11. Nakhaei, F.; Rahimi, S.; Fathi, M. Prediction of Sulfur Removal from Iron Concentrate Using Column Flotation Froth Features: Comparison of k-Means Clustering, Regression, Backpropagation Neural Network, and Convolutional Neural Network. *Minerals* **2022**, *12*, 1434. [CrossRef]
12. Zhang, W.; Liu, D.; Wang, C.; Liu, R.; Wang, D.; Yu, L.; Wen, S. An Improved Python-Based Image Processing Algorithm for Flotation Foam Analysis. *Minerals* **2022**, *12*, 1126. [CrossRef]
13. Gholami, A.; Movahedifar, M.; Khoshdast, H.; Hassanzadeh, A. Hybrid Serving of DOE and RNN-Based Methods to Optimize and Simulate a Copper Flotation Circuit. *Minerals* **2022**, *12*, 857. [CrossRef]
14. Mamani-Quiñonez, O.; Cisternas, L.A.; Lopez-Arenas, T.; Lucay, F.A. Control Structure Design Using Global Sensitivity Analysis for Mineral Processes under Uncertainties. *Minerals* **2022**, *12*, 736. [CrossRef]
15. Nakhaei, F.; Irannajad, M.; Yousefikhoshbakht, M. Simultaneous optimization of flotation column performance using genetic evolutionary algorithm. *Physicochem. Probl. Miner. Process.* **2016**, *52*, 874–893.
16. Riquelme, A.; Desbiens, A.; del Villar, R.; Maldonado, M. Predictive control of the bubble size distribution in a two-phase pilot flotation column. *Miner. Eng.* **2016**, *89*, 71–76. [CrossRef]

minerals

MDPI

Article

Prediction of Sulfur Removal from Iron Concentrate Using Column Flotation Froth Features: Comparison of k-Means Clustering, Regression, Backpropagation Neural Network, and Convolutional Neural Network

Fardis Nakhaei [1,*], Samira Rahimi [2] and Mohammadbagher Fathi [3]

[1] School of Chemical and Minerals Engineering, North West University, Potchefstroom 2520, South Africa
[2] Civil Hydraulic Structures, Azad University, South Tehran Branch, Tehran 1584743311, Iran
[3] Mining Department, Engineering Faculty, Urmia University, Urmia 5756151818, Iran
* Correspondence: 36598704@mynwu.ac.za

Abstract: Froth feature extraction plays a significant role in the monitoring and control of the flotation process. Image-based soft sensors have received a great deal of interest in the flotation process due to their low-cost and non-intrusive properties. This study proposes data-driven soft sensor models based on froth images to predict the key performance indicators of the flotation process. The ability of multiple linear regression (MLR), the backpropagation neural network (BPNN), the k-means clustering algorithm, and the convolutional neural network (CNN) to predict the amount of sulfur removal from iron ore concentrate in the column flotation process was examined. A total of 99 experimental results were used to develop the predictive models. Extracted froth features including color, bubble shape and size, texture, stability, and velocity were used to train the traditional predictive models, whereas in the CNN model the froth images were directly fed into the model. The results comparison indicated that the three-layered feedforward NN model (17-10-1 topology) and CNN model provided better predictions than the MLR and k-means algorithm. The BPNN model displayed a correlation coefficient of 0.97 and a root mean square error of 4.84% between the actual data and network output for both training and the testing datasets. The error percentages of the CNN, BPNN, MLR and k-means models were 10, 11, 15 and 18%, respectively. This study can become a key technical support for the application of intelligent models in the control of the operational variables for the flotation process used to desulfurize iron concentrate.

Keywords: column flotation; sulfur removal; iron ore; prediction; multiple linear regression; neural network; k-means clustering; convolutional neural network

Citation: Nakhaei, F.; Rahimi, S.; Fathi, M. Prediction of Sulfur Removal from Iron Concentrate Using Column Flotation Froth Features: Comparison of k-Means Clustering, Regression, Backpropagation Neural Network, and Convolutional Neural Network. *Minerals* **2022**, *12*, 1434. https://doi.org/10.3390/min12111434

Academic Editors: Dave Deglon and Luís Marcelo Tavares

Received: 26 September 2022
Accepted: 7 November 2022
Published: 12 November 2022

1. Introduction

Flotation is a significant industrial technology for separating valuable minerals from tailings. It is a complicated physical-chemical separation method that takes advantage of differences in the surface properties of valuable and gangue minerals [1]. Due to the depletion of high-grade iron reserves and the need for very fine grinding to improve liberation, flotation is the most effective way to remove impurities from iron ore concentrate [2,3].

Although the flotation process has been widely used over a long period of time, how to evaluate the operational conditions, the automatic control and robust model of the process are challenging issues and still have not been fully considered by academic scholars. Furthermore, the weakness of sufficiently accurate and reliable process measurements amplifies this difficulty [4]. The ultimate goal in the flotation process is to maximize the separation between valuable minerals (concentrate) and gangue minerals (tailings) [5]. Technically, the performance of the flotation process is described by the concentrate characteristics (grade and recovery). These are important economic and technical indexes needed for process control and optimization [6]. Accurate values of these parameters are obtained only after sampling, filtration, drying, preparation and chemical analysis of samples, which are time-consuming operations [7]. In industrial plants, the grade can be continuously measured using an XRF analyzer and the recovery can only be calculated from a mass balancing [8,9]. The online measurement of these parameters with a relatively low accuracy using an X-ray analyzer generally requires the purchase and maintenance of costly and sophisticated equipment [5,10,11], which justifies the execution of models for the prediction of key performance indicators based on secondary variables.

Due to the nonlinear and complex nature of the flotation process, the large number of variables involved, and the very limited understanding of the physicochemical rules of the flotation process, accurate prediction of grade and recovery parameters is a difficult task [12–15]. Therefore, data-driven approaches requiring less prior knowledge about the system state variables (the input/output) are viewed as alternative strategies for modeling the flotation process. In the past few decades, modelling and predicting flotation behavior based on operational variables (such as reagent dosages) using numerical techniques has been extensively studied by researchers [16–22], although there is no universally accepted forecasting model. Most of these models only used relatively few operational parameters as the input variables so that they failed to enable the capture of the strong nonlinear relationship between the variables in a wide range of operating conditions. These models still have challenges, such as needing a large quantity of data, and have a complex configuration that limits their application.

In the last ten years, several studies have shown that the froth's appearance contains a lot of valuable information, which plays a pivotal role in raising the flotation process efficiency [23]. The froth's surface appearance reflects the changes in the flotation process induced by those affecting the operational variables, such as air flow rate and dosage of reagents [24]. Then, it can be directly used for flotation metallurgical performance estimation [25,26]. Traditionally, in the flotation plant the froth state is observed by experienced workers in order to control and make a decision for the working conditions [7]. However, decision-making based on the monitoring of froth surface changes just by the naked eye is inaccurate; that is, a time-consuming operation that limits the real-time control of the process. Therefore, the combination of froth imaging analysis techniques and the predictive mathematical models could provide the soft sensors for modelling the effects of operating variables on flotation process performance.

In recent years, there has been a rapid development in computing technology with an increased interest in machine vision applications for monitoring, analyzing and controlling the output of processes for most engineering purposes [27,28]. Machine vision is an automated, non-destructive and cost-effective alternative method used for the on-stream estimation of the efficiency of the flotation process. Image processing methods have been developed for the extraction and interpretation of froth appearance features such as color and texture [29–32], velocity [33,34], mineral loading rate [35], stability [11,36], and bubble size [25,37]. These studies showed that many methods can be used to extract and analyze the froth features; thus, several extracted features should be addressed at the same time for a better interpretation of froth behavior.

The performance of the reverse flotation of iron ores is largely governed by the interactions between the operational variables, which are complex. The prediction of the amount of sulfur removal in the reverse flotation of iron ore concentrate is an important issue to enhance the process efficiency. When the sulfur content of iron concentrate increases, the sulfur recovery and efficiency will reduce. In this situation, the operators at concentrators try to identify the problem and find a way to fix it based on the froth surface appearance. Visual assessments by operators are time-consuming and difficult tasks that can be carried out only at a given time. An image-based soft sensor with a view to monitor and control the operating conditions could be a great help for operators at flotation plants. Due to a dearth of reports on the properties of flotation froth during the desulfurization of iron concentrate [38], this research opted to combine the machine vision system and different statistical and intelligent techniques with a view to investigating the relationship between column flotation performance (sulfur recovery) and the froth features under different operating conditions.

For this, a dataset of images was first captured from froth surfaces at defined time intervals. Afterward, the bubble size and shape, froth color, texture and dynamic features (froth velocity and stability) were considered as good supplements for operating condition recognition prior to an offline extraction of the images from the desulfurization of iron ore concentrate using the in-column flotation. Finally, a comparative study was carried out using multiple linear regression (MLR), the k-means algorithm, the backpropagation neural network (BPNN) and the convolutional neural network (CNN) to predict the flotation performance based on froth features.

The machine learning models, such as NN, are widely used as a powerful approach for classification and prediction because of their non-linear learning ability. The NN is a computational method developed by copying human brain behavior. The NN is able to extract the complex nonlinear relationships existing between input and output variables through a highly interconnected system of simple processing elements (neurons or nodes) [39]. Recently, the models based on BPNN have been widely used as rapid and reliable tools for forecasting flotation performance [40,41]. Many studies in the literature specialized in the integration of machine vision and machine learning systems for improving the flotation process control. The proposed soft sensors were capable of extracting and analyzing the froth image features and using them as the inputs to the machine learning models [42,43].

MLR is another estimation tool which can help to predict the flotation performance based on a number of froth image features. The MLR model is presented based on fitting a linear equation to observed data. The main advantage of MLR is its simple form and easily interpretable mathematical expression [44]. The MLR is easy to formulate and has been used in flotation research to obtain models as an alternative to other mathematical methods [4,19].

The use of clustering algorithms has been widely used in diverse fields of study to offer a favorable alternative to traditional prediction methods [45,46]. Clustering is a process of organizing a set of objects into groups of similar objects, based on their characteristics. The k-means algorithm is one of the well-known unsupervised clustering techniques which is mainly applied in image and signal processing, pattern recognition and data mining [47]. K-means clustering is an iterative algorithm that tries to minimize variation within the clusters, and maximize variation between clusters [48,49]. In this research, the k-means algorithm is proposed to classify the froth characteristics to estimate the sulfur recovery in the column flotation process at different operating conditions.

The above-mentioned models concentrate on extracting specific froth features such as bubble size, color, and stability for classification of froth images to evaluate the flotation process performance. The big challenge of these models is that the prediction accuracy remarkably relies on the capability of feature extraction methods. Furthermore, these models are not sufficiently effective and reliable because they employ only low-level features without having enough mid-level and high-level features [50]. To overcome this problem and develop the classification accuracy, a deep learning method has been developed to classify image features under different conditions.

The convolutional neural network (CNN), one of the fastest-growing deep leaning algorithms, was developed for the classification of objects in the fields of engineering [51,52]. This method obviates the complexity of image modification and allows users to enter the original image directly to the model.

CNN was first implemented for froth image feature extraction in 2018 by Fu and Aldrich. They proved that the CNN model could generate more accurate results than the traditional models in the processing of flotation froth images. Recently, many researchers have employed CNN as a robust and efficient model for the evaluation of flotation influencing factors and the prediction of metallurgical performance parameters [5,53,54]. This article provides the application of CNN to categorize the collected froth images during the column flotation process. Then, its classification accuracy was compared to the traditional models.

Despite the fact that the above-mentioned methods were all individually used to forecast the mineral flotation performance, there is still a gap in identifying the best method with the highest prediction accuracy, especially the amount of sulfur removal from iron ore concentrate. Thus, the current study compared four methods of MLR, k-means, BPNN, and CNN to predict the amount of sulfur removal in the desulfurization of iron ore concentrate. The aim of this work was to elucidate if these predictive models could predict column flotation performance for the desulfurization of the iron ore concentrate with high accuracy based on the analysis of froth image characteristics. Such studies can make a remarkable contribution towards the improvement of soft sensors on the basis of froth image analysis for the real-time monitoring and control of flotation process.

In the following section, the reverse column flotation process of iron ore concentrate and the image-acquisition device are presented. Then, the techniques for extraction of froth features and the predictive models are briefly introduced. In the third part, the prediction results obtained by k-means, MLR, BPNN, and CNN methods are discussed. In the Section 4, the performances of these methods are compared.

2. Materials and Methods

2.1. Iron Ore Concentrate

The sample used in the present study was taken from the feed stream of the mechanical flotation cells in an iron ore concentrate plant in Iran. In this plant, the iron ore after comminution, gravity and magnetic separation, is fed to the flotation cells. The feed flotation has high sulfur (pyrite) content. In general, pyrite is considered as a gangue mineral creating problems during the steelmaking processes [55]. Sulfur causes brittleness and frangibility of steel at high temperatures, reduces weldability, increases corrosion, air pollution during the pellet-firing process, increases limestone consumption and slag production in the steelmaking procedure [56]. Pyrite and pyrrhotite are paramagnetic minerals and are attracted by the magnetic field in the vicinity of iron ores (magnetite and hematite). Therefore, they are not completely recovered by magnetic separators [57]. In this plant, the concentrate from the magnetic separation generally presents a sulfur content of 0.4 to 0.6%, which value surpasses the allowable limit for steel production (0.1%). The reverse flotation conducted at the end of the processing circuit enables a reduction of the sulfur content in the magnetic separation concentrate.

The representative flotation feed sample consisted of 63.3% Fe, 1.48% FeO with a significant sulfur content 0.50%. The particle size analysis of the representative sample 188 gave the d80 equal to 95 µm.

Mineralogical and liberation degree studies of the sample were conducted using optical microscope (Axio Plan 2, Zeiss, Germany) and wild zoom stereo microscope (Zeiss, Germany), respectively. The main iron minerals were hematite, magnetite and goethite. Calcite, quartz, dickite, chlorite and pyrite were the main gangue minerals. The results of X-ray diffraction pattern (XRD) analysis of the sample are shown in Figure 1. The XRD analysis confirmed the microscopic findings. Mineral liberation studies showed that more than 90% of pyrite particles were liberated at -105 µm.

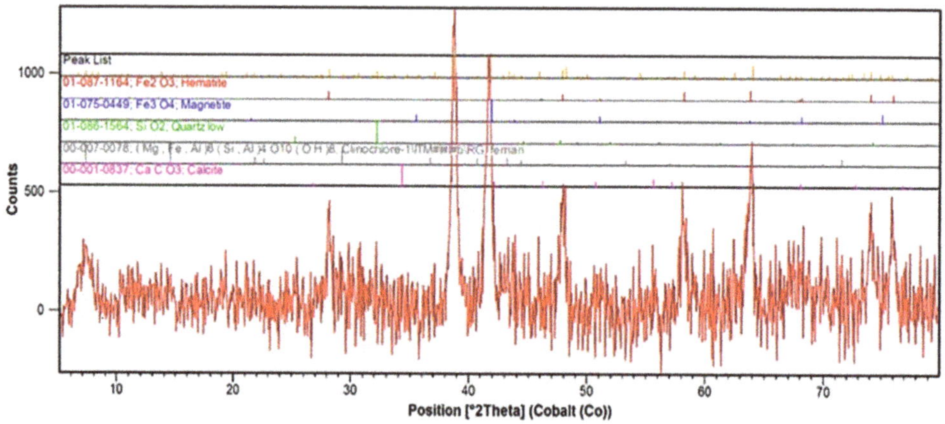

Figure 1. X-ray diffraction pattern of sample used in this work.

2.2. Column Flotation Test

The tests were carried out in a plexiglass cell column of 400 cm in height, and 10 cm inside diameter in order to desulfurize the iron ore concentrate. To provide the initial feed, a tank equipped with a stirrer was used to prevent the material from settling while being prepared. Two peristaltic pumps were used: one for sending the feed to the middle of the cell and the other for withdrawing the pulp at the bottom of the cell. During the experiments, the pulp pH was continuously measured by a digital laboratory pH meter and adjusted to the desired value by adding sulfuric acid or NaOH.

Since the purpose of the in-column flotation operation was to reduce the sulfur content in the final product, potassium amyl xanthate (PAX) as a collector and MIBC as a frother were used. After adjusting the pH and sufficient mixing with chemical reagents (five minutes), the pulp was entered into the cell at a flow rate of 1.5–2 L/min and a defined air flow rate provided by a compressor. No wash water was used during the experiments. The gas flow rate was measured by a flow meter and manually regulated by a needle valve. The froth height was determined by direct observation as well as pressure transducers. The control system of the froth height changed the speed of the discharging pump based on the position of the pulp–froth interface relative to the set point. Prior to the sampling of concentrate and tailing, the pulp–froth interface was kept constant to ensure that column flotation operated under steady-state conditions. The froth overflowed from the top of the cell as tailing product, while iron concentrate was discharged from the bottom of the column. The froth and concentrate were filtered and dried and their sulfur content determined. Figure 2 shows a schematic diagram of the experimental apparatus.

The image-acquisition system included a fixed device, a 100 W LED lamp, a protective cover, and a digital camera, as shown in Figure 2. The camera was placed inside a fixed chamber and positioned 250 mm away vertically from the top edge of the cell so that the camera lens center matched that of the cell. The target area was illuminated by the light. The color rendering index and color temperature of the light source were greater than 80 and 5000 K, respectively.

The froth surface appearance was indirectly varied by changing the process operation conditions. The data acquisition was performed for 2 min during each test. In this study, the most important froth features including bubble size and shape, color, texture, bubble burst rate and velocity were primarily extracted offline from the froth images in each run through image processing techniques. Then they were fed into the k-means algorithm, regression, and NN models. Since the froth appearance might frequently change, a single image could not fully elucidate the froth characteristics during the sampling period, thus eight random images were individually analyzed, and the mean value of each feature was measured. In the CNN algorithm the froth images were directly fed into the model.

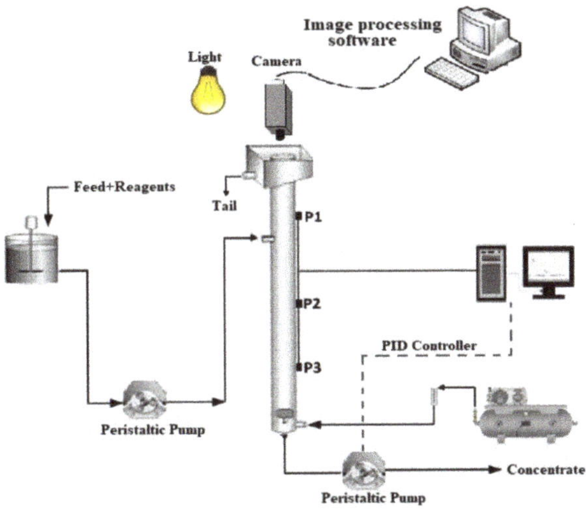

Figure 2. The flotation column and froth image-acquisition system.

The average of the gray, RGB, HSB (hue, saturation and brightness) and lab values were extracted from images to describe the froth color descriptors. Furthermore, based on each gray level co-occurrence matrix (GLCM), four texture features (entropy, contrast, inverse different moment (IDM), and angular second moment (ASM) or energy) were selected to describe the froth surface textural characteristics. A watershed technique was applied to measure the bubble size distribution, roundness (circularity) and aspect ratio (AR) (Figure 3a). The bubble burst rate (stability) and froth velocity (transport rate) which are dynamic descriptors were extracted from an image pair. The pixel tracing algorithm was applied to quantify the froth speed (pixel/s) between the consecutive frames (Figure 3b). The bubble burst rate was computed by finding the difference between two consecutive frames (Figure 3c). The particular implementation of the image processing techniques to extract the froth features offline was presented by Nakhaei et al. [38].

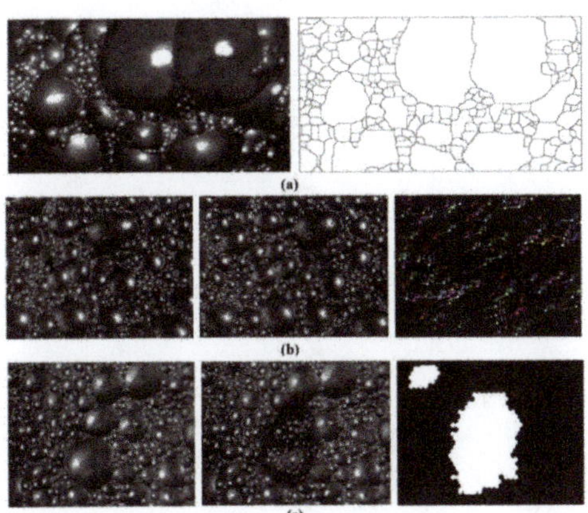

Figure 3. (**a**) The initial image after contrast enhancement and watershed algorithm; (**b**) froth transport rate measurement by the pixel tracing technique; (**c**) bubble burst rate measurement.

Since the purpose of this study was to find a valid model to describe the relationship between sulfur recovery and froth surface characteristics, the flotation experiments were performed on a wide range of operational variables. Therefore, the most important operational variables including aeration rate, solid percentage, froth height, chemicals dosage and pulp pH were changed for each run. A total of 99 experiments were designed and performed. The technical parameters of the reverse in-column flotation experiments are given in Table 1. The values of sulfur recovery spread out over a large range, as shown by the high values of the standard deviations, due to the different conditions applied in the flotation process (Table 2).

Table 1. Variables and levels used in column flotation experiments.

Variable	Range
Collector concentration (g/t)	40–220
Frother concentration (g/t)	60–230
Froth height (cm)	10–40
Air flowrate (cm/s)	1.3–1.9
Solids (%)	10–35
Pulp pH	2.5–8

Table 2. The statistical measures of sulfur recovery (the output variable).

Metallurgical Factor	Maximum	Minimum	Average	SD
Recovery (%)	85.08	6.53	49.28	18.56

Table 3 summarizes the values of the input dataset (froth characteristics) for the k-means algorithm, regression, and BPNN models. This dataset was extracted offline by applying a wide range of operational conditions to form a robust predictive model for sulfur recovery estimation.

Table 3. Descriptive statistics of the froth features obtained at different operating conditions as inputs for the k-means algorithm, regression and BPNN models.

Features	Maximum	Minimum	Average	Features	Maximum	Minimum	Average
Gray level	112	74	86.36	Energy (0°)	1×10^{-3}	2.5×10^{-4}	5.2×10^{-4}
R	136	79	95.3	Contrast (0°)	260	45.81	128.5
G	111	78	89.03	IDM (0°)	0.39	0.15	0.23
B	101	63	76.01	Entropy (0°)	10.05	7.75	8.50
Hue	0.36	0.11	0.25	Bubble size	50.97	5.6	30.80
Saturation	0.41	0.19	0.28	Aspect ratio	1.75	1.10	1.43
Brightness	0.53	0.33	0.39	Circularity	0.81	0.34	0.51
L	45.4	33.66	37.68	Speed	26.01	10.77	16.41
a	10.44	−6.3	−0.17	Stability	85.02	57.88	74.31
b	15.71	4.62	9.2				

According to the expert experiences, the process conditions (sulfur removal) can be sorted into six groups: A–F (Table 4). Each group stands for a specific process condition. A large number of froth images under various operating conditions were collected and directly fed into the CNN model. The training and testing of the CNN model were performed using 2079 and 891 froth images, respectively.

Table 4. Different froth classes in CNN model under various operating conditions.

Type	Sulfur Recovery (%)	Class Label	Number of Images
Very low	≤20	A	180
Low	20–30	B	300
Low–medium	30–40	C	390
Medium	40–50	D	480
Medium–high	50–65	E	1140
High	≥65	F	480

2.3. Models

In a flotation plant, the sulfur content of the concentrate needs to be sampled in the laboratory for analysis and this takes several hours. Consequently, when unexpected events happen, the sulfur content in the iron concentrate may increase with a decrease in the process efficiency. The late detection may lead this phenomenon to last for a long time resulting in economic losses. A soft sensor based on mixed image analysis and mathematical methods can predict the amount of sulfur removal in real-time and thus enable workers to quickly tackle these problems. In this study, the applicability of k-means, MLR, BPNN, and CNN methods for predicting the sulfur removal was compared. The general purpose of using different methods is to find which model most accurately discovers the relationship between inputs and output.

2.3.1. K-Means Clustering

Clustering is an unsupervised learning technique in the area of machine learning, which categorizes objects with high similarity into the same cluster according to a certain distance metric [45,58]. Thus, a cluster is a collection of similar data which are not similar to the data located in other clusters [59,60]. The more similar the objects in the cluster, the better the clustering effect [45]. Among various types of clustering methods, k-means is one of the most widely used clustering algorithms. In this work, the k-means algorithm was employed because of its high efficiency and simplicity in pattern recognition of massive data. The k-means algorithm splits a dataset into K discrete non-overlapping clusters. The clustering is performed by minimizing the sum of squares of distances between the center and the data in each cluster [61]. Details of the k-means algorithm procedure are presented in [47,61].

The simplest procedure of the k-means algorithm can be described as follows [48,62]: (1) Choose random initial clusters centroids. (2) Compute the Euclidean distance between each object to each cluster centroid according to Equation (1):

$$D = \sqrt{\sum_{i=1}^{n} \sum_{j=1}^{m} |X_{ij} - C_j|^2} \quad (1)$$

where D is the Euclidean distance; n is the number of data; m is the number of dimensions; X_{ij} is stated as the j dimensionality of the i-th data; Cj is the j dimensionality of the cluster center [48].

(3) Assign data points to one of the clusters on the basis of the proximity to the centers. (4) Assign new centroids for each cluster by averaging the data of each cluster. (5) Return to step 2 and repeat the process until convergence is obtained.

In the current study, the k-means algorithm was applied to cluster the froth features' dataset into specific groups (based on a given k value) depending on the flotation recovery (low to high). The algorithm described above was implemented based on the code in Python software (3.1, Python Software Foundation manufacturer).

The efficiency of each cluster is evaluated using the accuracy criterion. The accuracy of each cluster is the percentage of observations of the dataset that is correctly classified by the model used. This can be written as:

$$Accuracy = \frac{CN}{CN + FN} \times 100 \quad (2)$$

where CN is the number of correct responses, FN is the number of false responses.

2.3.2. Multiple Linear Regression

MLR is one of the mathematical methods used to describe the linear relationship between the independent variables and dependent variables [63]. MLR is used when the value of a specific variable can be estimated based on the values of other variables. In MLR, the model is fit by minimizing the sum of squares of the difference between the observed and fitted values. The linear equation is as follows [64]:

$$Y = C_0 + C_1x_1 + C_2x_2 + \cdots + C_nx_n + \varepsilon \quad (3)$$

where Y corresponds to the output variable, X_i is the independent variables, C_i is the regression coefficients, and ε represents the error term.

The froth features were taken as input variables for the MLR model to predict the sulfur recovery. In the MLR modeling, 85% of data was randomly used for computing the equation (84 runs) and 15% for testing (15 runs). The MLR analysis was performed using the SPSS 27 software (Property of IBM Corp.).

2.3.3. Backpropagation Neural Network

NNs are nonlinear computational methods that have been successfully used in various fields of science and technology over the past decade [65–67]. The main feature of the NN is its ability to learn the complex relationships between input and output data. The main advantages of the NN models are as follows: (1) It only considers the input and output data without referring to the process phenomenology; and (2) it has a generalization ability to accurately estimate the outputs corresponding to a new dataset that were not applied for training of the model [39,68].

NN is a complex network structure consisting of three or more layers to learn the nonlinear relationship between input and output data. The layers are processed with a large number of interconnected neurons. Each neuron, has an associated weight and bias. Figure 4 shows the NN structure, where X_1 and X_n are the inputs. W is the corresponding connection weight, b is the bias and Y is the output.

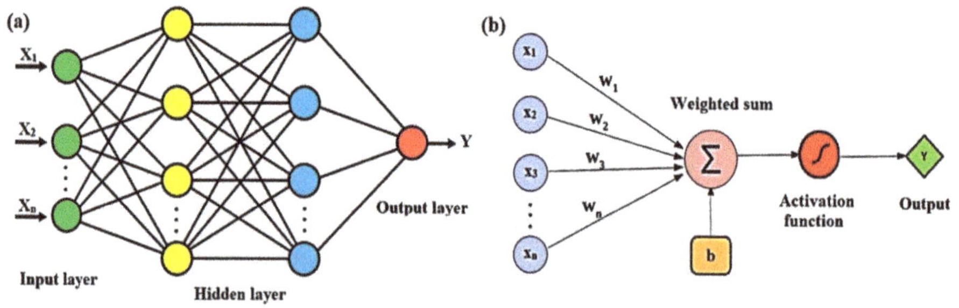

Figure 4. (**a**) The schematic of a three-layer NN; (**b**) signal-flow graph of a perceptron.

The modelling process is as follows: The first stage consists of choosing the network configuration and the number of nodes in the layers, and giving the random value of the weight and the bias matrixes. In the second stage, the input and the output matrixes are entered. In the third stage, the hidden layer output matrix and the output matrix of the last layer are calculated; The next step is to compute the difference between the output of the model and the actual output; if the error value is within the specified range, the training ends; otherwise, the fifth stage is continued. In the fifth stage, the error is back-propagated and the data are forwarded and the weight and bias matrixes are updated.

The dataset of the flotation tailing froth features is applied to train the network for predicting the amount of sulfur removal from iron ore concentrate in the column flotation process. In this study, 74 of 99 (75%) items in the dataset were used in training and the rest 25 (25%) in the test and validation of the network. The training process was carried out by applying the backpropagation algorithm with the Levenberg–Marquardt training method. The neural net fitting toolbox from MATLAB program, version R2022, was used to make the code for the NN. The tansig, and purelin activation functions were used in the hidden layer and the output layer, respectively. These complex functions offer the NN model the ability to learn both linear and nonlinear relationships between the inputs and the outputs.

The proper choice of the number of neurons in the hidden layer is an important task. According to the literature, as long as the number of neurons in the hidden layer is appropriately specified within reasonable limits, a three-layer BPNN can be effectively applied to model a wide range of complex problems [69]. Too many neurons may result in increasing the computational time or overlearning so that the NN loses its ability to generalize the patterns present in the training dataset. On the other hand, very few numbers of neurons may cause underfitting so that the model is not complex enough to capture patterns in the data [70,71].

In general, it is not possible to comment definitively on the number of neurons required for a proper network execution. The best number of hidden neurons was determined by trial and error, based on the evaluation of the mean square error (MSE). An empirical formula has been proposed [72] to find the number of hidden neurons:

$$n_h = \sqrt{(n_i + n_o)} + c \tag{4}$$

where n_h, n_i and n_o are the number of neurons in hidden, input and output layers, respectively. c is an adjustment constant ranging from 1 to 10. In this article, based on Equation (4), the number of hidden neurons was examined from 6 to 16 to determine the optimal value.

Performance Evaluation of the NN and the MLR Models

Three statistical indexes, including correlation coefficient (R^2), root mean square error (RMSE) and error percentage (E), were applied to assess the performance of the developed NN and MLR models in order to find whether there was any significant difference in their performance. These three performance indicators are calculated as follows [73,74]:

$$R^2 = \left[\frac{n(\sum XY) - (\sum X)(\sum Y)}{\sqrt{(n\sum X^2 - (\sum X)^2)(n\sum Y^2 - (\sum Y)^2)}}\right]^2 \tag{5}$$

$$RMS = \left[\frac{1}{n}\sum (X - Y)^2\right]^{\frac{1}{2}} \tag{6}$$

$$E = \frac{1}{n}\sum\left|\frac{Y - X}{X}\right| \tag{7}$$

In these equations, X represents the measured values, Y is the estimated values, and n is the number of data. The correlation coefficient can determine how the network output changes according to the actual values, and when it is equal to one, there is a complete correlation between the estimated and the measured values. The lower RMSE and error percentage values represent the more accurate estimation results.

2.3.4. Convolutional Neural Network

CNN is considered to be one of the most popular NNs for image classification problems. Basically, a CNN is a kind of feedforward NN which comprises two main sections: The first section (feature extraction) comprises input layer, convolution and pooling layers and the second section (classification) consists of a dense (flattened) layer, fully connected layers and an output layer [75]. Figure 5 illustrates the architecture of a typical CNN.

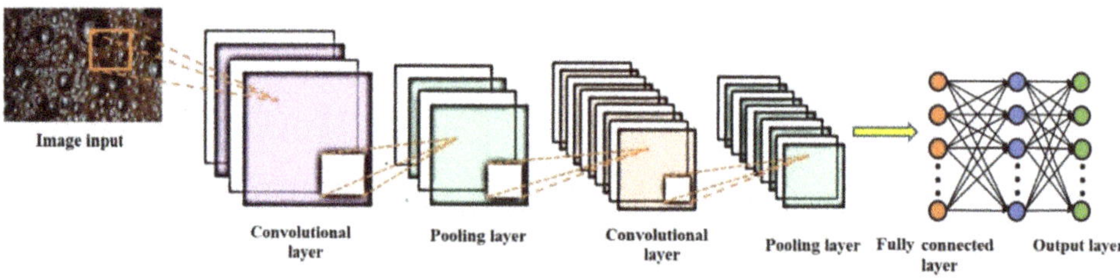

Figure 5. The structure of the CNN model.

The convolutional layer (CL) is the main part of a CNN which extracts features by convoluting the kernel (filter) over an original image (matrix). The filter works by moving a window from left to right and top to bottom to multiply and sum each position of the source pixel (Figure 6). Each convolution layer adopts a 3 × 3 filter with a stride of 1. ReLU (rectified linear unit) is used as an activation function after each CL. ReLU is always levelled as 0 and 1. The pooling layer decreases the dimension of feature maps while keeping the most important information to avoid overfitting [76]. It is generally situated between consecutive CLs. The maximum pooling function adopts a matrix of size 2 × 2 (Figure 7).

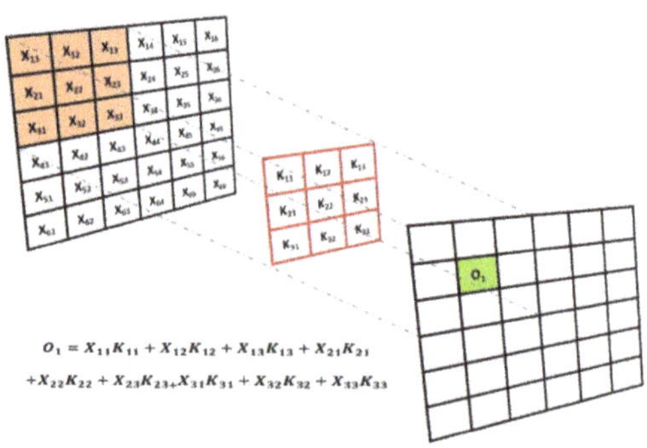

Figure 6. Schematic diagram of image convolution operation.

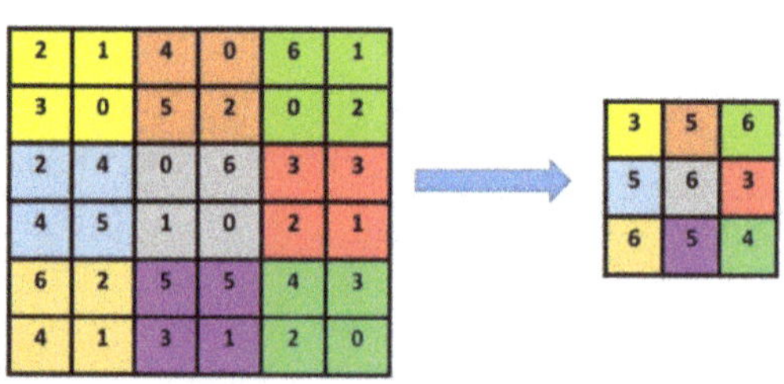

Figure 7. The maximum pooling function operation (2 × 2 filter and stride 2).

After the pooling layer, the output feature maps are flattened in the dense layer by transforming the input from multidimensional space to a one-dimensional array of numbers to smooth the connection of nodes in the fully connected layer. The classification of the input image is performed in the classification layer with the ability to adjust weights. The number of nodes in the classification layer corresponds to the number of classes in the output.

Many popular CNN architectures have been developed such as LeNet, GoogleNet, VGG16, ResNet, and AlexNet to solve machine vision-based problems [65]. In this study, ResNet18 was selected. MATLAB 2021a (The MathWorks, Inc. protected by U.S and international patents), with Machine toolbox deep neural network designer, was used in this experiment.

3. Results and Discussion

3.1. Correlation between Froth Features and the Process Metallurgical Performance

The predictive model performance depends on the data nature, and the variable selection [77]. Feature selection should be considered to find the best set of variables to build useful predictive models.

The correlation coefficient is a statistical measure that is commonly used to calculate the strength of the relationship between two variables [78]. In order to select the appropriate variables and minimize the required dataset, the Pearson correlation coefficients between the froth features and the sulfur recovery were implemented. The values of the correlation matrix between the froth images features and the recovery are shown in Figure 8. As shown in the heatmap, color features obtained from the froth images except H and b were strongly associated with the amount of sulfur removal, showing almost similar correlation coefficients. As a result, these parameters were suitable for the sulfur recovery estimation. Furthermore, there were very weak correlations between the H, b and the recovery. Therefore, these two features were excluded from the model predictors due to having a low contribution rate.

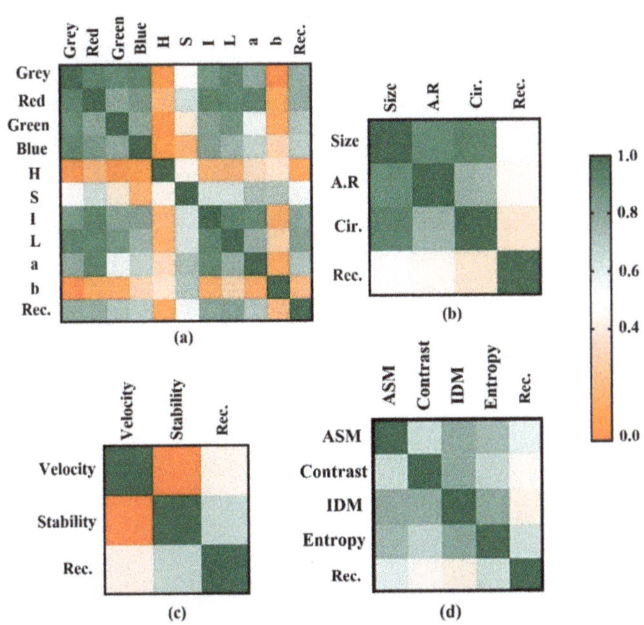

Figure 8. Pearson correlation matrix between the inputs and output. (**a**) color; (**b**) size and shape; (**c**) dynamic; (**d**) textural properties.

The color characteristics of the froth images are more related to the process performance than to the geometric properties of the bubbles. Figure 8d shows the correlation values between the textural properties of the froth images and the process performance factor. There was a significant correlation between the textural features and the recovery. Among the textural properties, the entropy and energy had a higher correlation with the sulfur removal values. According to the heatmap, both geometric features and dynamic factors were most associated with the recovery. The froth stability relative to the velocity had a higher correlation with the process output variable.

Although the results of the correlation matrix between the amount of sulfur removal and the froth visual properties showed that these features are somewhat in line with expectations, the interaction between the factors may obscure the results. Therefore, the use of nonlinear models for process modelling is justifiable. The input and output variables of the proposed models for prediction of the output parameter are shown in Figure 9.

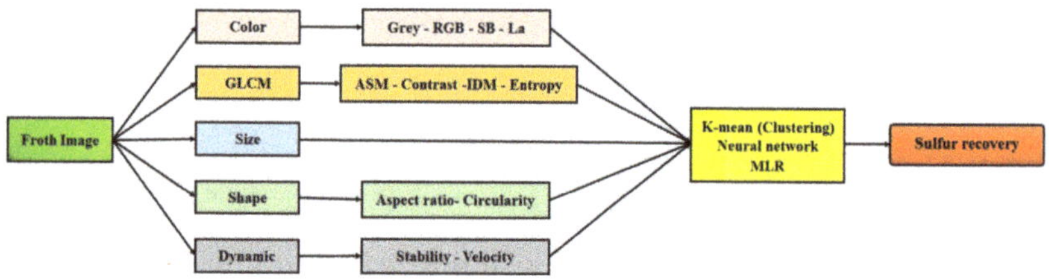

Figure 9. The structure of the inputs and output of the proposed models.

3.2. *Sulfur Removal Estimation Based on K-Means Algorithm*

In this study, the k-means algorithm was applied to cluster the froth image features using Python software. The effect of the number of clusters (five and six) was examined. According to the results of this study, a model with six clusters was optimal. The clusters of samples based on k-means algorithm are visualized in Figure 10. The samples were classified into the different clusters with different colors. It is clear that the k-means algorithm was able to classify the dataset. The number of samples in each cluster is given in Table 5. It can be seen that clusters 3 and 2 had the highest and lowest populations, respectively.

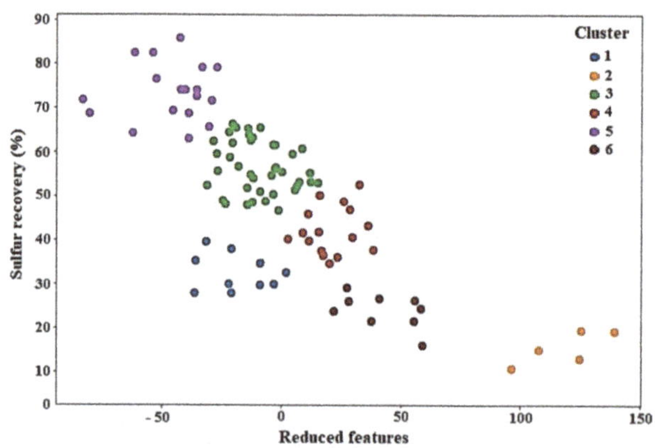

Figure 10. K-means clustering (K = 6) applied to froth feature dataset in order to group the sulfur removal value.

Table 5. The number of samples in each cluster.

Cluster	Number of Samples in Cluster
1	10
2	5
3	41
4	16
5	18
6	9

The statistical results obtained by applying the k-means method are shown in Table 6. According to the results of the flotation experiments, the sulfur recovery was split into six clusters, which were too low, low, low–medium, medium, medium–high and high, as shown in Table 6. For example, cluster 5 was characterized by the highest level of sulfur recovery (average = 73.74%). Clusters 3 and 2 were characterized by the medium–high and the very low level of recovery, respectively. In other words, the average of sulfur recovery in cluster 3 was higher than the average of the corresponding recovery in cluster 2.

Table 6. Statistical results of predicted values in each cluster.

Cluster	Max.	Min.	Average	SD	Type	Sulfur Recovery (%)
1	37.97	27.91	32.00	3.39	Low–medium	30–40
2	20.57	10.92	15.88	4.14	Very low	≤20
3	66.42	44.06	56.40	6.06	Medium–high	50–65
4	52.62	35.05	42.05	5.42	Medium	40–50
5	87.28	63.73	73.74	6.56	High	≥65
6	29.22	17.08	24.34	3.65	Low	20–30

The accuracy measure clarifies how well the k-means algorithm was able to group the samples with similar features in the same cluster. To evaluate the errors, it was necessary to check the members of each cluster to see if they were properly classified. The efficiency of each cluster was evaluated using the accuracy criterion (Equation (1)). As shown in Table 7, the results of the k-means algorithm almost matched the results of the actual classification. Results demonstrated that the relationships between the froth characteristics and recovery were successfully modelled using the k-means method with rational errors. In general, the clustering results confirmed the principle that sulfur recovery is greatly affected by froth features. The froth features classification could estimate the sulfur recovery more accurately than human workers, avoiding the large fluctuation caused by the personal decisions of different workers. The proposed classifier could be used to cluster other data whose class label is not specified.

Table 7. The accuracy of each cluster.

Metallurgical Factor	Cluster					
	1	2	3	4	5	6
Misclassification	2	0	7	6	2	1
Classification accuracy (%)	80	100	83	63	89	89

3.3. Sulfur Removal Estimation Based on MLR

To predict the amount of sulfur removal from iron ore concentrate, the MLR model was obtained by processing the full data, as shown in Figure 9. The data were split 85:15 into training (to constitute the regression equation) and test data sets. The results of ANOVA for the regression model are shown in Table 8. The model is shown in the following equation:

$$\begin{aligned} SE = 212.04 + 0.204(Grey) - 1.25(Red) - 1.23(Green) + 0.30(Blue) - 106.71(S) \\ +361.62(I) + 0.66(L) - 1.35(a) + 62,409.34(ASM) - 0.133(Contrast) \\ -164.25(IDM) - 5.99(Entropy) - 0.58(Buble\ size) - 13.34(AR) \end{aligned} \quad (8)$$

Table 8. Results of ANOVA for regression model.

Sums of Squares Regress	df Regress	Sums of Squares Residual	df Residual	Mean Squares Regress	Mean Squares Residual	F	P-Level
25,256.54	17	3772.41	66	1485.69	57.16	25.99	0.00

The relationship between measured and calculated values using the proposed mathematical model in Equation (8) is presented in Figure 11. The multivariable regression equation predicted the sulfur recovery with a correlation coefficient of 0.92. The comparison of the actual values and the estimated values for sulfur recovery in the validation stage is presented graphically in Figure 12.

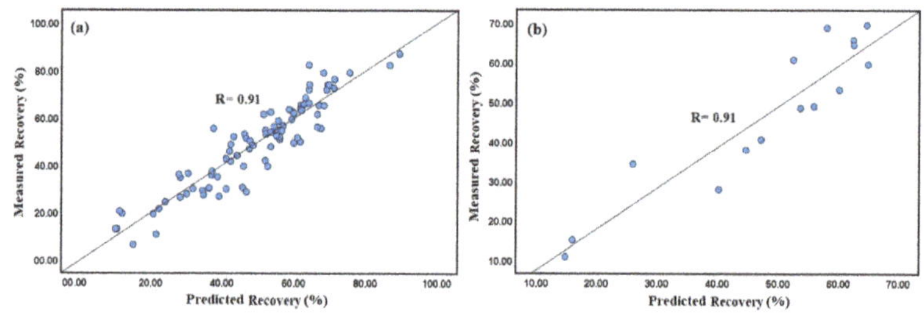

Figure 11. The correlation between the experimental values and predicted values obtained using the MLR (**a**) training (**b**) validation.

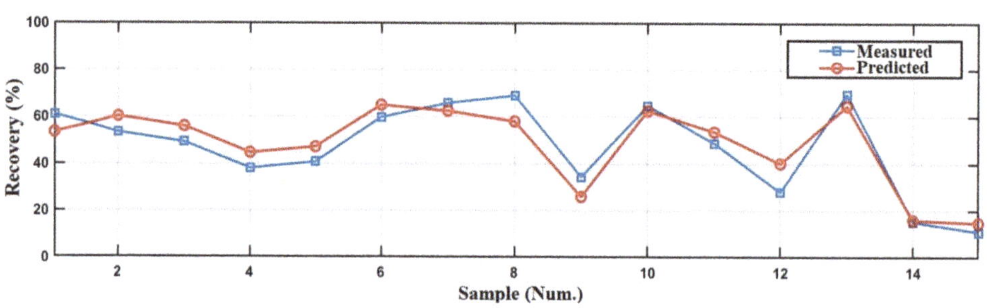

Figure 12. The comparison of measured values and predicted values for sulfur recovery using MLR.

According to Figure 11, the R values in the training and validation data set for sulfur recovery were 0.91. The descriptive statistics of the differences between the measured and estimated values for the evaluation data are given in Table 9. Although the MLR has been stated as a capable tool for data modelling, it could not accurately describe the sulfur recovery within the entire process variables space (RMSE = 6.82). This may be due to the complexity of the selected data covering a wide range of operational conditions. Thus, the ANN model was examined as an alternative method and is discussed in the next section.

Table 9. Descriptive statistics of errors between actual and BNN predicted values for the test data.

Variables	Minimum	Maximum	Mean	RMSE	Error Percentage
Prediction error	−11.02	12.14	1.04	6.82	15.4

3.4. Sulfur Removal Estimation Based on BPNN Model

The desulfurization of iron ore concentrate by the column flotation process is highly nonlinear and complicated [34]. The main objective of the current study was to combine image analysis methods with an intelligent predictive model such as NN to forecast the amount of sulfur removal from the iron ore concentrate.

As stated, the visual properties of the flotation froth were included in the input layer and the sulfur recovery in the output layer. The total 99 experimental data were randomly divided into three subsets as training, testing and validation. The training data (74 runs) were used by BPNN to learn how to map the inputs to the output by updating the network weights. The validation data (10 runs) were applied to assess the quality of the model. The ultimate check on the performance and generalization ability of the trained model was performed using testing data (15 runs).

BPNNs with at least one hidden layer are able to effectively estimate any function with the appropriate approximation, as long as the number of hidden neurons is appropriately determined. In the hidden layer, only the number of neurons determines the structure of the network and plays a major role in a network capability. Thus, if the number of neurons is small, the model does not accurately reflect the nonlinear mapping between inputs and outputs. On the other hand, if the number of middle layer neurons is too large, the model becomes overtrained and loses its generalizability [79,80]. To choose the best structure, the number of neurons in the hidden layer was changed to achieve minimum MSE.

The training settings of the BPNN model in this study are summarized as follows: number of input nodes: 17, number of hidden neurons: from 6 to 16, number of output nodes: 1, number of epochs: 1000. As stated in the literature [81], unlike changing the number of hidden nodes, changing the activation function does not have a significant effect on the model performance and results with similar MSE and R. The tangent sigmoid and linear activation functions were used in the hidden and output layers, respectively.

The BPNN model consisting of 10 neurons in the hidden layer gave the lowest MSE error among all the models studied. The network's performance in the training stage is shown in Figure 13. In the training stage, Mu first increased and then dropped, then fell to 0.01, and remained stationary, which shows that the model had reached its optimum state. The number of validation checks, which describes the number of consecutive iterations that the validation performance fails to reduce, was equal to six. In other words, if the number of validation checks reaches six, the training phase will stop.

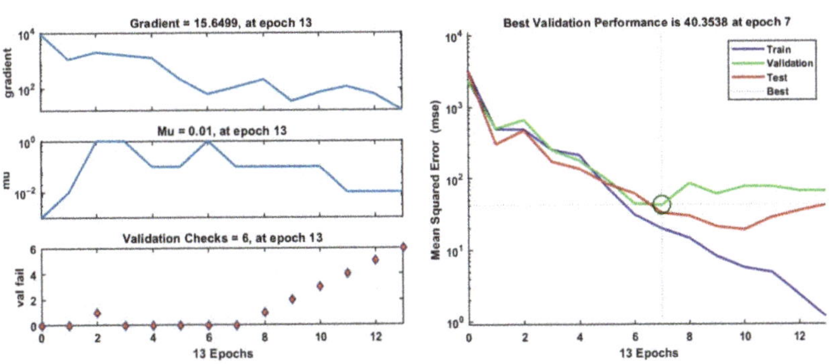

Figure 13. Training state and performance of the generated BPNN model.

Figure 13 shows the MSE variation of the training, validation and testing stages versus the iteration number. As shown in this figure, the large values for the MSE gradually reduced to a smaller value as the weights were updated. Training stage stopped at the seventh epoch, i.e., after epoch seven, there was no significant improvement in the performance of the model. The best validation performance was 48.98 at epoch seven, and after six error iterations (validation checks), the process stopped at epoch thirteen. In epoch seven, the MSE values for training and testing phases were 18.87 and 31.83, respectively, implying a good stable network behavior.

The comparison between the actual and predicted values of the sulfur recovery during training, validation, and testing stages is displayed in Figure 14. The dashed line shows the perfect result, i.e., outputs are equal to targets, while the solid line represents the best fit linear regression.

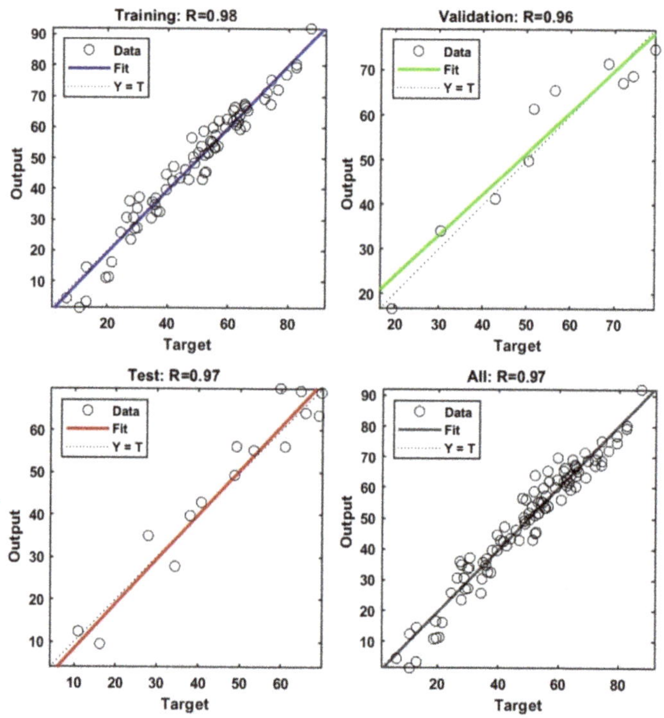

Figure 14. The regression plot of the data used in the training, validation and testing phases.

In the training stage, the predicted data obtained from the BPNN modeling were close to the actual data with a correlation of R = 0.98. Figure 14 shows the good fit attained in both the validation and testing of the proposed model (R values above 0.96). Despite the high variability and wide range of the output data, the overall R of the model was 0.97, which showed the appropriateness of the training, testing, and validity. From these comparison plots, it can be concluded that the BPNN is appropriately trained and shows consistency in forecasting sulfur recovery. The R values obtained from this BPNN model are higher than those reported in other studies in the literature for the prediction of metallurgical performance of the flotation process [19,36,82]. The novelty and superiority of the proposed BPNN model are its highest accuracy compared to the previously reported models to predict the metallurgical performance parameters of the flotation process.

The BPNN error histogram (differences between actual and predicted values) for the prediction of sulfur removal is illustrated in Figure 15. It can be observed that most of the errors were distributed between −1.43 and 2.54. Furthermore, the validation set and test set had similar behavior with no occurrence of overfitting. The comparison of the predicted recoveries with the measured values in the testing phase is shown in Figure 16. The values estimated by utilizing the BPNN model were very close to the actual measurements. The descriptive statistics of the errors in the testing stage (Table 10) confirmed that the BPNN model based on selected froth features was able to estimate the sulfur recovery quite precisely and satisfactorily. In the testing phase, the R, RMSE, and percent error values were obtained as 0.97, 4.84 and 11.29%, respectively, which confirmed a precise and robust prediction of the experimental data.

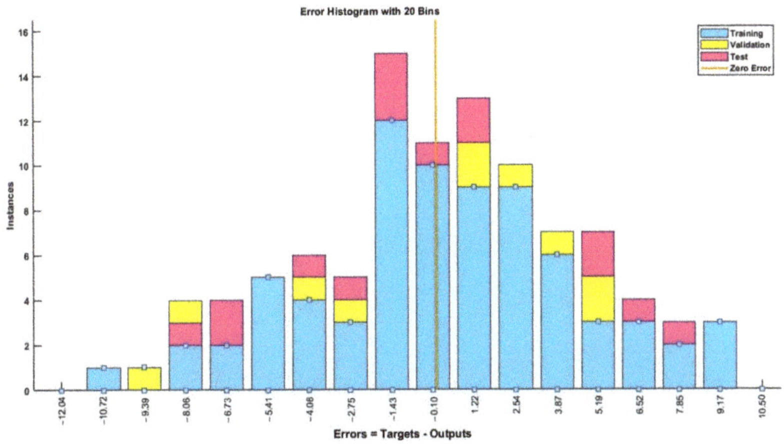

Figure 15. Error histogram with 20 bins for the training, validation and testing of BPNN for sulfur recovery prediction.

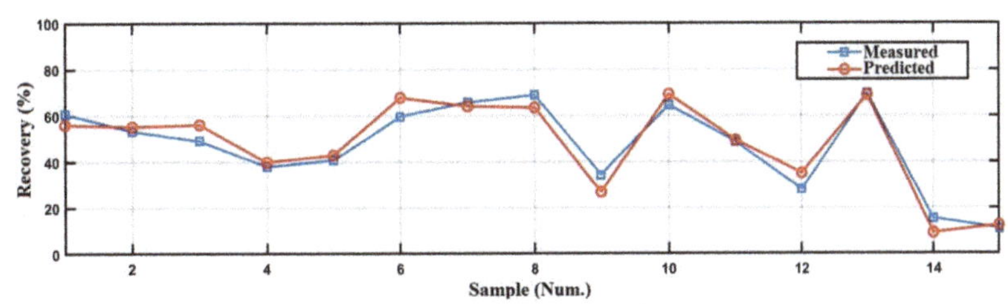

Figure 16. Comparison of measured and predicted values in the BPNN testing phase.

Table 10. Descriptive statistics of errors between actual and BPNN predicted values for the test data.

Variables	Minimum	Maximum	Mean	RMSE	Error Percentage
Prediction error	8.06	−7.59	0.57	4.84	11.29

It was demonstrated that a well-trained BPNN model based on froth features could be used to predict sulfur recovery from iron ore concentrate by column flotation without needing more experimental study requiring much time and high experiment costs.

3.5. *Sulfur Removal Estimation Based on CNN*

To classify the six groups of froth data, the ResNet18 deep learning classification method was used. The CNN has 71 layers containing input, convolutional, pooling, fully connected, ReLU, batch normalization, softmax and output layers. The layer structure of the proposed CNN is given in Figure 17.

	Name	Type	Activations	Learnable Properties
1	imageinput 256×512×3 images with 'zerocenter' nor...	Image Input	256(S) × 512(S) × 3(C) × 1(B)	-
2	conv1 64 7×7×3 convolutions with stride [2 2] a...	Convolution	128(S) × 256(S) × 64(C) × 1(B)	Weights 7 × 7 × 3 × 64 Bias 1 × 1 × 64
3	bn_conv1 Batch normalization with 64 channels	Batch Normalization	128(S) × 256(S) × 64(C) × 1(B)	Offset 1 × 1 × 64 Scale 1 × 1 × 64
4	conv1_relu ReLU	ReLU	128(S) × 256(S) × 64(C) × 1(B)	-
5	pool1 3×3 max pooling with stride [2 2] and pa...	Max Pooling	64(S) × 128(S) × 64(C) × 1(B)	-
⋮				
20	res3a_branch2a 128 3×3×64 convolutions with stride [2 2...	Convolution	32(S) × 64(S) × 128(C) × 1(B)	Weights 3 × 3 × 64 × 128 Bias 1 × 1 × 128
21	bn3a_branch2a Batch normalization with 128 channels	Batch Normalization	32(S) × 64(S) × 128(C) × 1(B)	Offset 1 × 1 × 128 Scale 1 × 1 × 128
22	res3a_branch2a_relu ReLU	ReLU	32(S) × 64(S) × 128(C) × 1(B)	-
⋮				
36	res4a_branch2a 256 3×3×128 convolutions with stride [2...	Convolution	16(S) × 32(S) × 256(C) × 1(B)	Weights 3 × 3 × 128 × 256 Bias 1 × 1 × 256
37	bn4a_branch2a Batch normalization with 256 channels	Batch Normalization	16(S) × 32(S) × 256(C) × 1(B)	Offset 1 × 1 × 256 Scale 1 × 1 × 256
38	res4a_branch2a_relu ReLU	ReLU	16(S) × 32(S) × 256(C) × 1(B)	-
⋮				
52	res5a_branch2a 512 3×3×256 convolutions with stride [2...	Convolution	8(S) × 16(S) × 512(C) × 1(B)	Weights 3 × 3 × 256 × 512 Bias 1 × 1 × 512
53	bn5a_branch2a Batch normalization with 512 channels	Batch Normalization	8(S) × 16(S) × 512(C) × 1(B)	Offset 1 × 1 × 512 Scale 1 × 1 × 512
54	res5a_branch2a_relu ReLU	ReLU	8(S) × 16(S) × 512(C) × 1(B)	-
⋮				
68	pool5 2-D global average pooling	2-D Global Average...	1(S) × 1(S) × 512(C) × 1(B)	-
69	fc 6 fully connected layer	Fully Connected	1(S) × 1(S) × 6(C) × 1(B)	Weights 6 × 512 Bias 6 × 1
70	prob softmax	Softmax	1(S) × 1(S) × 6(C) × 1(B)	-
71	classoutput crossentropyex	Classification Output	1(S) × 1(S) × 6(C) × 1(B)	-

Figure 17. Layer structure analysis for the CNN model training.

Figure 18 illustrates the confusion matrices of the CNN model in the classification process. A confusion matrix explains the estimation results of a classifier. The diagonal elements represent the number of correct predictions, while the off-diagonal elements represent incorrect estimations. This figure shows that the classification accuracy of the CNN for six froth classes was high, indicating that the model could provide an accurate prediction of the flotation performance (the amount of sulfur removal). The bottom and right values display the overall accuracy of the classification model. The prediction accuracy using this deep learning model was 90% which was much higher than the k-means algorithm.

A	5.72	0.17	0.17	0.00	0.00	0.00	94.44
B	0.00	9.60	0.34	0.17	0.00	0.00	95.00
C	0.00	0.17	12.29	0.34	0.34	0.00	93.59
D	0.00	0.17	0.34	14.14	1.01	0.51	87.50
E	0.00	0.00	0.34	2.19	33.67	2.19	87.72
F	0.00	0.00	0.00	0.67	0.84	14.65	90.63
	100.00	95.00	91.25	80.77	93.90	84.47	90.07
	A	B	C	D	E	F	

Figure 18. Confusion matrix of the CNN classifier.

The results showed that the CNN model was suitable for finding the process conditions of the column flotation in desulfurization of iron ore concentrate with high accuracy and efficiency. For example, if the froth images are categorized in Class A, Class B, or Class C, it is a sign that the current condition is abnormal. In this circumstance, the plant workers need to regulate the operating factors to achieve the desired performance.

3.6. Comparison of MLR, K-Means, BPNN and CNN Models

The results showed that all the proposed models, k-means, MLR, BPNN, and CNN, could be considered for sulfur removal prediction in the desulfurization of iron ore concentrate using flotation froth images. However, when different statistical results such as R, RSME, and percent error were examined, it was clear that the BPNN and CNN predicted values were more accurate than those of the MLR and k-means models.

The total accuracy of the CNN and k-means algorithms were 91% and 82%, respectively. In the k-means algorithm, regression and BPNN models, the froth features were firstly extracted offline through image analysis techniques and then they were entered into the models, whereas in the CNN model the froth images were directly fed into the model. Since, the traditional predictive models were built based on off-line feature extraction techniques, the training time for evaluation of the models was ignored. So, it can be concluded that the CNN and BPNN were able to categorize the froth images with high performance.

The BPNN and CNN models were found to be more successful where nonlinear and complex relationships were involved in the system, such as the desulfurization process of iron ore concentrate using the column flotation process. Thus, it was possible to establish a complex relationship between sulfur recovery and froth flotation features using the BPNN and CNN models with an excellent accuracy level.

It is important to note that sometimes BPNN does not represent the most cost-effective solution. In some cases, the interpretability of the network and weights may be difficult, the determination of the optimum structure and parameters may be troublesome, and the convergence of the training algorithm may be endless [81]. In these situations, MLR models can have advantages in terms of accuracy, variability, and model creation; therefore, its choice is preferred [83].

4. Conclusions

In the iron reverse flotation process, the amount of sulfur removal and the quality of flotation concentrate are usually judged according to the froth's surface appearance. Visual assessments by workers are challenging tasks and often inaccurate; in practice, they can only be performed at a given time. A machine vision system for monitoring and control of the operating conditions of the flotation process can be a great help for workers and iron and steel producers. Therefore, to forecast the amount of sulfur removal from the iron concentrate in the flotation process, soft-sensor models based on extracted froth features were proposed.

A total of 99 flotation column experiments were conducted in a wide range of different operating conditions and the froth features were extracted for each run. In this study, the abilities of four algorithms, including multiple linear regression (MLR), the k-means algorithm, the backpropagation neural network (BPNN), and the convolutional neural network (CNN) to predict the sulfur recovery, were examined and compared. In the first three models, the different froth features such as color, bubble shape and size, texture, froth stability, and velocity were primarily extracted offline through image processing techniques and then they were fed into the models, whereas in the CNN model the froth images were directly fed into the model.

The results showed that among the four algorithms implemented, the three-layer BPNN and the CNN models outperformed the MLR and k-means techniques when predicting sulfur recovery. The best NN model was a backpropagation network (topology 17-10-1) using a Levenberg–Marquardt algorithm for the training step, showing R = 0.97, RMSE = 4.84 and percent error = 11.29%. The total accuracy of the CNN and k-means algorithms were 91% and

80%, respectively. Therefore, the sulfur recovery, which is the most important parameter for control purposes of the flotation process, can be accurately predicted from the froth surface appearance by the BPNN and CNN models. Considering the accuracy and time consumed in measuring the metallurgical parameters of the flotation process, with the use of BPNN and CNN models, satisfactory results can be predicted rather than measured in the laboratory which thereby reduces the testing time and cost. As a result, it can be confirmed that such modelling studies pave the way to find an alternative to the on-stream analyzer, such as XRF, which is costly and not widely available.

In the future, the performance of the proposed models can be improved if more machine learning techniques are tested on larger datasets. Finally, future studies should concentrate on the validation and generalization of the proposed models in an industrial flotation process.

Author Contributions: Conceptualization, F.N.; methodology, F.N. and S.R.; software, F.N. and S.R.; data analysis, F.N.; writing—original draft preparation, F.N. and M.F.; writing—review and editing, F.N.; visualization, F.N.; supervision, F.N. All authors have read and agreed to the published version of the manuscript.

Funding: This research received no external funding.

Data Availability Statement: Not applicable.

Conflicts of Interest: The authors declare no conflict of interest.

References

1. Corin, K.C.; McFadzean, B.J.; Shackleton, N.J.; O'Connor, C.T. Challenges related to the processing of fines in the recovery of platinum group minerals (PGMs). *Minerals* **2021**, *11*, 533. [CrossRef]
2. Nakhaei, F.; Irannajad, M. Sulphur removal of iron ore tailings by flotation. *J. Dispers. Sci. Technol.* **2017**, *38*, 1755–1763. [CrossRef]
3. Nakhaei, F.; Irannajad, M.; Mohammadnejad, S. Comparison of conventional and column flotation performance for desulfurization of iron ore concentrate. *New Find. Appl. Geol.* **2019**, *13*, 55–66.
4. Jovanovíc, I.; Miljanovíc, I. Contemporary advanced control techniques for flotation plants with mechanical flotation cells—A review. *Miner. Eng.* **2015**, *70*, 228–249. [CrossRef]
5. Pu, Y.; Szmigiel, A.; Apel, D.B. Purities prediction in a manufacturing froth flotation plant: The deep learning techniques. *Neural Comput. Applic.* **2020**, *32*, 13639–13649. [CrossRef]
6. Jing, Z.; Jinling, N.; Yong, Z. Prediction of concentrate grade and recovery rate of tailings in the process of production based on chaotic ant colony algorithm. In Proceedings of the 2018 Chinese Control and Decision Conference (CCDC), Shenyang, China, 9–11 June 2018; pp. 5308–5313. [CrossRef]
7. Tang, M.; Zhou, C.; Zhang, N.; Liu, C.; Pan, J.; Cao, S. Prediction of the ash content of flotation concentrate based on froth image processing and bp neural network modeling. *Int. J. Coal Prep. Util.* **2021**, *41*, 191–202. [CrossRef]
8. Nakhaei, F.; Sam, A.; Mosavi, M.R.; Nakhaei, A. Prediction of XRF analyzers error for elements on-line assaying using Kalman Filter. *Int. J. Min. Sci. Technol.* **2012**, *22*, 595–601. [CrossRef]
9. Nakhaei, F.; Abdollahzade, A.; Irannejad, M. Presenting a method for examining the measurement system construction of mineral processing complicated circuits: Extracting observable variables and conducting mass balance. *J. Anal. Numer. Methods Min. Eng.* **2013**, *3*, 77–88.
10. Brooks, K.S.; Koorts, R. Model predictive control of a zinc flotation bank using online x-ray fluorescence analysers. *IFAC-Papers OnLine* **2017**, *50*, 10214–10219. [CrossRef]
11. Zhang, J.; Tang, Z.; Xie, Y.; Ai, M.; Gui, W. Convolutional memory network based flotation performance monitoring. *Miner. Eng.* **2020**, *151*, 106332. [CrossRef]
12. Jahedsaravani, A.; Marhaban, M.H.; Massinaei, M. Application of statistical and intelligent techniques for modeling of metallurgical performance of a batch flotation process. *Chem. Eng. Commun.* **2016**, *203*, 151–160. [CrossRef]
13. Vieira, S.; Sousa, J.; Durão, F. Fuzzy modelling strategies applied to a column flotation process. *Miner. Eng.* **2005**, *18*, 725–729. [CrossRef]
14. Bergh, L.G.; Yianatos, J.B. The long way toward multivariate predictive control of flotation processes. *J. Process Control* **2011**, *21*, 226–234. [CrossRef]
15. Shean, B.J.; Cilliers, J.J. A review of froth flotation control. *Int. J. Miner. Processing* **2011**, *100*, 57–71. [CrossRef]
16. Nakhaei, F.; Irannajad, M. Application and comparison of RNN, RBFNN and MNLR approaches on prediction of flotation column performance. *Int. J. Min. Sci. Technol.* **2015**, *25*, 983–990. [CrossRef]
17. Merma, A.G.; Olivera, C.A.C.; Hacha, R.R.; Torem, M.L.; Santos, B.F. optimization of hematite and quartz bioflotation by an artificial neural network (ANN). *J. Mater. Res. Technol.* **2019**, *8*, 3076–3087. [CrossRef]

18. Geng, Z.; Chai, T.; Yue, H. A method of hybrid intelligent optimal setting control for flotation process. In Proceedings of the Intelligent Control and Automation, 7th World Congress on IEEE, Chongqing, China, 25–27 June 2008; pp. 4713–4718.
19. Nakhaei, F.; Mosavi, M.R.; Sam, A.; Vaghei, Y. Recovery and grade accurate prediction of pilot plant flotation column concentrate: Neural network and statistical techniques. *Int. J. Miner. Processing* **2012**, *110–111*, 140–154.
20. Wang, J.S.; Han, S. Feed forward neural network soft-sensor modeling of flotation process based on particle swarm optimization and gravitational search algorithm. *Comput. Intel. Neurosc.* **2015**, *2015*, 1–10. [CrossRef]
21. Massinaei, M.; Sedaghati, M.R.; Rezvani, R.; Mohammadzadeh, A.A. Using data mining to assess and model the metallurgical efficiency of a copper concentrator. *Chem. Eng. Commun.* **2014**, *201*, 1314–1326. [CrossRef]
22. Al-Thyabat, S. Investigating the effect of some operating parameters on phosphate flotation kinetics by neural network. *Adv. Powder. Technol.* **2009**, *20*, 355–360. [CrossRef]
23. Nakhaei, F.; Irannajad, M.; Mohammadnejad, S. Application of image analysis systems in flotation process. *Soft Comput. J.* **2021**, *5*, 66–83.
24. Irannajad, M.; Soltanpour, R.; Nakhaei, F. Evaluation of factors affecting carrying capacity of laboratory flotation column treating copper sulfides. *Amirkabir J. Civ. Eng.* **2019**, *51*, 725–732.
25. Tan, J.; Liang, L.; Peng, Y.; Xie, G. The concentrate ash content analysis of coal flotation based on froth images. *Miner. Eng.* **2016**, *92*, 9–20. [CrossRef]
26. Li, Z.; Gui, W. The method of reagent control based on time series distribution of bubble size in a gold-antimony flotation process. *Asian J. Control* **2017**, *20*, 2223–2236. [CrossRef]
27. Nakhaei, F.; Irannajad, M.; Mohammadnejad, S.; Hajizadeh Omran, A. Sulfur content reduction of iron concentrate by reverse flotation. *Energy Sources Part A Recover. Util. Environ. Eff.* **2019**, 1–17. [CrossRef]
28. Morar, S.H.; Harris, M.C.; Bradshaw, D.J. The use of machine vision to predict flotation performance. *Miner. Eng.* **2012**, *36–38*, 31–36. [CrossRef]
29. Nakhaei, F.; Irannajad, M.; Mohammadnejad, S. A comprehensive review of froth surface monitoring as an aid for grade and recovery prediction of flotation process. Part A: Structural features. *Energy Sources Part A Recovery Util. Environ. Eff.* **2019**, 1–23. [CrossRef]
30. Bartolacci, G.; Pelletier, P.; Tessier, J.; Duchesne, C.; Bossé, P.A.; Fournier, J. Application of numerical image analysis to process diagnosis and physical parameter measurement in mineral processes—Part I: Flotation control based on froth textural characteristics. *Miner. Eng.* **2006**, *19*, 734–747. [CrossRef]
31. Ren, C.C.; Yang, J.G.; Liang, C. Estimation of copper concentrate grade based on color features and least-squares support vector regression. *Physicochem. Probl. Miner. Processing* **2015**, *51*, 163–172.
32. Xu, D.; Chen, X.; Xie, Y.; Yang, C.; Gui, W. Complex networks-based texture extraction and classification method for mineral flotation froth images. *Miner. Eng.* **2015**, *83*, 105–116. [CrossRef]
33. Núñez, F.; Cipriano, A. Visual information model-based predictor for froth speed control in flotation process. *Miner. Eng.* **2009**, *22*, 366–371. [CrossRef]
34. Nakhaei, F.; HajizadehOmran, A.; Irannajad, M.; Mohammadnejad, S. Relationship between froth behavior and operating parameters of flotation column. In Proceedings of the XXIX International Mineral Processing Congress (IMPC 2018), Moscow, Russia, 17–21 September 2018.
35. Hemmati Chegeni, M.; Abdollahy, M.; Khalesi, M.R. Bubble loading measurement in a continuous flotation column. *Miner. Eng.* **2016**, *85*, 49–54. [CrossRef]
36. Nakhaei, F.; Irannajad, M.; Mohammadnejad, S. Column flotation performance prediction: PCA, ANN and image analysis-based approaches. *Physicochem. Probl. Miner. Processing* **2019**, *55*, 1298–1310.
37. Lin, B.; Recke, B.; Knudsen, J.K.H.; Jørgensen, S.B. Bubble size estimation for flotation processes. *Miner. Eng.* **2008**, *21*, 539–548. [CrossRef]
38. Nakhaei, F.; Iranajad, M.; Mohammadnejad, S. Evaluation of column flotation froth behavior by image analysis: Effects of operational factors in desulfurization of iron ore concentrate. *Energy Sources Part A Recovery Util. Environ. Eff.* **2018**, *40*, 2286–2306. [CrossRef]
39. Kalyani, V.K.; Pallavika; Chaudhuri, S.; Charan, T.G.; Haldar, D.D.; Kamal, K.P.; Badhe, Y.P.; Tambe, S.S.; Kulkarni, B.D. Study of a laboratory-scale froth flotation process using artificial neural networks. *Miner. Process. Extr. Metall. Rev.* **2008**, *29*, 130–142.
40. Nakhaei, F.; Irannajad, M.; Yousefikhoshbakht, M. Flotation column performance optimisation based on imperialist competitive algorithm. *Int. J. Min. Miner. Eng.* **2016**, *7*, 1–17. [CrossRef]
41. Alsafasfeh, A.H.; Alagha, L.; Alzidaneen, A.; Nadendla, V.S.S. Optimization of flotation efficiency of phosphate minerals in mine tailings using polymeric depressants: Experiments and machine learning. *Physicochem. Probl. Miner. Processing* **2022**, *58*, 150477. [CrossRef]
42. Bu, X.; Zhou, S.; Danstan, J.K.; Bilal, M.; Hassan, F.U.; Chao, N. Prediction of coal flotation performance using a modified deep neural network model including three input parameters from feed. *Energy Sources Part A Recover. Util. Environ. Eff.* **2022**, 1–13. [CrossRef]
43. Wang, X.; Song, C.; Yang, C.; Xie, Y. Process working condition recognition based on the fusion of morphological and pixel set features of froth for froth flotation. *Min. Eng.* **2018**, *128*, 17–26. [CrossRef]

44. Stangierski, J.; Weiss, D.; Kaczmarek, A. Multiple regression models and Artificial Neural Network (ANN) as prediction tools of changes in overall quality during the storage of spreadable processed Gouda cheese. *Eur. Food Res. Technol.* **2019**, *245*, 2539–2547. [CrossRef]
45. Du, X.; Niu, D.; Chen, Y.; Wang, X.; Bi, Z. City classification for municipal solid waste prediction in mainland China based on K-means clustering. *Waste Manag.* **2022**, *144*, 445–453. [CrossRef] [PubMed]
46. Zhao, Y.; Peng, T.; Li, K. An improved K-means algorithm based recognition method for working condition of flotation process. *Int. J. Syst. Control Inf. Processing* **2017**, *2*, 113–126. [CrossRef]
47. Yassin, S.; Pooja, S. Road accident prediction and model interpretation using a hybrid K-means and random forest algorithm approach. *SN Appl. Sci.* **2020**, *2*, 1576. [CrossRef]
48. Abo-Elnaga, Y.; Nasr, S. K-means cluster interactive algorithm-based evolutionary approach for solving bilevel multi-objective programming problems. *Alex. Eng. J.* **2022**, *61*, 811–827. [CrossRef]
49. Jansson, N.F.; Allen, R.L.; Skogsmo, G.; Tavakoli, S. Principal component analysis and K-means clustering as tools during exploration for Zn skarn deposits and industrial carbonates, Sala area, Sweden. *J. Geochem. Explor.* **2022**, *233*, 106909. [CrossRef]
50. Li, Z.; Gui, W.; Zhu, J. Fault detection in flotation processes based on deep learning and support vector machine. *J. Cent. South Univ.* **2019**, *26*, 2504–2515. [CrossRef]
51. Zhang, W.; Jin, L.; Song, E.; Xu, X. Removal of impulse noise in color images based on convolutional neural network. *Appl. Soft Comput.* **2019**, *82*, 105558. [CrossRef]
52. Shi, C.; Tan, C.; Wang, T.; Wang, L.A. Waste classification method based on a multilayer hybrid convolution neural network. *Appl. Sci.* **2021**, *11*, 8572. [CrossRef]
53. Fu, Y.; Aldrich, C. Froth image analysis by use of transfer learning and convolutional neural networks. *Miner. Eng.* **2018**, *115*, 68–78. [CrossRef]
54. Wen, Z.; Zhou, C.; Pan, J.; Nie, T.; Zhou, C.; Lu, Z. Deep learning-based ash content prediction of coal flotation concentrate using convolutional neural network. *Miner. Eng.* **2021**, *174*, 107251. [CrossRef]
55. Nakhaei, F.; Irannajad, M.; Mohammadnejad, S. The effect of column flotation operational variables on desulfurization of iron ore concentrate. *J. Anal. Numer. Methods Min. Eng.* **2020**, *9*, 43–58.
56. Nakhaei, F.; Irannajad, M. Reagents types in flotation of iron oxide minerals: A review. *Miner. Processing Extr. Metall. Rev.* **2018**, *39*, 89–124. [CrossRef]
57. Liu, X.H.; Liao, Z.H.; Yan, X.H.; Chen, W. Comprehensive recovery of magnetite and pyrrhotite from a low-grade iron ore. *Min. Metall. Eng.* **2014**, *34*, 47–51.
58. Xu, R.; Wunsch, D. Survey of clustering algorithms. *IEEE Trans. Neural Netw.* **2005**, *16*, 645–678. [CrossRef]
59. Miri, S.E.; Azadi, M.; Pakdel, S. Development of a duty cycle with K-means clustering technique for hydraulic steering in an instrumented TIBA vehicle. *Transp. Eng.* **2022**, *8*, 100114. [CrossRef]
60. Rodriguez, A.; Laio, A. Clustering by fast search and find of density peaks. *Science* **2014**, *344*, 1492–1496. [CrossRef]
61. Minh, H.L.; To, T.S.; Wahab, M.A.; Le, T.C. A new metaheuristic optimization based on K-means clustering algorithm and its application to structural damage identification. *Knowl. -Based Syst.* **2022**, *251*, 109189. [CrossRef]
62. Yuan, K.; Chi, G.; Zhou, Y.; Yin, H. A novel two-stage hybrid default prediction model with k-means clustering and support vector domain description. *Res. Int. Bus. Financ.* **2022**, *59*, 101536. [CrossRef]
63. Agha, S.R.; Alnahhal, M.J. Neural network and multiple linear regression to predict school children dimensions for ergonomic school furniture design. *Appl. Ergon.* **2012**, *43*, 979–984. [CrossRef] [PubMed]
64. Abrougui, K.; Gabsi, K.; Mercatoris, B.; Khemis, C.; Amami, R.; Chehaibi, S. Prediction of organic potato yield using tillage systems and soil properties by artificial neural network (ANN) and multiple linear regressions (MLR). *Soil Tillage Res.* **2019**, *190*, 202–208. [CrossRef]
65. Zhang, D.; Gao, X. Soft sensor of flotation froth grade classification based on hybrid deep neural network. *Int. J. Prod. Res.* **2021**, *59*, 4794–4810. [CrossRef]
66. Fu, Y.; Aldrich, C. Flotation Froth Image Recognition with Convolutional Neural Networks. *Miner. Eng.* **2019**, *132*, 183–190. [CrossRef]
67. Horn, Z.C.; Auret, L.; McCoy, J.T.; Aldrich, C.; Herbst, B.M. Performance of convolutional neural networks for feature extraction in froth flotation sensing. *IFAC-Pap.* **2017**, *50*, 13–18. [CrossRef]
68. Nandi, S.; Ghosh, S.; Tambe, S.S.; Kulkarni, B.D. Artificial neural–network-assisted stochastic process optimization strategies. *AICHE J.* **2001**, *47*, 126–141. [CrossRef]
69. Zhang, N.N.; Zhou, C.C.; Pan, J.H.; Xia, W.; Liu, C.; Tang, M.C.; Cao, S.S. The response of diasporic-bauxite flotation to particle size based on flotation kinetic study and neural network simulation. *Powder Technol.* **2017**, *318*, 272–281. [CrossRef]
70. Çakman, G.; Gheni, S.; Ceylan, S. Prediction of higher heating value of biochars using proximate analysis by artificial neural network. *Biomass Conv. Bioref.* **2021**, 1–9. [CrossRef]
71. Lee, K.M.; Zanil, M.F.; Chan, K.K.; Chin, Z.P.; Liu, Y.C.; Lim, S. Synergistic ultrasound-assisted organosolv pretreatment of oil palm empty fruit bunches for enhanced enzymatic saccharification: An optimization study using artificial neural networks. *Biomass Bioenergy* **2020**, *139*, 105621. [CrossRef]

72. Bu, X.; Xu, Y.; Zhao, M.; Li, D.; Zou, J.; Wang, L.; Bai, J.; Yang, Y. Simultaneous extraction of polysaccharides and polyphenols from blackcurrant fruits: Comparison between response surface methodology and artificial neural networks. *Ind. Crops Prod.* **2021**, *170*, 113682. [CrossRef]
73. Samani, B.H.; Samani, M.A.; Shirneshan, A.; Fayyazi, E.; Najafi, G.; Rostami, S. Evaluation of an enhanced ultrasonic-assisted biodiesel synthesized using safflower oil in a diesel power generator. *Biofuels* **2020**, *11*, 523–532. [CrossRef]
74. Wang, W.; Liu, Y.; Bai, F.; Xue, G. Capture power prediction of the frustum of a cone shaped floating body based on BP neural network. *J. Mar. Sci. Eng.* **2021**, *9*, 656. [CrossRef]
75. Lu, F.; Liang, Y.; Wang, X.; Gao, T.; Chen, Q.; Liu, Y.; Zhou, Y.; Yuan, Y.; Liu, Y. Prediction of amorphous forming ability based on artificial neural network and convolutional neural network. *Comput. Mater. Sci.* **2022**, *210*, 111464. [CrossRef]
76. Tegegne, A.M. Applications of Convolutional Neural Network for Classification of Land Cover and Groundwater Potentiality Zones. *J. Eng.* **2022**, *2022*, 6372089. [CrossRef]
77. Asim, Y.; Shahid, A.R.; Malik, A.K.; Raza, B. Significance of machine learning algorithms in professional blogger's classification. *Comput. Electr. Eng.* **2018**, *65*, 461–473. [CrossRef]
78. Batabyal, A.K. Correlation and multiple linear regression analysis of groundwater quality data of Bardhaman District, West Bengal, India. *Int. J. Res. Chem. Environ.* **2014**, *4*, 42–51.
79. Charan, T.G.; Kalyani, V.K.; Sharma, K.K.; Kumar, L.; Kumari, S.; Sinha, A. Use of an artificial neural network to evaluate the oleo-flotation process to treat coal fines. *Int. J. Coal Prep. Util.* **2014**, *34*, 229–238. [CrossRef]
80. Pan, X.; Lee, B.; Zhang, C. A comparison of neural network backpropagation algorithms for electricity load forecasting. In Proceedings of the IEEE International Workshop on Intelligent Energy Systems (IWIES), Vienna, Austria, 14 November 2013.
81. Pereira, A.A.C.; Olivera, C.A.C.; Merma, A.G.; Hacha, R.R.; Santos, B.F.; Torem, M.L. Mineral bioflotation optimization: Comparison between artificial neural networks and response surface methodology. *Miner. Eng.* **2021**, *169*, 106983. [CrossRef]
82. Jahedsaravani, A.; Marhaban, M.H.; Massinaei, M. Prediction of the metallurgical performances of a batch flotation system by image analysis and neural networks. *Miner. Eng.* **2014**, *69*, 137–145. [CrossRef]
83. Paliwal, M.; Kumar, U.A. Neural networks and statistical techniques: A review of applications. *Expert Syst. Appl.* **2009**, *36*, 2–17. [CrossRef]

minerals

Article

Comparison of Fuzzy and Neural Network Computing Techniques for Performance Prediction of an Industrial Copper Flotation Circuit

Ivana Jovanović [1,*], Fardis Nakhaei [2,*], Daniel Kržanović [1], Vesna Conić [1] and Daniela Urošević [1]

1 Mining and Metallurgy Institute Bor, Zeleni Bulevar 35, 19210 Bor, Serbia
2 School of Chemical and Minerals Engineering, North West University, Potchefstroom 2531, South Africa
* Correspondence: ivana.jovanovic@irmbor.co.rs (I.J.); 36598704@mynwu.ac.za (F.N.)

Abstract: This paper presents the development and validation of five different soft computing methods for flotation performance prediction: (1) two models based on fuzzy logic (Mamdani and Takagi-Sugeno fuzzy inference system) and (2) three models based on artificial neural networks. Copper content in the ore feed, collector dosage in the rougher and the scavenger flotation circuits, slurry pH in the rougher flotation circuit and frother consumption were selected as input parameters to estimate the copper grade and recovery of final concentrate, as well as the copper content in the final tailings of the flotation plant. The training and evaluation of the proposed models were performed on the basis of real process data collected by the multiannual monitoring of industrial flotation plant operation in "Veliki Krivelj Mine". The results showed that the proposed soft computing-based models well describe the behavior of the industrial flotation plant in a wide range of circumstances. Among the proposed algorithms, artificial neural networks gave the most accurate predictions for the final copper concentrate grade and recovery (R^2 = 0.98 and R^2 = 0.99, respectively) and copper content in final tailings (R^2 = 0.87). At some points, fuzzy logic models are almost equally efficient, but artificial neural networks had lower values for all error functions.

Keywords: copper flotation; fuzzy logic; artificial neural network; mathematical modeling

Citation: Jovanović, I.; Nakhaei, F.; Kržanović, D.; Conić, V.; Urošević, D. Comparison of Fuzzy and Neural Network Computing Techniques for Performance Prediction of an Industrial Copper Flotation Circuit. *Minerals* **2022**, *12*, 1493. https://doi.org/10.3390/min12121493

Academic Editor: Hyunjung Kim

Received: 15 September 2022
Accepted: 16 November 2022
Published: 23 November 2022

1. Introduction

In the mining industry, the separation of valuable minerals from raw ores is carried out using the flotation process to obtain qualified concentrate and to eliminate tailings. Recovery and grade of the concentrate are the important metallurgical factors of the flotation process. In an industrial flotation plant, the online prediction of the metallurgical factors is costly and time-consuming. Since the flotation is a nonlinear process, modeling, automatic monitoring and control of the industrial plants based on the metallurgical factors have met with limited success at the present time. Furthermore, the process models based on classical experimental methods do not provide sufficiently efficient results when it comes to such complex systems [1]. To overcome these difficulties, the use of statistical and artificial intelligence methods for control and monitoring purposes of the flotation process has been developed [2]. Therefore, soft sensors based on problem-solving technologies such as fuzzy logic and neural networks (NNs) emerged as perspective alternatives to the classical modeling approaches. These methods, unlike the conventional mathematical techniques, exhibit a certain tolerance to the imprecision and uncertainty of technological parameters in the description of real systems [3–6].

The most commonly used soft computing methods in the modeling of flotation process are different types of NNs due to their acceptable accuracy, robustness, simplicity and nonlinearity [7–9]. NNs use very powerful computational techniques for modeling the complex nonlinear processes. Among them, a frequently used NN is a multilayer perceptron (MLP) based on which models for predicting various flotation parameters have been developed. For example, Al-Thyabat [10] examined the suitability of different MLP architectures by considering the influence of three input parameters (mean particle size in the feed slurry, collector dosage and impeller rotation speed) on the grade and recovery of the concentrate obtained by the phosphate flotation process. Similarly, Farghaly et al. [11] used an artificial neural network (ANN) model to investigate the effect of flotation time, collector dosage, frother dosage and impeller speed on the grade and recovery of coal concentrate. Jorjani et al. [12] formed two MLP architectures (7-10-3-1) and (5-30-2-1), for predicting the recovery and grade of the coal concentrate, respectively.

Nakhaei et al. [13–15] proposed several ANN models for predicting the contents and recoveries of copper and molybdenum in the column flotation concentrate. The authors concluded that NNs proved to be a superior technique compared to the regression models. The same conclusion was obtained by other researchers comparing the statistical methods and techniques based on MLP for the modeling of copper minerals' flotation process [16,17]. More recently, the researchers proposed ANN models for the prediction of the phosphate concentrate grade [18], and removal Cu (II) ion in flotation systems [19].

Some authors have also considered the application of other types of NNs. For example, Gholami et al. [20] used two types of recurrent neural networks (RNN)—Long short-term memory network (LSTM) and Gated recurrent unit network (GRU)—to predict the grade and recovery of copper concentrate within different operating conditions. They compared the results obtained by the RNN with models based on classical NN and random forest methods. They demonstrated that the RNN models have a better prediction ability. The potential of RNNs was also considered by Nakhaei and Irannajad [15], where they demonstrated good abilities to predict the content and recovery of Cu and Mo in the final concentrate obtained in the flotation column (correlation coefficient greater than 0.8).

There are also numerous studies for the application of NN in the models related to the identification, categorization and interpretation of flotation froth images [21–30]. These models typically use extracted features from flotation froth images. In recent times, convolutional neural networks are increasingly being used to model the flotation process, especially when it comes to the classification and feature recognition of froth images [31–35].

Fuzzy logic supports the ability of the human mind to effectively express a way of reasoning that is more approximate than exact. It is a computation and reasoning system where the objects of computation and reasoning are classes with fuzzy limitations. Fuzzy logic system allows analysis for modeling complex and poorly defined systems in which linguistic expressions are used rather than numerical variables [36].

Fuzzy inference is the process of formulating the mapping from a given input to an output using fuzzy logic. The fuzzy logic generally presents a simple knowledge of the complex and non-linear process in terms of if-then precise rules with different matching degrees for a given operational situation [37,38]. This is especially valuable where models are developed based on expert's knowledge and individuals without a mathematical background are involved. There are two kinds of fuzzy inference systems that can be implemented in the Fuzzy Logic Toolbox: Mamdani and Sugeno types.

Fuzzy logic has been increasingly used in different mineral processing fields with a variety purpose, such as crushing [39], grinding [40], flotation [4,41] and sample preparation [42] systems. Attempts to model flotation processes by a fuzzy logic approach are numerous, whether it is about models related to the entire flotation system or to specific subsystems that are its integral part.

Jahedsaravani et al. [16] used NN, adaptive neuro-fuzzy and regression methods for modeling the metallurgical factors in the batch flotation process. Their results indicated that intelligent techniques outperform statistical approaches. Carvalho and Durão [43] proposed a model based on Mamdani's fuzzy inference system within the control system for stabilizing the operation of the flotation column in a two-phase system (air and water). In another study, they more fully described the implementation of this model in the control system of a pilot flotation plant [44]. Vieira et al. [45] employed a multiple-input and multiple-output (MIMO) model to identify the performance of a flotation column using the fuzzy modeling technique, where the Takagi-Sugeno inference system was applied.

Liao et al. [46] created a fuzzy inference system within a slurry level control system in a flotation column. They considered the position of the valve that regulates the slurry level, depending on the error between the reference and measured slurry level and the change rate of that error. Jahedsaravani et al. [47] developed a fuzzy model that simulates the relationship between process conditions (air velocity, slurry solids content, frother dosage and type) and metallurgical process parameters (grade and recovery) in an industrial copper flotation column. This model was incorporated into an intelligent system for controlling the flotation process. The simulation results demonstrated that the proposed model was able to maintain the process performance at the desired values during an acceptable period of time.

Recently, Zhou and Zhou [48] proposed a fuzzy logic-based methodology for bubble edge detection in flotation froth images. Liang et al. [49] used a fuzzy logic model to optimize the configuration of flotation circuit. Gao et al. [35] applied several soft computing methodologies to develop a system that recognizes and classifies flotation froth images, including fuzzy logic.

The importance of process modeling is reflected in several technological and production aspects of flotation concentration. For example, models can be used for data analysis, which, from the aspect of the complexity of the technological process, would not be easy to achieve. Ali [50] provides several advantages that can be obtained by mathematical modeling, optimization and flotation process simulation, such as: improvement of knowledge management, better understanding of current problems, enhancement of technology transfers, decision-making support systems for plant personnel, improvements in plant work conditions, improvement in product quality, the reduction of potentially hazardous experiments, reduction of waste in process development, etc. Proper prediction of metallurgical performance factors of the flotation circuit leads to optimum production and high profit margin for industrial mineral companies.

Despite the popularity of the flotation process, and the numerous published studies, comprehensive and robust model that can represent the relationship between the operation variables (such as grinding fineness, reagent regime, slurry pH and density, residence time and similar) and performance parameters (usually concentrate quality and recovery of useful components) of the whole industrial flotation circuit is a challenging task.

Hence, finding a robust and more accurate prediction method for estimating the metallurgical performance parameters of an industrial flotation circuit is still necessary. Therefore, this work aims to assess and compare the performance of the two predictive methods that are ANN and fuzzy logic and to reveal the relevant approach for predicting the metallurgical parameters in an industrial copper flotation plant. These models are applied to a large set of industrial flotation circuit data and the results are compared in terms of accuracy, complexity and suitability for control of the flotation process. For this purpose, a real large dataset (obtained by multiannual monitoring) of the industrial copper flotation plant (Veliki Krivelj) is used. The final copper grade and recovery of the flotation circuit concentrate along with copper content in final tailings are predicted depending on the five effective variables, which are copper content in the ore feed, collector dosage in the rougher and the scavenger flotation lines, pulp pH in the rougher line and frother consumption.

The purpose of modeling the production process in Veliki Krivelj is the possibility of implementing the obtained models into an automatic control system of this process. This system would include the application of controllers based on soft computing methods, as a form of decision support for operators (fuzzy logic controller) or as an independent control unit (ANN controller). Classic PID controllers, in this case, would be used at lower hierarchical levels of control to regulate certain parameters such as: pulp level in flotation cells, air flow, pH value, value of electrochemical potential (Eh) and the like.

2. Materials and Methods

2.1. Input and Output Variables

2.1.1. Brief Description of the Veliki Krivelj Flotation Plant

The flow sheet of the Veliki Krivelj flotation circuit is shown in Figure 1. In the comminution plant, the output of three-stage crushing is fed to the two-stage grinding trains (rod mill and ball mill) in the closed circuit with cyclones to produce 58% of the product finer than 74 μm. The product of the grinding circuit is separately fed to two lines of roughing flotation cells. The first roughing flotation bank consists of 2 × 16 cells, and the second one of 3 × 21 cells. The final copper concentrate is produced by three stages of flotation cleaning. The first, second and third cleaning banks have 2 × 9, 1 × 8 and 1 × 18 cells, respectively. There is a scavenging flotation stage after the first cleaning. The scavenging bank consists of 8 cells. The combined concentrates of the roughing and scavenging lines are ground by using a regrinding mill. The tailings of the roughing and scavenging cells are discarded as the final tails. In this flotation circuit, Potassium ethyl xanthate (PEX) is used as collector and Dowfroth 250 (D-250) is used as frother. After the flotation stages, a copper concentrate was produced with an average grade and recovery of 18%–22% and 80%–90%, respectively.

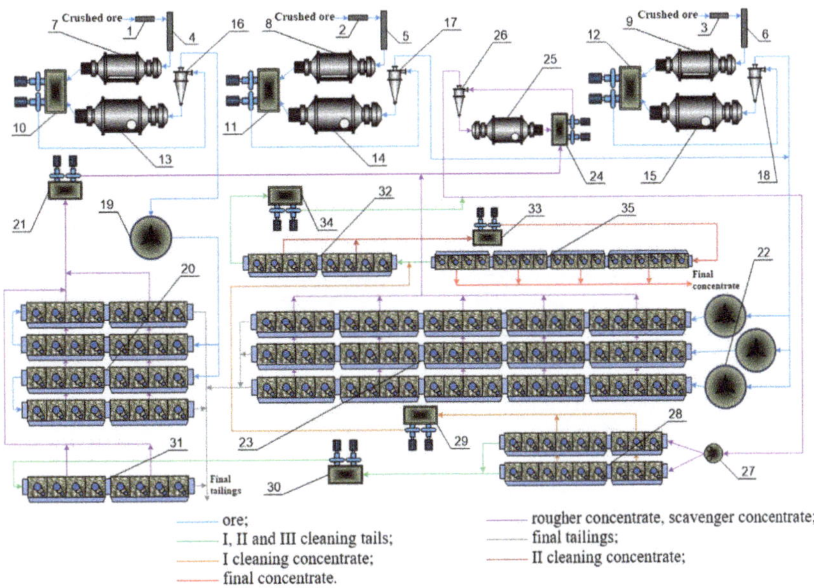

Figure 1. Process flow diagram of Veliki Krivelj flotation plant: 1–6—conveyor belts; 7–9—rod mills; 10–12, 21, 24, 29, 30, 33, 34—centrifugal slurry pumps; 13–15, 25—ball mills; 16–18, 26—hydrocyclones; 19, 22—agitators; 20, 23—roughing flotation cells; 27—divider; 28—first cleaning cells; 31—scavenging cells; 32—second cleaning cells; 35—third cleaning cells (Adapted from [4]).

2.1.2. Data Collection

The flotation process is characterized by a large number of influential parameters, i.e., input variables that influence the course of the process. All data (the input and output parameters of the model) were collected by daily/shiftly process monitoring during the multiannual plant's operation. The samples on which the copper content is determined—feed ore, final concentrate and final tailings—were taken at every hour. From the hourly increments, shift composite samples are formed (one shift lasts 8 h). On these, composites chemical analyses for copper are performed. The average consumption of reagents as well as the pulp pH value during one shift are read from the control room. More details about process parameters measurement and data collection are given in Appendix A.

Furthermore, Appendix B contains information about all datasets that were used in the process of model development and evaluation.

2.1.3. Selection of Variables

For the purposes of developing the process models, during the selection of the appropriate influential parameters, their importance was taken into account on the basis of expert analysis, as well as the availability of appropriate data. The key input variables are shown in Table 1. Other variables are considered mainly at their constant values as shown in Table 2. The output parameters of the models are shown in Table 3.

Table 1. Input (independent) parameters.

Type of Variable	Label in the Model	Unit of Measurement
Copper content in the feed	FCU	%
Collector consumption at rougher flotation circuit	PXR	g/t
Frother consumption	FRT	g/t
pH value of slurry at rougher flotation circuit	PHR	-
Collector consumption at scavenger flotation circuit	PXS	g/t

Table 2. Input variables that are considered constant.

Parameter	Value	Unit of Measurement
Grinding fineness	58	% of class $-74 + 0$ μm
Regrinding fineness	85	% of class $-74 + 0$ μm
Pulp density in rougher flotation circuit	1190	g/L
Pulp density in 1st, 2nd and 3rd cleaning	1150, 1130 and 1125	g/L
Pulp density in scavenger flotation circuit	1120	g/L
pH value in 1st, 2nd and 3rd cleaning	11.5, 11.8 and 12	-
pH value in scavenger flotation circuit	11.5	-
Residence time of rougher flotation circuit	21	minutes
Residence time of 1st, 2nd and 3rd cleaning	10, 20 and 19	minutes
Residence time of scavenger flotation circuit	10	minutes

Table 3. Output (dependent) parameters.

Type of Variable	Label in the Model	Unit of Measure
Copper content in the final concentrate	CCU	%
Copper content in the final tailings	TCU	%
Copper recovery in the final concentrate	RCU	%

The reasons for choosing constant variables are varied. For example, the liberation of the mineral raw material, i.e., its particle size distribution after grinding and regrinding is an important influencing parameter in the process. However, the analysis of the particle size distribution on the sieves is not performed daily, but as needed, every few days. The monitoring of the particle size distribution of the raw material mainly takes place on the mining pan, so there was not enough information about the changes of this parameter during each shift. Furthermore, when it comes to pulp density, there are shiftly data on the density of hydrocyclone overflows at grinding and regrinding. However, it should be taken into account that water is added to the roughing flotation cells, as well as to each cleaning stage, with the aim of correcting the pulp density. Since there is no continuous monitoring of pulp density in flotation cells, data on changes in pulp density in any of the stages of flotation concentration were not available.

By analysis of data about pH values of the hydrocyclone overflow at the regrinding, it was observed that they generally range between 9 and 10. Given that the first cleaning takes place at a pulp pH of approximately 11.5, it is clear that in the stage of the first cleaning, with addition of lime milk, was almost always necessary to achieve the desired pH value. Consequently, although there is regular monitoring and record keeping of pH values of the hydrocyclone overflow at the regrinding, these data cannot be considered confident for process model development. Choosing the consumption of the titrant in the cleaning stages as input variables makes more sense; however, from the automatic process control and regulation point of view, it is much more efficient to monitor pulp pH values (because they can be directly and continuously measured). Bearing in mind this fact, it was decided not to take the titration values into account in the modeling, but to consider the pH values of the pulp at cleaning and scavenging as constants.

It should also be noted that one of the criteria when choosing the number of input variables was the complexity of the fuzzy logic model. Specifically, when the number of input parameter increases, the number of fuzzy rules will increase exponentially [51]. This significantly increases the number of calculation and response time of fuzzy system. Therefore, the decision was to adopt an optimally small number of input parameters in the model.

In order to gain a better insight into the range and characteristics of the input and output variables, Table 4 presents some of the statistical indicators of the associated data sets.

Table 4. Statistical indicators of the input and output variables in the flotation system.

Statistical Indicator	Input Variables					Output Variables		
	FCU	PXR	FRT	PHR	PXS	CCU	TCU	RCU
Minimum	0.12	10.00	1.02	8.44	2.50	7.91	0.009	40.78
Maximum	0.51	49.98	16.97	11.97	7.90	28.09	0.149	96.48
Mean	0.26	32.27	6.19	10.40	5.38	19.24	0.041	84.24
Mod	0.26	33.50	3.80	9.85	5.00	18.48	0.035	90.77
Median	0.26	33.30	5.87	10.45	5.25	19.34	0.038	85.18
Standard deviation	0.046	5.046	2.319	0.645	0.774	3.129	0.017	6.751
Confidence interval	0.002	0.226	0.104	0.029	0.035	0.140	0.001	0.303

2.2. Methodology of Fuzzy System

Fuzzy logic is a mathematical rule-based system in which two human capabilities, which are the reasoning ability and the ability to fulfill different mental tasks, are tried to be mechanized with IF–THEN rules [52,53]. Fuzzy modeling uses linguistic expressions instead of numerical variables. The typical fuzzy logic architecture comprises four parts: 1. fuzzification, 2. fuzzy rule base, 3. fuzzy inference system (FIS) and 4. defuzzification, as it is illustrated in Figure 2 [54].

To implement a fuzzy rule-based model, the following stages are needed. First, the input and output variables are determined. Second, the fuzzy sets are determined for all variables. Within the fuzzification process, membership functions were determined for each of the input and output variable. In practice, different types of membership functions are used (such as triangular, trapezoidal, Gaussian, sigmoidal, polynominal, etc.), and the selection of the form of the membership functions will rely on the suitability of representing the element belonging to a given set and expert's decision. In this research, several membership functions—triangular, trapezoidal, Gaussian and sigmoidal right—are taken into consideration, and the most appropriate membership functions: Gaussian, trapezoidal and constant (Sugeno) were chosen. According to expert analysis, these functions gave the response surfaces most corresponding to the real process.

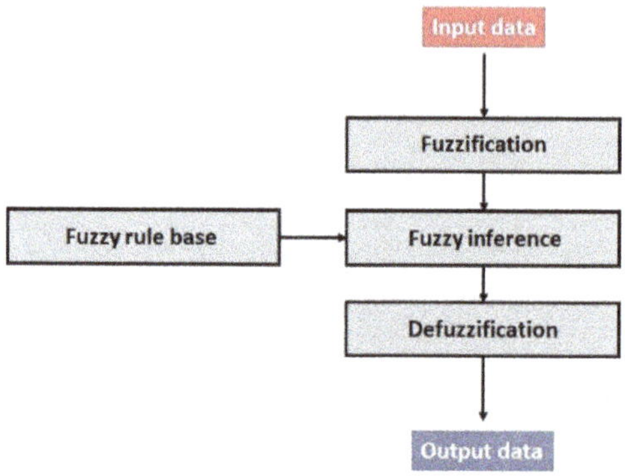

Figure 2. Fuzzy logic system [54].

FIS is the heart of the fuzzy logic system in which the classical numeric values are translated to linguistic forms. These linguistic forms are related to membership function which assigns them a membership degree (always in the interval between 0 and 1) [55]. FIS is known as a decision-making platform, which syndicates fuzzification facts with a rule base and performs the fuzzy reasoning process [56]. FIS includes membership functions, IF-THEN rules and other logical operations. IF-THEN rules are made to describe the association between input and output variables [53]. In this work, a set of the linguistic variables of a fuzzy model are specified as Low, Medium and High (see Table 5).

For the purpose of the current research, the two most common FIS types were established: the Mamdani fuzzy inference system marked as PMM and the other based on the Takagi Sugeno fuzzy inference system, marked as PSM. There are some differences between them. The PSM model is made by the appropriate transformation of the PMM model. The output of the Sugeno is linear or constant, but the output of Mamdani is the membership function. The final stage is the defuzzification in which the fuzzy results are translated to crispy form values. Some of the defuzzification techniques are the mean of the maximum, centroid, smallest of the maximum, and largest of the maximum.

In this study, the built fuzzy models are established, trained and authenticated to predict the flotation performance parameters. The input parameters are FCU, PXR, FRT, PHR, PXS, while CCU, TCU, RCU are considered as the output responses. The fuzzy models are constructed by using a fuzzy toolbox of MATLAB®. The fuzzification of the variables is performed based on the experiential knowledge of the flotation process as well as the analysis of the collected data. During the training stage, the values of the fuzzy numbers within the membership functions were adjusted. For each membership function, an appropriate type and range are determined and assigned.

In order to see more clearly the type and range of the used membership functions, in the following text, the mathematical formulas describing trapezoidal and Gaussian functions are given. Trapezoidal membership function (μ_T) is specified by four parameters (a, b, c, d) as follows:

$$\mu_T(x; a,\ b,\ c, d) = \left\{ \begin{matrix} 0,\ x \le a \\ \frac{x-a}{b-a},\ a \le x \le b \\ 1,\ b \le x \le c \\ \frac{d-x}{d-c},\ c \le x \le d \\ 0,\ d \le x \end{matrix} \right\} \quad (1)$$

where the parameters a, b, c and d (with $a < b \le c < d$) determine the x coordinates of the four corners of the underlying trapezoidal membership function. An illustration of this membership function is given in Figure 3a.

Table 5. Fuzzification of the range of variables in the PMM and PSM.

Variable		Membership Function	Range of Values	Range Fuzzification (Linguistic Value of the Variable)
Input (independent)	FCU	Gaussian	σ/2 = 0.04778; c = 0.112 σ/2 = 0.07976; c = 0.3008 σ/2 = 0.02871; c = 0.492	Low Medium High
	PXR	Gaussian	σ/2 = 4.743; c = 10.5 σ/2 = 6.37; c = 30.0 σ/2 = 4.678; c = 49.0	Low Medium High
	FRT	Gaussian	σ/2 = 1.362; c = 0.325 σ/2 = 1.486; c = 6.5 σ/2 = 5.335; c = 17.4	Low Medium High
	PHR	Gaussian	σ/2 = 0.779; c = 8.08 σ/2 = 0.598; c = 9.989 σ/2 = 1.071; c = 11.88	Low Medium High
	PXS	Gaussian	σ/2 = 1.225; c = 2.2 σ/2 = 0.7887; c = 6.0 σ/2 = 0.3802; c = 7.9	Low Medium High
Output PMM	CCU	Trapezoidal	a = 3.439; b = 6.845; c = 8.015; d = 13.98 a = 10.16; b = 12.36; c = 15.28; d = 21.66 a = 19.78; b = 24.18; c = 25.38; d = 28.08	Low Medium High
	RCU	Trapezoidal	a = 34.32; b = 43.92; c = 54.12; d = 65.73 a = 56.05; b = 71.45; c = 76.15; d = 94.35 a = 61.90; b = 82.32; c = 89.90; d = 100.0	Low Medium High
	TCU	Trapezoidal	a = 0.0124; b = 0.0394; c = 0.0461; d = 0.0731 a = 0.0609; b = 0.0878; c = 0.0946; d = 0.1216 a = 0.1227; b = 0.1497; c = 0.1565; d = 0.1834	Low Medium High
Output PSM	CCU	Constant	z = 8.8 z = 15.08 z = 24.25	Low Medium High
	RCU	Constant	z = 49.61 z = 74.69 z = 83.07	Low Medium High
	TCU	Constant	z = 0.0427 z = 0.0912 z = 0.153	Low Medium High

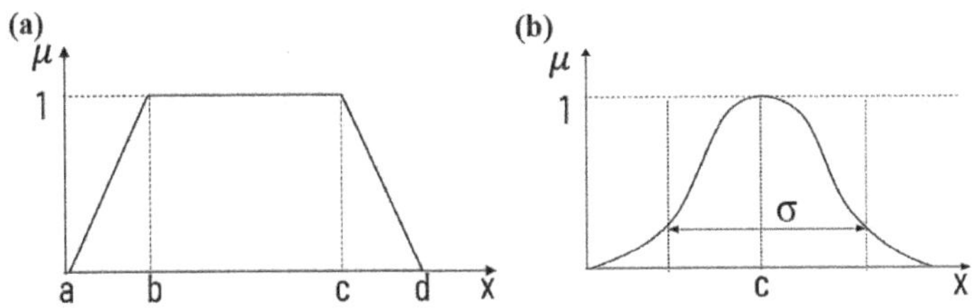

Figure 3. Membership functions: (**a**) Trapezoidal; (**b**) Gaussian.

Gaussian membership function (μ_G) is specified by two parameters (c, σ) as follows:

$$\mu_G(x; c, \sigma) = e^{-\frac{1}{2}\left(\frac{x-c}{\sigma}\right)^2} \tag{2}$$

where c represents the membership functions center and σ determines the membership function width. An illustration is given in Figure 3b.

Table 5 shows the fuzzified linguistic values of the input and output variables. The parameter labels given in the column entitled "Range of values" correspond to those in Equations (1) and (2). Figures 4 and 5 show the examples of defined membership functions.

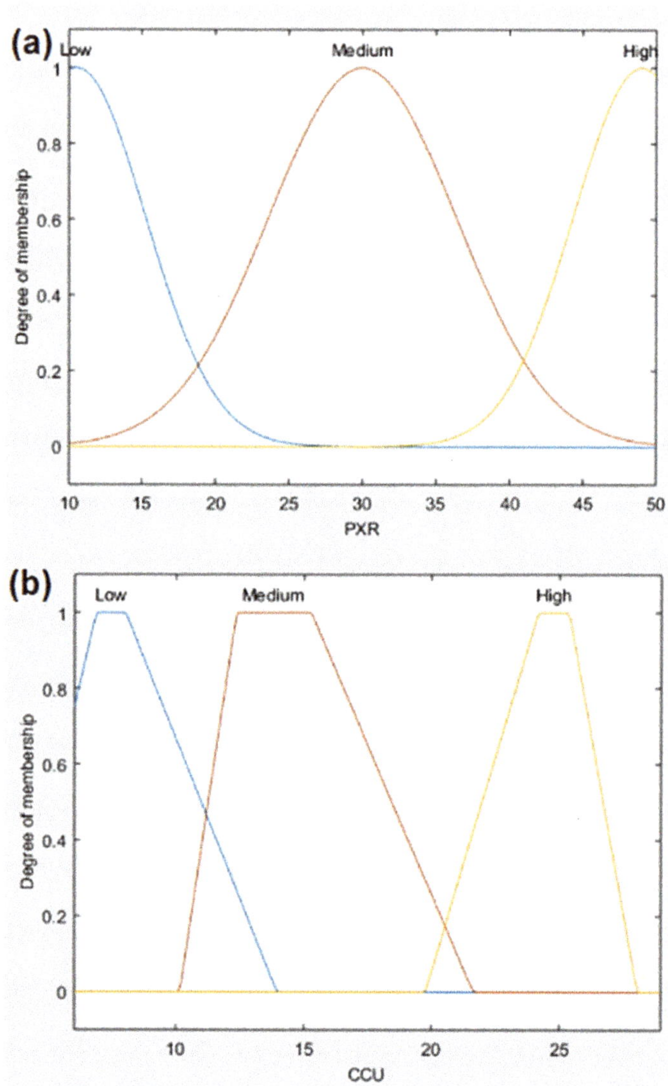

Figure 4. Example of defining the parameters of membership functions in the PMM: (**a**) collector consumption at rougher flotation circuit (PXR); (**b**) copper content in final concentrate (CCU).

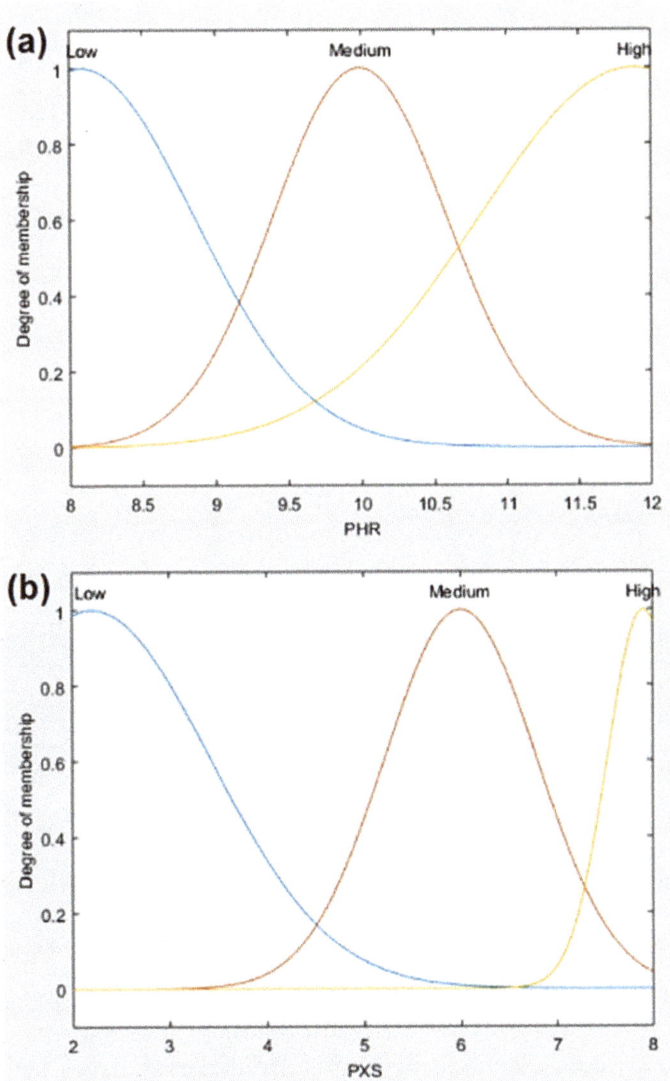

Figure 5. Example of defining the parameters of membership functions in the PSM: (**a**) pH value at rougher flotation circuit (PHR); (**b**) collector consumption in scavenging circuit (PXS).

As a starting point for defining fuzzy rules, Table 6 shows the interdependence matrix of process parameters of the flotation process. The action, that is, the behavior of dependent variables in the process, was considered for the case of an increase in the value of independent variable factors (understood within the established range, with no influence of the remained parameters).

By combining the linguistic values of the input and output variables using the logical operators of conjunction (AND) or disjunction (OR), fuzzy rules of the following type were formed (Table 7):

Fuzzy rules were formed based on literature review and expert knowledge of the flotation process. In this process, 753 rules were constructed by using a rule editor in a fuzzy toolbox environment. Figure 6 shows the details of rule bases for PMM and PSM models, where the rule constituents are expressed through membership functions. This display of the rule base allows the possibility of setting the values of the input parameters and insight into the resulting values of the output variables.

Table 6. Interdependence matrix of parameters in the flotation process.

Independent Variable	**Action** [1]		
	CCU	**RCU**	**TCU**
FCU	↑	↓	↑
PXR	↑	↑	↓
FRT	↓	↑	↓
PHR	↑	↓	↑
PXS	↓	↑	↓

[1] ↑—increase; ↓—decrease.

Table 7. Fuzzy rules formation.

IF FCU is *"low"* **AND** PXR is *"high"* **AND** FRT is *"medium"* **AND** PHR is *"medium"* **AND** PXS is *"high"*
THEN CCU is *"medium"* **AND** TCU is *"low"* **AND** RCU is *"high"*
IF FCU is *"medium"* **AND** PXR is *"medium"* **AND** FRT is *"high"* **AND** PHR is *"low"* **AND** PXS is *"medium"*
THEN CCU is *"medium"* **AND** TCU is *"medium"* **AND** RCU is *"high"*
IF FCU is *"high"* **AND** PXR is *"medium"* **AND** FRT is *"medium"* **AND** PHR is *"high"* **AND** PXS is *"low"*
THEN CCU is *"high"* **AND** TCU is *"medium"* **AND** RCU is *"medium"*
⋮
etc.

2.2.1. Fuzzy Logic Model Based on Mamdani Inference System—PMM

The elementary characteristics of the PMM model are as following:

- Mamdani inference system;
- Applied AND operator in all rules;
- Implication by the minimum method;
- Aggregation by the maximum method; and
- Defuzzification by the centroid method.

PMM model contains a base of 753 rules. The resulting surfaces are shown in Figures 7–9.

By observing the obtained surfaces, and taking into account predefined boundaries, it can generally be concluded that these surfaces follow real dependencies of the process parameters to an acceptable extent. Due to the existence of a large number of resulting surfaces, a detailed analysis of each surface would require a too extensive textual presentation. Therefore, the authors decided to analyze two resulting surfaces of the PMM model:

(1) The surface that shows the dependence of the final concentrate grade (CCU) on copper content in the feed (FCU) and collector consumption at a rougher flotation circuit (PXR), Figure 7c;
(2) The surface that demonstrates dependence of the final grade of tailings (TCU) on the copper content in the feed (FCU) and frother consumption (FRT), Figure 9a.

By visual analysis of the first chosen surface, Figure 7c, it can be observed that with increasing the collector dosage to a certain extent, the copper grade of the final concentrate increases and then decreases with a further increase in collector dosage. Similarly, with increasing the copper content in the feed ore, the quality of the final concentrate also increases. When we consider the second chosen surface, Figure 9a, it can be observed that an increase in the frother dosage leads to the reduction of the copper content in the tailings. These characteristics are in alignment with the literature [57–59].

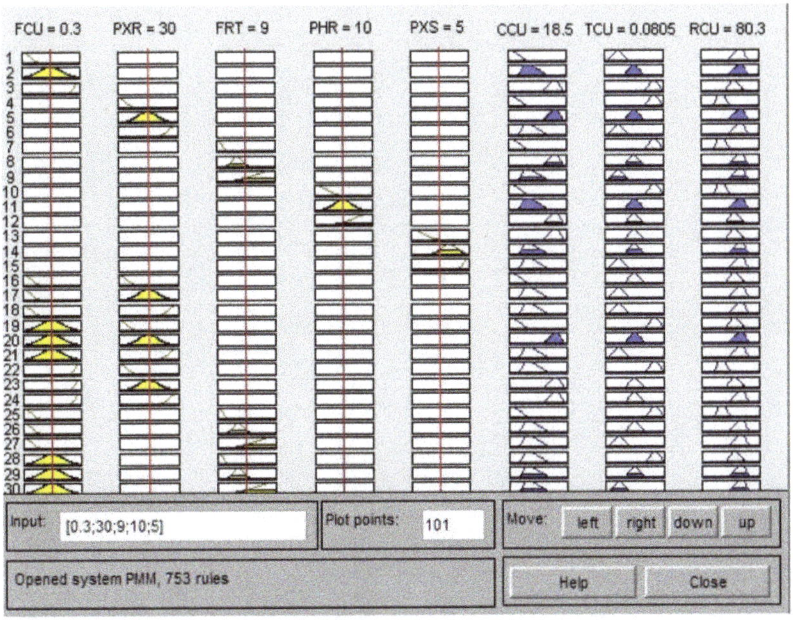

(a)

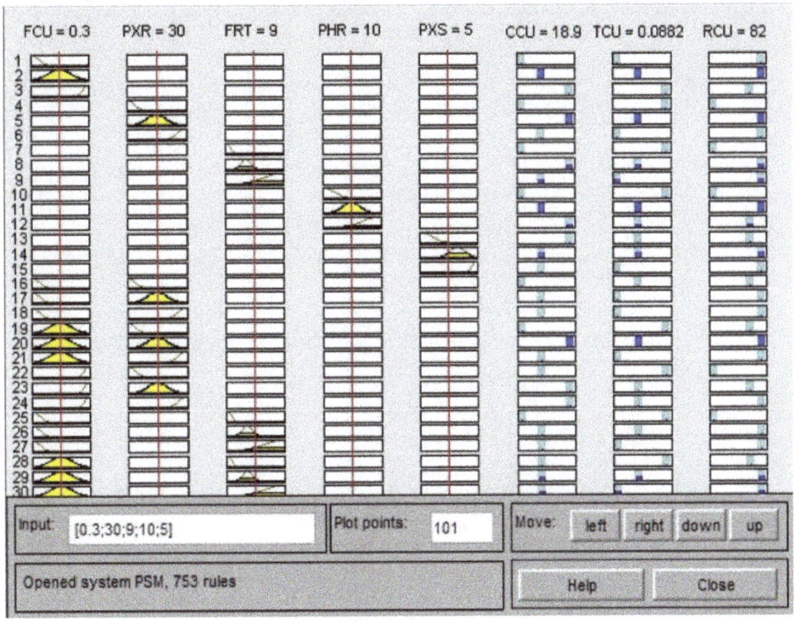

(b)

Figure 6. Display of the detail of the rule base: (**a**) PMM and (**b**) PSM model.

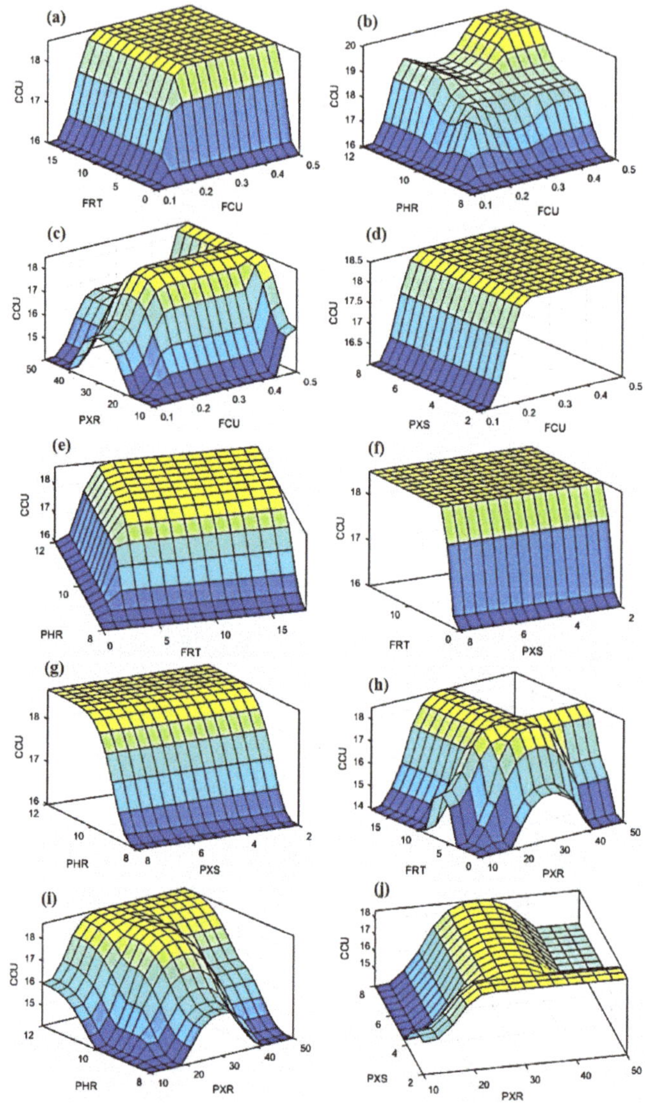

Figure 7. Resulting response surfaces of the PMM model—dependence of final concentrate quality (CCU) on input parameters. (**a**) copper content in the feed (FCU) and frother consumption (FRT); (**b**) copper content in the feed (FCU) and pH value of slurry at rougher flotation circuit (PHR); (**c**) copper content in the feed (FCU) and collector consumption at rougher flotation circuit (PXR); (**d**) copper content in the feed (FCU) and collector consumption at scavenger flotation ciruit (PXS); (**e**) frother consumption (FRT) and pH value of slurry at rougher flotation circuit (PHR); (**f**) frother consumption (FRT) and collector consumption at scavenger flotation ciruit (PXS); (**g**) pH value of slurry at rougher flotation circuit (PHR) and collector consumption at scavenger flotation ciruit (PXS); (**h**) frother consumption (FRT) and collector consumption at rougher flotation circuit (PXR); (**i**) pH value of slurry at rougher flotation circuit (PHR) and collector consumption at rougher flotation circuit (PXR); (**j**) collector consumption at scavenger flotation ciruit (PXS) and collector consumption at rougher flotation circuit (PXR).

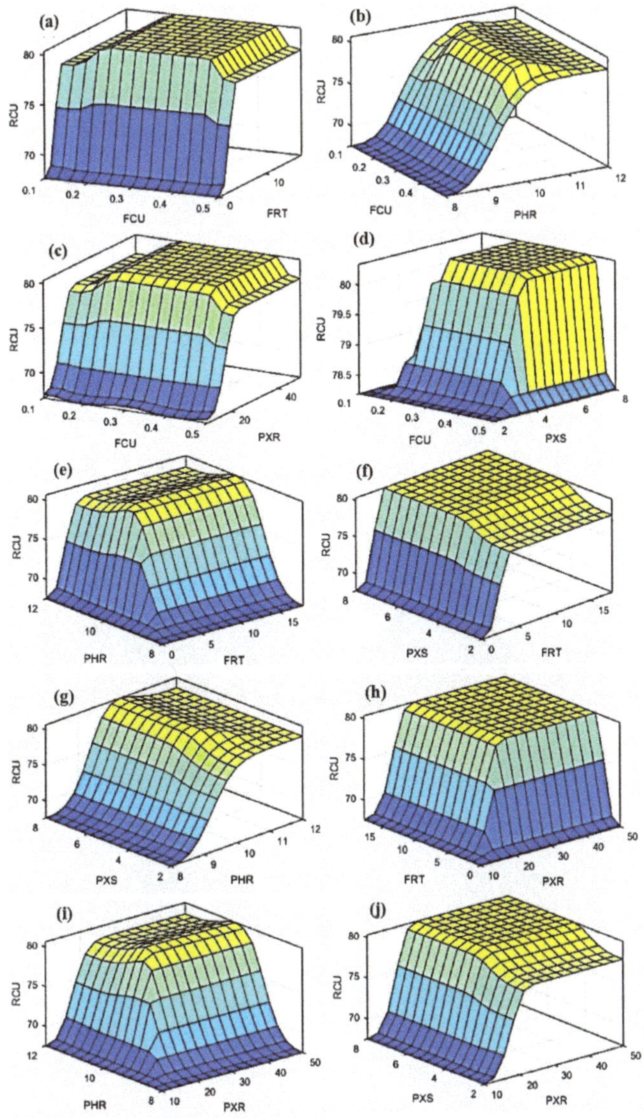

Figure 8. Resulting response surfaces of the PMM model—dependence of copper recovery in final concentrate (RCU) on input parameters. (**a**) copper content in the feed (FCU) and frother consumption (FRT); (**b**) copper content in the feed (FCU) and pH value of slurry at rougher flotation circuit (PHR); (**c**) copper content in the feed (FCU) and collector consumption at rougher flotation circuit (PXR); (**d**) copper content in the feed (FCU) and collector consumption at scavenger flotation ciruit (PXS); (**e**) frother consumption (FRT) and pH value of slurry at rougher flotation circuit (PHR); (**f**) frother consumption (FRT) and collector consumption at scavenger flotation ciruit (PXS); (**g**) pH value of slurry at rougher flotation circuit (PHR) and collector consumption at scavenger flotation ciruit (PXS); (**h**) frother consumption (FRT) and collector consumption at rougher flotation circuit (PXR); (**i**) pH value of slurry at rougher flotation circuit (PHR) and collector consumption at rougher flotation circuit (PXR); (**j**) collector consumption at scavenger flotation ciruit (PXS) and collector consumption at rougher flotation circuit (PXR).

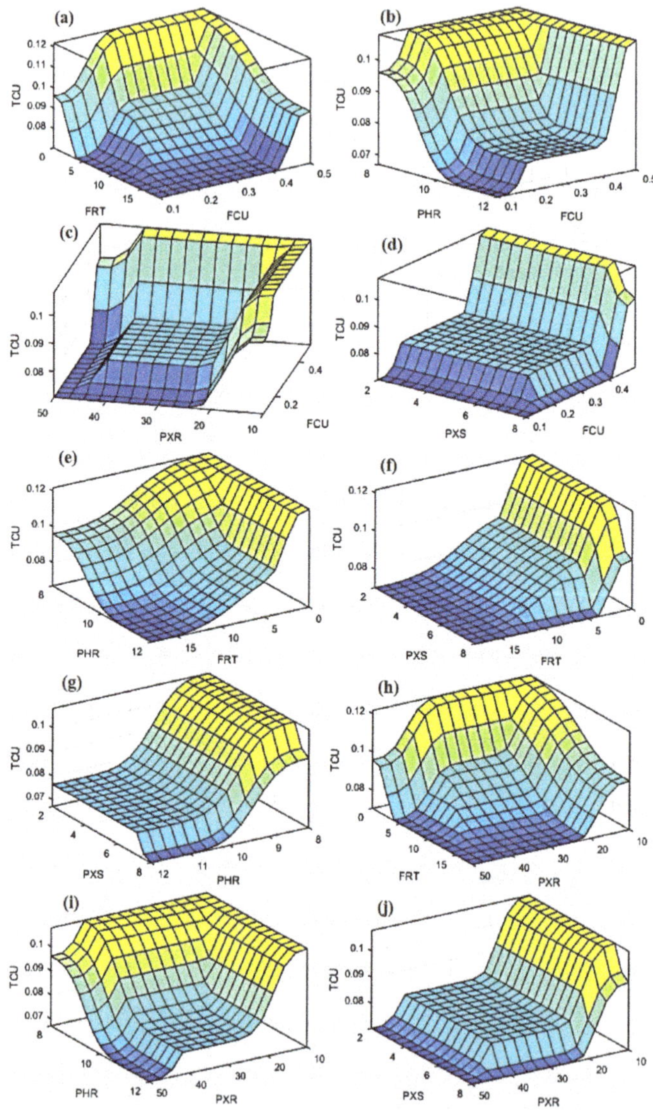

Figure 9. Resulting response surfaces of the PMM model—dependence of copper content in final tailings (TCU) on input parameters. (**a**) copper content in the feed (FCU) and frother consumption (FRT); (**b**) copper content in the feed (FCU) and pH value of slurry at rougher flotation circuit (PHR); (**c**) copper content in the feed (FCU) and collector consumption at rougher flotation circuit (PXR); (**d**) copper content in the feed (FCU) and collector consumption at scavenger flotation ciruit (PXS); (**e**) frother consumption (FRT) and pH value of slurry at rougher flotation circuit (PHR); (**f**) frother consumption (FRT) and collector consumption at scavenger flotation ciruit (PXS); (**g**) pH value of slurry at rougher flotation circuit (PHR) and collector consumption at scavenger flotation ciruit (PXS); (**h**) frother consumption (FRT) and collector consumption at rougher flotation circuit (PXR); (**i**) pH value of slurry at rougher flotation circuit (PHR) and collector consumption at rougher flotation circuit (PXR); (**j**) collector consumption at scavenger flotation ciruit (PXS) and collector consumption at rougher flotation circuit (PXR).

2.2.2. Fuzzy Logic Model Based on Takagi Sugeno Inference System—PSM

The fuzzy logic model was formed by applying a suitable transformation of the PSM model. Its basic methodological characteristics are as follows:

- Takagi-Sugeno inference system; applied AND operator in all rules;
- Implication by the product method;
- Aggregation by the sum method and
- Defuzzification by the weighted average method.

The surfaces obtained by the Takagi Sugeno methodology were smoother than those obtained by the Mamdani fuzzy inference system, which is expected [60]. In general, it can be concluded that these surfaces follow the real dependencies of the process parameters somewhat better than the PMM model. As in the previous section, a detailed analysis of all surfaces requires a substantial textual representation. Because of that, the analyses of two selected resulting surfaces are given in this section:

(1) The surface that shows the dependence of the final concentrate grade (CCU) on the copper content in the feed (FCU) and the pulp pH at the rougher circuit (PHR), Figure 10b; and
(2) The surface that displays the dependence of copper recovery in the final concentrate (RCU) on the copper content in the feed (FCU) and the consumption of the collector at the scavenger circuit (PXS), Figure 11d.

It is known from practice that an increase in the pulp pH leads to an increase in the copper grade of concentrate, which can also be observed from the first selected surface, Figure 10b. At first glance, the question of the regularity of the given surface can be imposed, considering that in this case only the pH value at the rough flotation is considered. However, we should not ignore the fact that the pH values in the other stages of flotation are considered optimal constants, and therefore this surface is justified from the aspect of the flotation concentration process as a whole.

The main reason for adding the collector at the scavenger stage is to increase the recovery of copper in the concentrate, which coincides with the appearance of the second selected surface, Figure 11d.

In industrial conditions, the increase in the copper content in the feed ore increase the concentrate grade. Moreover, if the feed is poor, recovery increases to a certain extent, and with a further increase in Cu in the feed, recovery decreases, because a part of the copper remains in the tailings. These technological actualities correspond to both observed surfaces.

2.3. *Artificial Neural Network*

NNs are used for solving complex and nonlinear engineering problems. NNs is a widely used model in mineral processing applications with the ability to recognize patterns among parameters and predict the key performance parameters of the processes. A typical NN comprises a large number of neurons or nodes, which are connected to each other. The neurons are grouped in layers and connected by weighted links and bias. The output of each neuron is transferred to the next layer as an input. Finally, the nonlinear basis function set is used to calculate the outputs of NN. The model learning is modifying the weights and biases to minimize the error, taking into account the targets. Details of the NN algorithm procedure are presented by Nakhaei et al. [61,62].

In the current study, three feed-forward back-propagation learning algorithms were employed for predicting the copper content in the final concentrate (NN1), copper recovery in the final concentrate (NN2), and copper content in the final tailings (NN3). A simple architecture for the model structures established in this study is shown in Figure 13.

The base also contains 753 rules, as well as the corresponding Mamdani model. The resulting surfaces of the PSM model are presented in Figures 10–12.

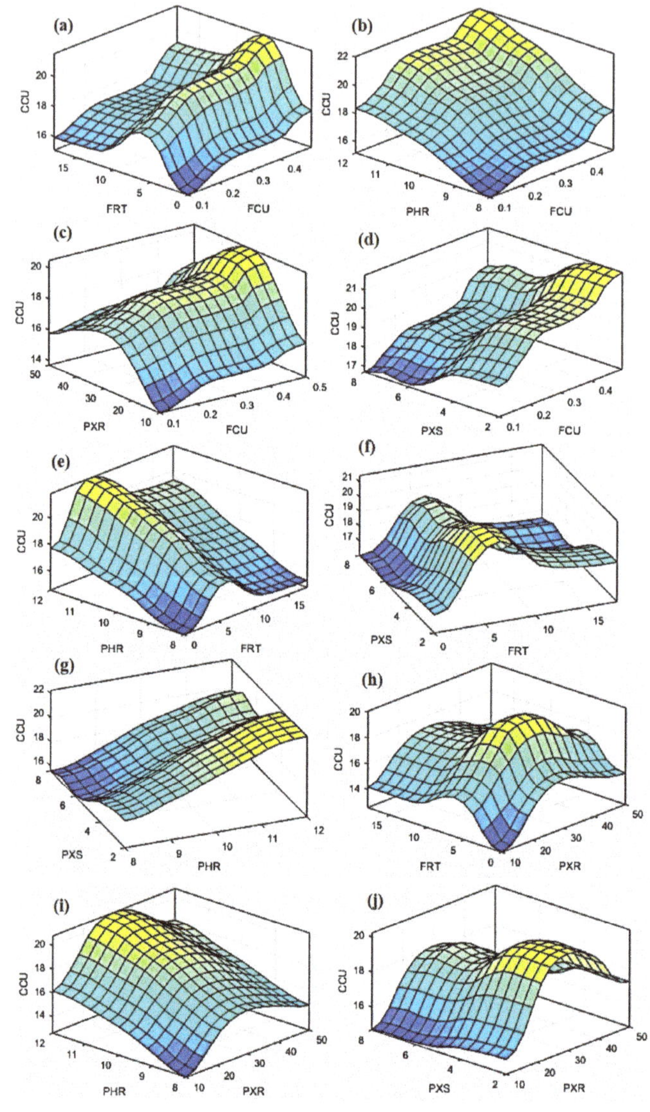

Figure 10. Resulting response surfaces of the PSM model—dependence of final concentrate quality (CCU) on input parameters. (**a**) copper content in the feed (FCU) and frother consumption (FRT); (**b**) copper content in the feed (FCU) and pH value of slurry at rougher flotation circuit (PHR); (**c**) copper content in the feed (FCU) and collector consumption at rougher flotation circuit (PXR); (**d**) copper content in the feed (FCU) and collector consumption at scavenger flotation ciruit (PXS); (**e**) frother consumption (FRT) and pH value of slurry at rougher flotation circuit (PHR); (**f**) frother consumption (FRT) and collector consumption at scavenger flotation ciruit (PXS); (**g**) pH value of slurry at rougher flotation circuit (PHR) and collector consumption at scavenger flotation ciruit (PXS); (**h**) frother consumption (FRT) and collector consumption at rougher flotation circuit (PXR); (**i**) pH value of slurry at rougher flotation circuit (PHR) and collector consumption at rougher flotation circuit (PXR); (**j**) collector consumption at scavenger flotation ciruit (PXS) and collector consumption at rougher flotation circuit (PXR).

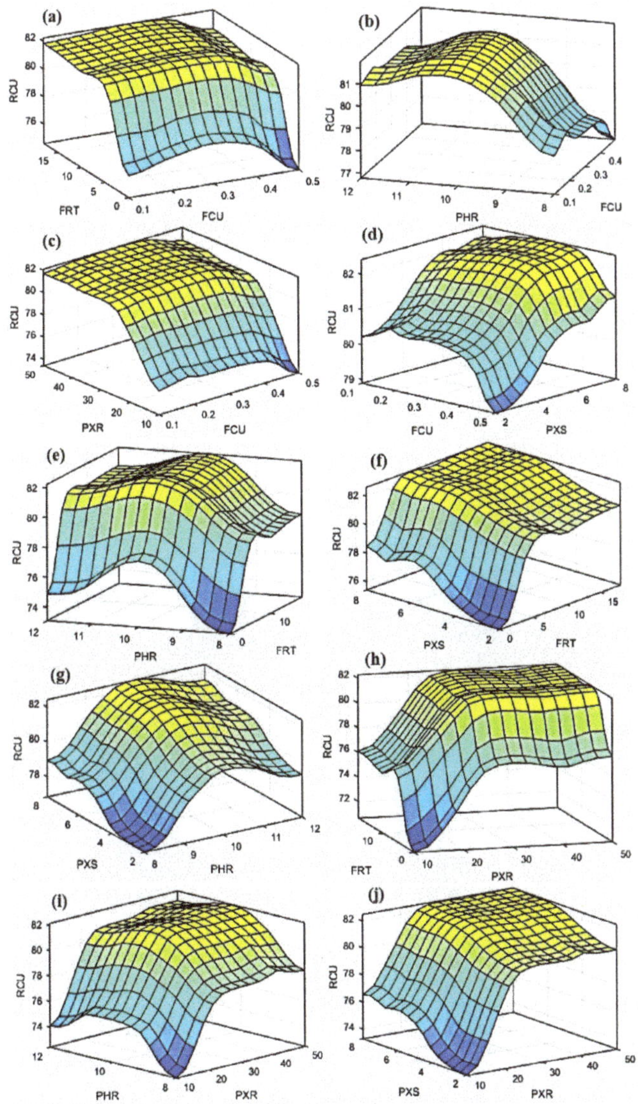

Figure 11. Resulting response surfaces of the PSM model—dependence of copper recovery (RCU) in final concentrate on input parameters. (**a**) copper content in the feed (FCU) and frother consumption (FRT); (**b**) copper content in the feed (FCU) and pH value of slurry at rougher flotation circuit (PHR); (**c**) copper content in the feed (FCU) and collector consumption at rougher flotation circuit (PXR); (**d**) copper content in the feed (FCU) and collector consumption at scavenger flotation ciruit (PXS); (**e**) frother consumption (FRT) and pH value of slurry at rougher flotation circuit (PHR); (**f**) frother consumption (FRT) and collector consumption at scavenger flotation ciruit (PXS); (**g**) pH value of slurry at rougher flotation circuit (PHR) and collector consumption at scavenger flotation ciruit (PXS); (**h**) frother consumption (FRT) and collector consumption at rougher flotation circuit (PXR); (**i**) pH value of slurry at rougher flotation circuit (PHR) and collector consumption at rougher flotation circuit (PXR); (**j**) collector consumption at scavenger flotation ciruit (PXS) and collector consumption at rougher flotation circuit (PXR).

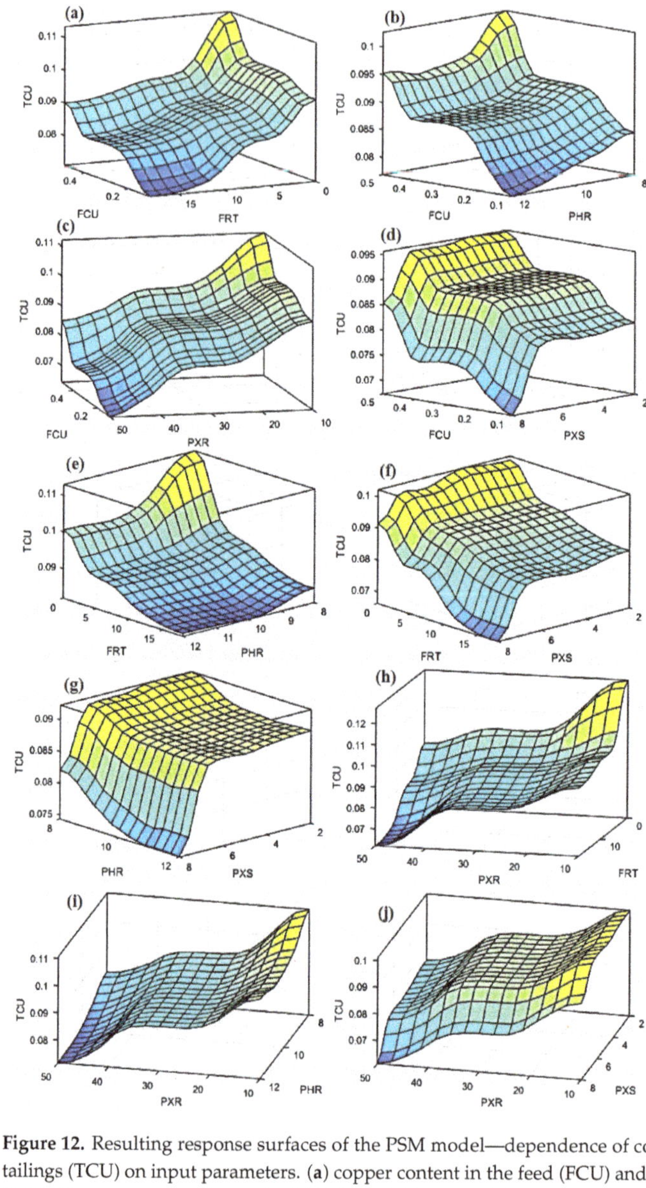

Figure 12. Resulting response surfaces of the PSM model—dependence of copper content in final tailings (TCU) on input parameters. (**a**) copper content in the feed (FCU) and frother consumption (FRT); (**b**) copper content in the feed (FCU) and pH value of slurry at rougher flotation circuit (PHR); (**c**) copper content in the feed (FCU) and collector consumption at rougher flotation circuit (PXR); (**d**) copper content in the feed (FCU) and collector consumption at scavenger flotation ciruit (PXS); (**e**) frother consumption (FRT) and pH value of slurry at rougher flotation circuit (PHR); (**f**) frother consumption (FRT) and collector consumption at scavenger flotation ciruit (PXS); (**g**) pH value of slurry at rougher flotation circuit (PHR) and collector consumption at scavenger flotation ciruit (PXS); (**h**) frother consumption (FRT) and collector consumption at rougher flotation circuit (PXR); (**i**) pH value of slurry at rougher flotation circuit (PHR) and collector consumption at rougher flotation circuit (PXR); (**j**) collector consumption at scavenger flotation ciruit (PXS) and collector consumption at rougher flotation circuit (PXR).

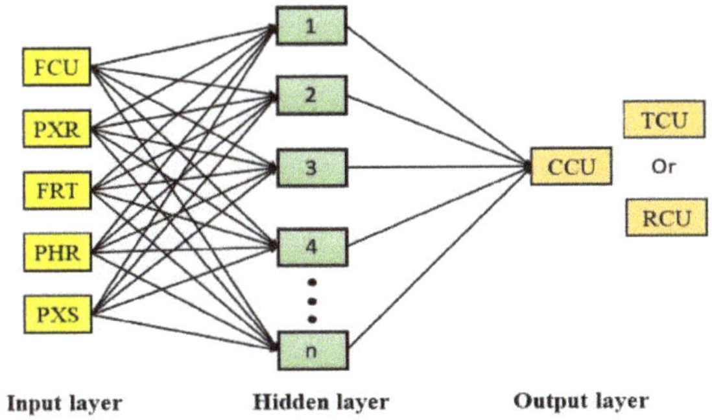

Figure 13. The structure of the NN models for the prediction of CCU, RCU and TCU from the industrial flotation plant.

The ANN models consist of one input layer with five elements (FCU, PXR, FRT, PHR and PXS), one hidden layer including the desired number of nodes, and the output layer where the CCU, RCU and TCU values are calculated. The logsigmoid and purelin functions (see Figure 14) are implemented as the activation functions in the hidden layer and the output layer. The optimum NN structure is selected using the "trial and error" method by adjusting the number of neurons in the hidden layer (from 10 to 100) in order to achieve the best model by minimizing the errors. The widely applied Levenberg Marquardt (LM) algorithm is used for the model training. For the purposes of designing the network, 1910 data were selected.

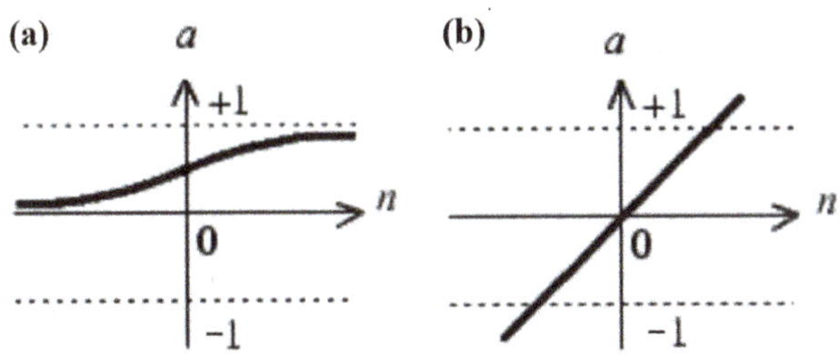

$a = logsig(n) = \frac{1}{1+e^{-n}}$ $\qquad$ $a = purelin(n) = n$

Figure 14. Activation functions: (**a**) logsigmoid; (**b**) purelin.

The first idea was to develop one ANN model consisting of 5 input and 3 output variables. However, the selected software tool considers the sets of output data as one set and performs all calculations and interpretations based on it. Since the output parameters are numerically independent and expressed in different units, this was not acceptable. Moreover, ANN models were developed in order to compare their performances with performances of fuzzy logic models. Given that fuzzy logic models were developed based on original data (without normalization), the same procedure was applied to neural networks. The building of new ANN models with 3 output variables, using more advanced software tools and normalized data sets, is a topic for further research.

3. Results and Discussion

3.1. *Performance Evaluation of the Fuzzy Models*

The evaluation of the proposed models was carried out in the MATLAB software package, by entering the real values of the input process variables from the industrial flotation plant "Veliki Krivelj" and generating the corresponding outputs predicted by the models. Several standard statistical criteria, including the determination coefficient (R^2), root mean square error (RMSE), and standard deviation of prediction error (SDE), are used to evaluate the performance of the developed models. R^2 is an indicator of how much changes in one variable are caused by changes in another variable, and is represented by Equation (3), while *RMSE* shows the accuracy of the model's predicted values versus the actual, and is represented by Equation (4):

$$R^2 = \left[\frac{n(\sum XY) - (\sum X)(\sum Y)}{\sqrt{\left(n\sum X^2 - (\sum X)^2\right)\left(n\sum Y^2 - (\sum Y)^2\right)}} \right]^2 \quad (3)$$

$$RMSE = \sqrt{\frac{\sum_{i=1}^{n}(X-Y)^2}{n}} \quad (4)$$

In these equations, X is the measured value, Y is the predicted value, and n is the number of samples.

SDE shows how much, on average, the elements of the dataset deviate from the arithmetic mean of that dataset and is expressed by Formula (5):

$$SDE = \sqrt{\frac{\sum_{i=1}^{n}(x-\mu)^2}{n}} \quad (5)$$

where n presents the number of elements in the dataset, μ is mean (average value) of the dataset and x is the i-th member of the dataset.

Standard deviation is determined for absolute (SDE) and relative prediction error (SDE_r).

The prediction error (ε) and relative prediction error (ε_r), which served as the criteria for evaluating the predictive properties of models, were calculated according to the Formulas (6) and (7):

$$\varepsilon = Y - X \quad (6)$$

$$\varepsilon_r = \frac{Y - X}{X} \quad (7)$$

where X, Y have the same meaning as in Equations (3) and (4).

The descriptive statistics of differences between measured and estimated values for the evaluation data are given in Tables 8 and 9. Figures 15–18 present the prediction errors of the fuzzy models.

Table 8. Statistical analysis of actual and predicted values using the PMM model.

Statistical Parameters	Technological Indicator of the Flotation Process		
	CCU	RCU	TCU
R^2	0.971	0.992	0.839
RMSE	3.126	7.007	0.034
Mean of prediction error, μ	−0.886	−2.666	0.044
SDE	3.306	7.227	0.019
Maximum positive error (maximum)	10.644	36.739	0.100
Minimum positive error	0.00069	0.00689	0.00015
Minimum negative error	−0.01968	−0.01318	−0.00001
Maximum negative error (minimum)	−10.743	−25.233	−0.047
Mean of relative prediction error, μ_r	−0.017	−0.024	1.485
SDE_r	0.190	0.098	1.238

Table 9. Statistical analysis of actual and predicted values using the PSM model.

Statistical Parameters	Technological Indicator of the Flotation Process		
	CCU	RCU	TCU
R^2	0.970	0.993	0.847
RMSE	3.291	6.739	0.034
Mean of prediction error	−0.288	−2.008	0.046
SDE	3.374	6.915	0.018
Maximum positive error (maximum)	10.961	39.619	0.091
Minimum positive error	0.00473	0.00232	0.00128
Minimum negative error	−0.01028	−0.01190	−0.00019
Maximum negative error (minimum)	−10.354	−17.502	−0.056
Mean of relative prediction error, μ_r	0.015	−0.0165	1.562
SDE_r	0.198	0.096	1.244

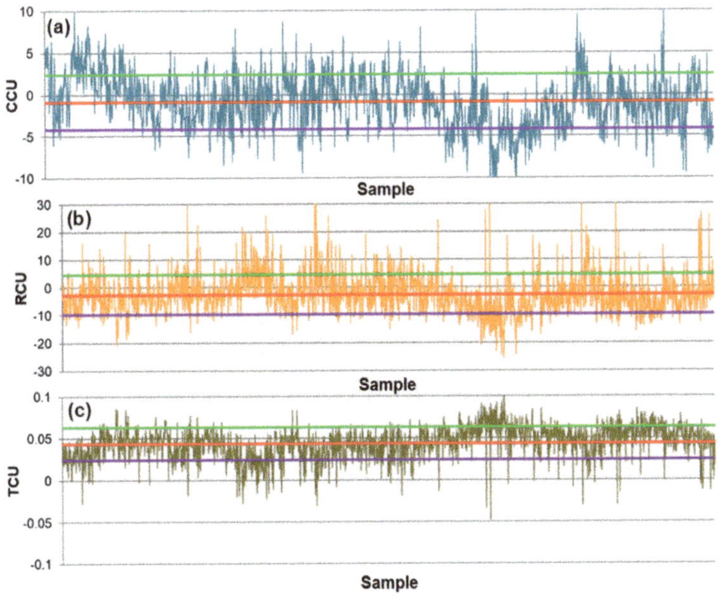

Figure 15. Prediction errors of technological indicators according to the PMM model: (**a**) CCU, (**b**) RCU and (**c**) TCU. The red line on the charts represents mean (μ), green line value μ + 1 SDE and purple line value μ − 1 SDE.

By considering the results of the regression analysis shown in Table 8, it can be observed that the values of the determination coefficients for predictions of the content and recovery of copper in the final concentrate are high. This means that a strong relationship has been established using the PMM model between the actual and the predicted values of the metallurgical indicators. In other words, the proposed model well describes the changes in the real values of the observed parameters related to their increase or decrease over time.

Yet, when the prediction error is taken into account (Figure 15), it is noted that there are insignificant and, in some occasions, relatively large deviations from the actual results. Although the maximum and minimum errors of the estimation (margin of error) are high, they occurred a few times and did not have a high frequency.

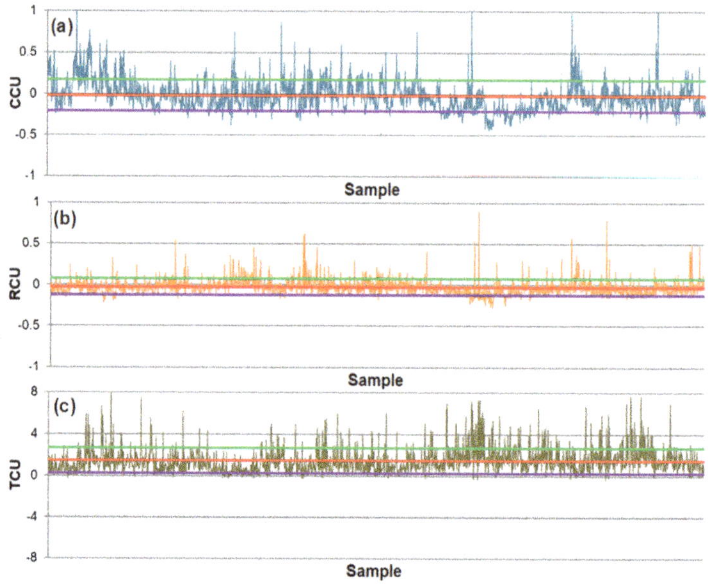

Figure 16. Relative prediction errors of technological indicators according to the PMM model: (**a**) CCU, (**b**) RCU and (**c**) TCU. The red line on the charts represents mean (μ_r), green line value $\mu_r + 1$ SDE_r and violet line value $\mu_r - 1$ SDE_r.

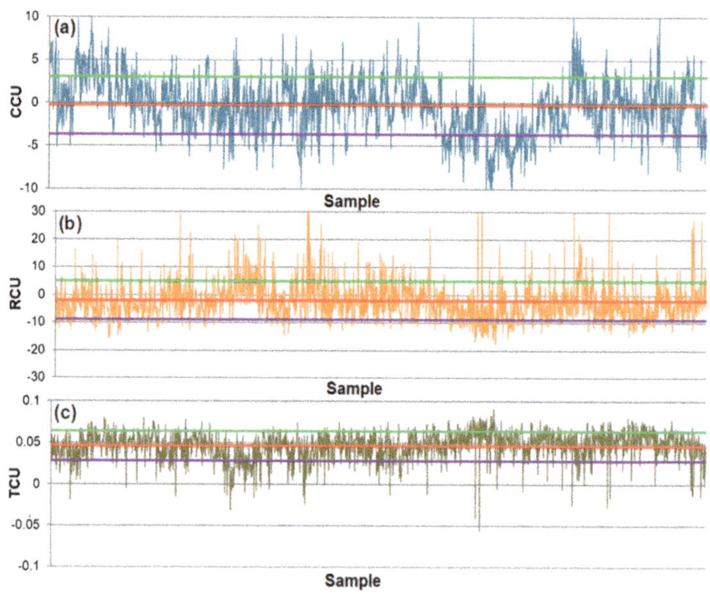

Figure 17. Prediction errors of technological indicators according to the PSM model: (**a**) CCU, (**b**) RCU and (**c**) TCU. The red line on the charts represents mean (μ), green line value $\mu + 1$ SDE and purple line value $\mu - 1$ SDE.

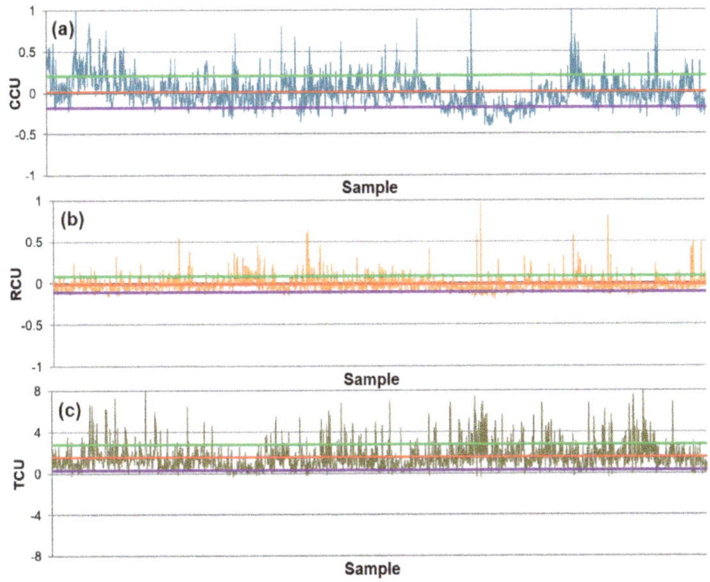

Figure 18. Relative prediction errors of technological indicators according to the PSM model: (**a**) CCU, (**b**) RCU and (**c**) TCU. The red line on the charts represents mean (μ_r), green line value $\mu_r + 1$ SDE_r and violet line value $\mu_r - 1$ SDE_r.

The red line on the charts represents mean (μ), green line value μ + 1 SDE and purple line value μ − 1 SDE. This is generally applied to all charts where absolute and relative prediction error is presented.

It should be noted that the complete regression analysis was based on a theoretical assumption that if the actual values of the metallurgical indicators were equal to zero, in that case the values predicted by the models would also be equal to zero. This statement was adopted for practical reasons, in order to mitigate illogical oscillations, extreme deviations and dispersion of values in real results, caused by imperfections in measurement, human factor errors and the like, which can significantly affect the results of the model validation. In this sense, this regression analysis should be given greater mathematical importance than the practical.

By observing the trend of the prediction errors of the final concentrate grade, it can be concluded that their values are mostly between ± 5% Cu, which is consistent with the RMSE value of 3.126 (Table 8). The only major deviation from this trend can be observed on the right half of the diagram, which corresponds to the beginning of the last third of the observed time period of the plant operation. In this period, prediction errors are mostly negative, which may indicate changes in the operation conditions of the plant. These large fluctuations can be caused by changing some factors that were not taken into account during modeling, such as changes in the feed ore hardness or changes in the quality of the reagents.

Furthermore, by looking at the prediction errors of recovery, it is observed that a positive prediction error for the recovery (predicted values are higher than the actual) is quite well associated to a negative prediction error for the tailings grade. The reverse is also true: a negative prediction error for the recovery is associated to a positive prediction error for the tailings grade. Therefore, the recovery of copper in the final concentrate and the copper content in the tailings are in alignment with each other. This matching indicates a good general setting of the model, as well as the potential influence of process factors that were considered constant during modeling.

It should also be noted that the TCU prediction error is mostly positive, which means that the values predicted by the model are, as a rule, slightly higher than the real ones (offset error). This result indicates that the additional fine tuning of the model is needed. Namely, the relatively small deviations of the predictive model results in relation to real results, with the tendency to "translatory skip" the real values, indicating that fuzzy values within the membership functions need to be additionally corrected. However, taking into account the good correlation between fuzzy rule base outputs with the real flotation system conditions, this is at the same time an indication that the performance of the model itself has "hit" its own limits. In this sense, hybridization of this fuzzy logic model with other soft computing methods is recommended.

By considering the data distribution in the range of ± 1 SDE, it was found that in this range lies 67.6% of the prediction error values for CCU, 74.7% for RCU and 71.2% for TCU.

Besides the absolute prediction error, one of the criteria for assessing model adequacy is the relative prediction error. This parameter can provide a clearer insight and additional data on the effectiveness of the predictive model, because it is used as a basis for comparing parameters that are expressed in different measurement units and different value ranges. Figure 16 presents the relative prediction errors of the CCU, RCU and TCU parameters of the PMM model.

By visual analysis of the charts in Figure 16, as well as consideration of the statistical parameters, it can be observed that the smallest relative prediction error is obtained when modeling RCU, and the highest when modeling TCU, which is consistent with their R^2.

The relative prediction error of RCU has the smallest standard deviation, and 82.1% of the results are in the range ± 1 SDE_r. When it comes to relative prediction errors of CCU and TCU, 73.6% and 79.0% of the results are within ± 1 SDE_r., respectively.

Table 9 shows that the results obtained by the PSM model are very similar to the PMM model results. The high values of determination coefficients indicate the existence of a strong relationship between the actual and predicted values of metallurgical indicators.

Prediction error diagrams (Figure 17) have shown small (more frequently) and larger (less frequently) deviations from the actual results. Margin errors (maximum and minimum) are also high, as in the previous case (Figure 15), but such extremes occur very rarely.

The prediction error trends of the copper content and recovery of the concentrate is almost identical to the trends shown in Figure 15.

Moreover, when it comes to the error of predicting the copper content in the tailings, the trend largely coincides with the corresponding trend of Figure 15. The only difference that is possible to observe is that a smaller number of error values are close to extremes. This leads to the conclusion that the deviations between the predicted and actual tailings grade are somewhat smaller, which also corresponds to the slightly higher value of the determination coefficient of the PSM model for the TCU variable.

Generally, for both models (PMM and PSM), the values of TCU prediction errors are mostly between ± 0.05% Cu, which is consistent with RMSE values (0.034 both, Table 9). It can be concluded that both models provided effective predictions, overall. As for the PMM model, the TCU prediction error is mostly positive, which indicates an offset error.

Similar to the PMM model, 67.7% of the prediction error values for CCU, 73.8% for RCU and 72.1% for TCU are in the range ± 1 SDE.

When considering the relative prediction error presented in Figure 18, it is also concluded that the results are very similar to the results of the PMM model.

In the range ± 1 SDE_r are 73.1% of the relative prediction error results for CCU, 82.1% for RCU and 79.4% for TCU (Figure 18).

3.2. *Neural Network-Based Models*

3.2.1. Model Performances

Three different models based on the principle of feed-forward backpropagation NN were trained. Model NN1 predicts copper content in the final concentrate, model NN2 predicts copper recovery in the final concentrate and model NN3 predicts copper content in the final tailings.

The architecture of the NN that gave the smallest Mean Square Error (MSE) of validation (best validation performance) was chosen as the most favorable predictive model. Mathematically, MSE can be defined as:

$$MSE = \frac{1}{n}\sum_{i=1}^{n}(X - Y)^2 \quad (8)$$

where X is the measured value, Y is the predicted value, and n is the number of samples.

The NN models consisting of 64, 43 and 24 neurons in the hidden layer gave the lowest mean squared error among all models studied for NN1, NN2 and NN3, respectively. The networks' performances in the training stage are shown in Figures 19–21.

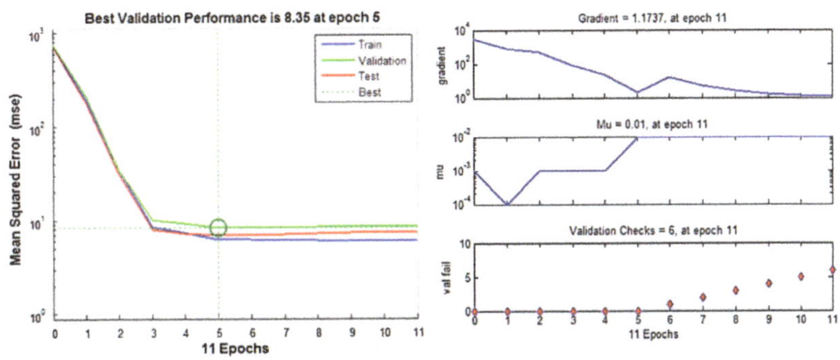

Figure 19. Training state and performance of the generated NN1 model for CCU.

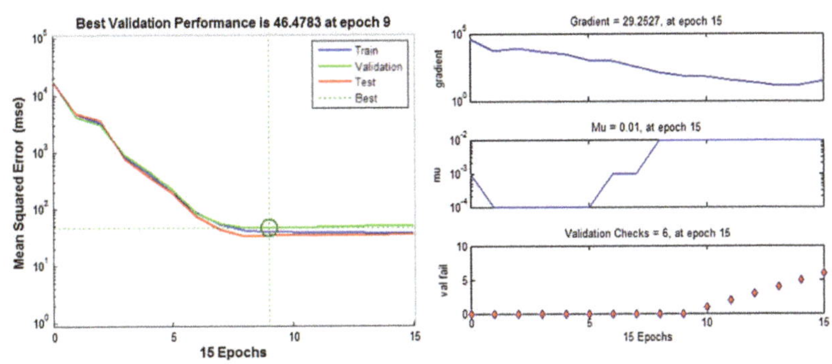

Figure 20. Training state and performance of the generated NN2 model for RCU.

Figure 19 demonstrates the MSE variation of the training, validation and testing stages versus the iteration number. As shown in this figure, the large values for the MSE were gradually reduced to a smaller value as the weights are updated. The training stage stopped at 5th epoch, i.e., after epochs 5, there was not a significant improvement in the performance of the model. The best validation performance was 8.35 at epoch 5, implying a good stable network behavior. Then, after six error iterations (validation checks), the process stopped at epoch 11. The similar explanation provided for Figure 19 can also be applied for Figures 20 and 21.

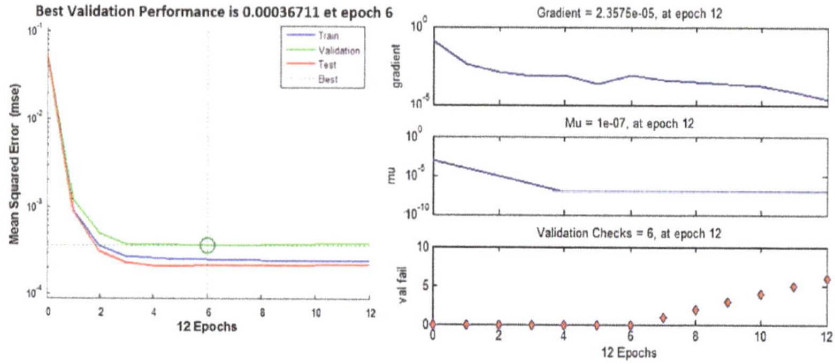

Figure 21. Training state and performance of the generated NN3 model for TCU.

One of the important phenomena that can be observed from the Figures 19–21 is that no overfitting occurred during training, testing and validation.

Overfitting can be spotted when the error on the training data decreases to a small value, but the error on the test/validation data increases to a large value. Some reasons for overfitting are: small size of the training dataset, very noisy dataset and complex ANN architecture [63].

When the network tries to learn from a small dataset, it will tend to have greater control over the dataset and to satisfy all the data points exactly. Therefore, the network is trying to memorize every single data point and failing to capture the general trend from the training dataset. In the observed case, a relatively large training dataset is applied, containing 1336 samples.

Overfitting also occurs when the model tries to make predictions on data that is very noisy, which results that the overfitted model is inaccurate, as the trend does not reflect the reality present in the data. By using the early stopping technique (small number of epochs), the network can be prevented from overfitting the noise in the data [63]. In the observed case, the number of epochs is relatively small, i.e., 11, 15 and 12 epochs for NN1, NN2 and NN3, respectively.

Increasing the number of hidden units and/or layers may lead to overfitting because it will make it easier for the neural network to memorize the training set, but to avoid generalization to unseen data. Therefore, a too large number of neurons in the hidden layer is undesirable. There is an empirical formula for the optimal number of hidden neurons (N_h) in the hidden layer [64]:

$$N_h = \frac{N_s}{\alpha(N_i + N_o)} \tag{9}$$

where N_i represents the number of input neurons, N_o is the number of output neurons, N_s is the number of samples in training dataset and α is the scaling factor, which takes values from 2 to 10.

If this formula is applied to the observed case, the number of hidden neurons should vary between 22 and 111. NN1, NN2 and NN3 models contain 64, 43 and 24 hidden neurons, respectively, which corresponds to the specified range. Since overfitting is not occurred, the numbers of hidden neurons are correctly chosen.

3.2.2. Predictive Abilities of NN-Based Models

The predicted results demonstrate the R^2 value 0.98 for CCU, 0.99 for RCU and 0.87 for TCU, demonstrating the success of the proposed NN models (Table 10). As it can be observed from Table 10, the RMSE between predicted and actual data by using NN models are very small, which means they can accurately predict the flotation performance indicators within the entire process variables' space.

Table 10. Statistical analysis of actual and predicted values using NN1, NN2 and NN3 models.

Statistical Parameters	Technological Indicator of the Flotation Process		
	CCU	RCU	TCU
R^2	0.982	0.994	0.867
RMSE	2.567	6.284	0.015
Mean of prediction error	−0.017	−0.338	0.0015
SDE	2.589	6.323	0.016
Maximum positive error (maximum)	10.933	44.501	0.043
Minimum positive error	0.00102	0.00694	0.000003
Minimum negative error	−0.00230	−0.01246	−0.00004
Maximum negative error (minimum)	−10.892	−16.341	−0.113
Mean of relative prediction error, μ_r	0.019	0.002	0.222
SDE_r	0.154	0.089	0.551

NN errors for the prediction of CCU, RCU and TCU are shown in Figure 22. It can be observed that these errors are mostly between ± 2.7% Cu and ± 5% for CCU and RCU, respectively. The prediction errors of the copper content and recovery in the final concentrate are very rarely out of range of ± 5% Cu and ± 10%, respectively. This is consistent with the RMSE values of 2.567 for NN1 and 6.284 for NN2 (Table 10).

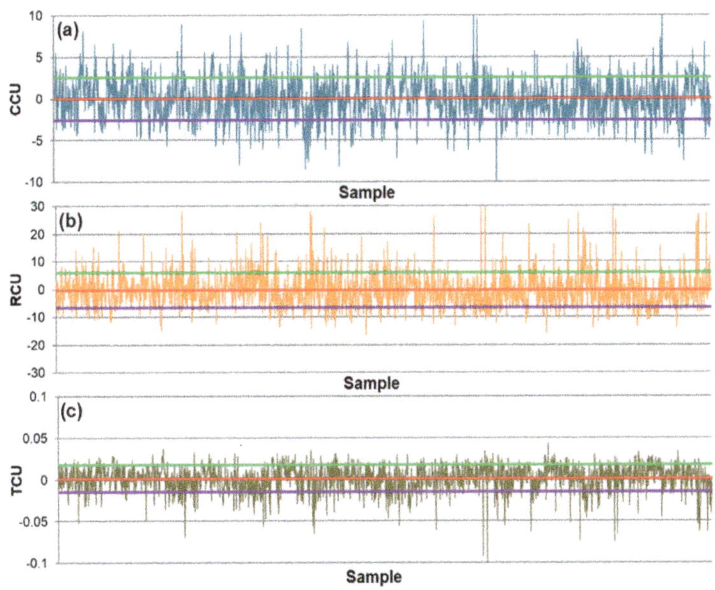

Figure 22. Prediction errors of technological indicators according to the (**a**) NN1, (**b**) NN2 and (**c**) NN3 models. The red line on the charts represents mean (μ), green line value μ + 1 SDE and purple line value μ − 1 SDE.

Analyzing the results obtained by the NN3 model from Figure 22, we explored that the prediction errors of the TCU mostly take values in the range of ± 0.016% Cu, which is consistent with its RMSE of 0.015 (Table 10). Examining the data distribution of the prediction error in the range of ± 1 SDE, it was found that 68.9%, 74.3% and 72.1% data lies within this range for CCU, RCU and TCU, respectively.

Figure 23 shows the relative error of the predictive models. Similar to the previous observations (for PMM and PSM), the smallest relative prediction error with the smallest standard deviation was obtained for the RCU variable. This is consistent with the highest coefficient of determination also achieved for the RCU (Table 10).

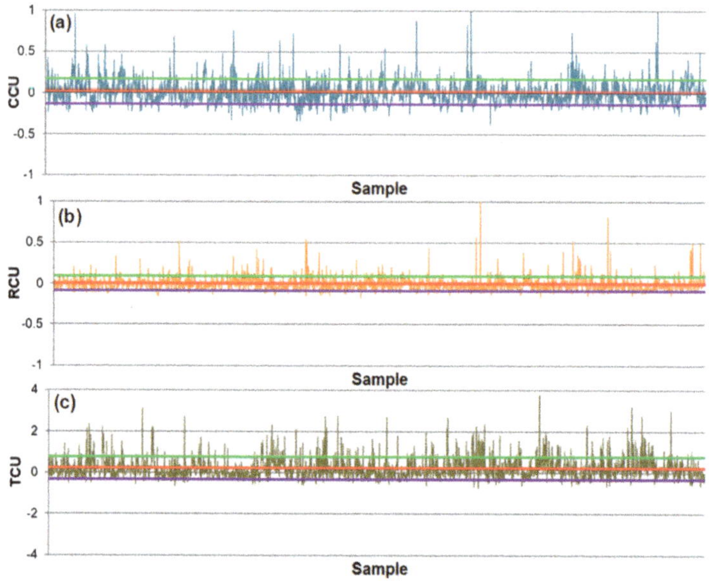

Figure 23. Relative prediction errors of technological indicators according to the (**a**) NN1, (**b**) NN2 and (**c**) NN3 models. The red line on the charts represents mean (μ_r), green line value μ_r + 1 SDE_r and violet line value μ_r − 1 SDE_r.

The number of datum lying in the ± 1 SDE_r range is similar to the previous two models, and is 75.1% for the NN1 model, 81.4% for the NN2 model and 78.7% for the NN3 model.

In general, from the obtained results, it was found that the NN models are well fitting the real data and have the ability to predict the output data. It can be concluded that the NN models have good predictive properties, because of significant fluctuations in real process data. The validation set and test set had similar behavior with no occurrence of overfitting.

3.3. Summary Discussion

Comparative presentation of statistical parameters of every model is given in Figures 24–26.

According to Figure 24, the accuracy of the NN models is higher than the accuracy of fuzzy models. The dependent variables (CCU, RCU and TCU) are better explained in the NN models than in the fuzzy models by the independent variables (FCU, PXR, FRT, PHR and PXS), since the determination coefficients are higher and the RMSE are significantly smaller. It can be clearly observed that models based on NN demonstrate the best predictive abilities, while both fuzzy models demonstrate very similar performances. Therefore, practically there is no difference in the utilization of either the Mamdani or Takagi Sugeno inference system under a wide range of operation conditions.

A similar conclusion can be made based on the results shown in Figures 25 and 26. The smallest standard deviations of predictive error (both absolute and relative) have neural networks when it comes to all output variables. A low standard deviation indicates that the values tend to be close to the mean (also called the expected value) of the set, while a high standard deviation indicates that the values are spread over a wider range.

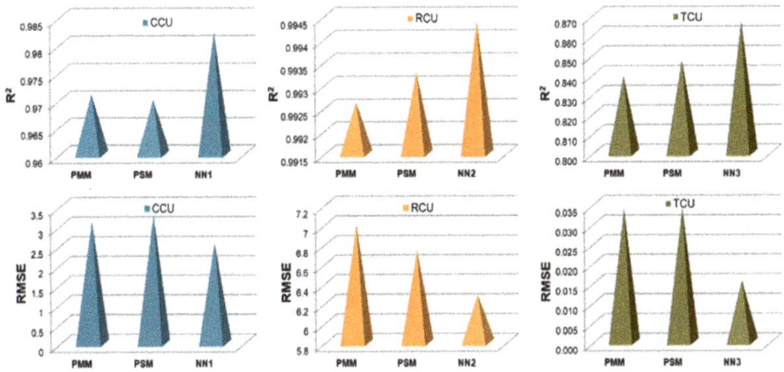

Figure 24. Comparison of fuzzy and ANN modeling results: R^2 and RMSE.

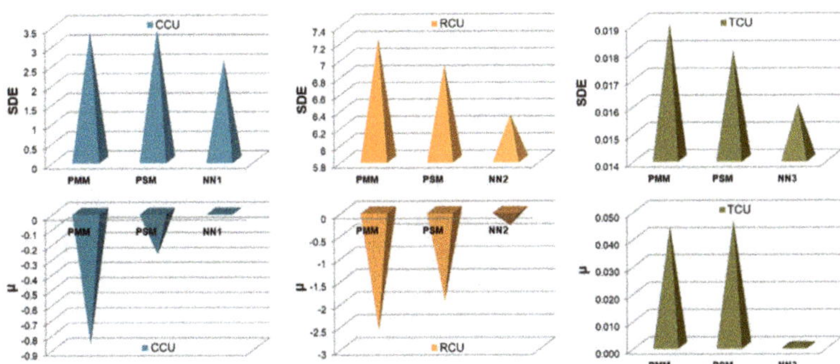

Figure 25. Comparison of fuzzy and ANN modeling results: SDE and μ.

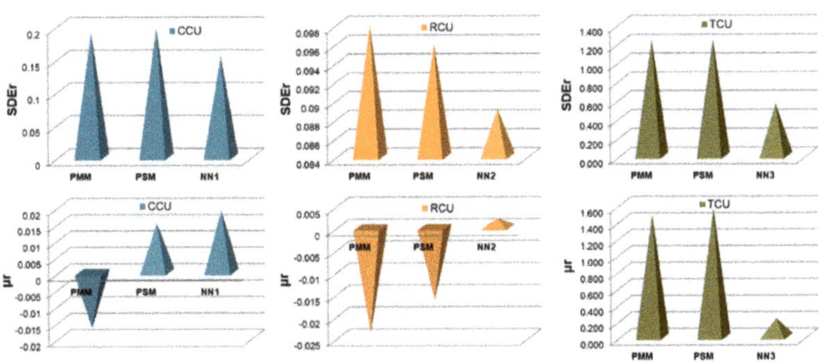

Figure 26. Comparison of fuzzy and ANN modeling results: SDE_r and μ_r.

By analyzing the statistical parameters of the relative error, it is observed that the RCU variable has the smallest standard deviations (SDE_r) in all models, which corresponds to the highest coefficients of determination for RCU variable in all models. Further, the variable TCU has the highest standard deviations of the relative prediction error, which also corresponds to the smallest values of R^2 in relation to CCU and RCU (Tables 8–10).

The means of the absolute errors for CCU and RCU variables are negative, which indicates that the models have a greater tendency to predict values that are less than the real ones. On the other hand, the means of errors for TCU are positive, which indicates that the models have a tendency to predict higher values from real ones. In accordance with the second observation are offset errors in the Figures 15c and 17c.

This study demonstrates that the developed models can be used as new tools to estimate the performance of the flotation circuit based on operational conditions, which are readily available at an industrial plant. Due to the dynamic nature and variations in real data, building an appropriate model for the prediction of flotation process in an industrial plant is a challenging task. More complex algorithms always require a larger amount of data. The sample size would depend on the number of input features and degree of nonlinearity between input and predicted variables (nature of the problem) [65]. In this paper, more than 1900 samples were used to capture the complexity of the process. The results indicate that the proposed NN and fuzzy models are able to predict the performance indicators of the flotation process in an industrial plant based on a large dataset with a high accuracy. The main reason why these models are more precise than some other presented models is that such models reported in the past were based on a small data set [2,28,66].

4. Conclusions

In this study, the metallurgical parameters of an industrial copper flotation plant were predicted by two fuzzy logic models (Mamdani and Takagi-Sugeno) and three back-propagation neural network (BPNN) models. Since the grade and recovery of concentrate are essential factors for decision making and control of the industrial flotation plant, their predictive model accuracy is very important. Therefore, a large dataset (1910 samples) was collected by daily/shiftly process monitoring during the multiannual plant's operation. The most effective parameters, namely copper grade in the ore feed, collector dosage in the rougher and the scavenger circuits, slurry pH in the rougher circuit and frother consumption, were selected and used in all models. The performance of each model was evaluated by well-known evaluation criteria, RMSE, R^2.and SD of prediction error. Based on the obtained results, the following can be concluded:

- The purpose of modeling the production process in the Veliki Krivelj plant is the possibility of implementing the obtained models into an automatic control system of this process. This system would include the application of controllers based on fuzzy logic or artificial neural networks.
- The NN and fuzzy logic models provided effective estimations due to the high R^2 and low RMSE values. However, the NN models were slightly more robust and accurate in predicting the values of metallurgical factors compared to the fuzzy logic models. The RMSE values of the NN models for the prediction of copper grade and recovery of the final concentrate were 2.567 and 6.284, respectively.
- Neural networks have the smallest standard deviations of the absolute and relative prediction error for all output variables.
- The differences between predicted and actual values are relatively small for all models. The significant deviations between actual and predicted values most likely occurred due to the fluctuations in real process data that can be caused by various factors, such as changing process dynamics due to the downtime of the plant, oscillations in the process parameters that were considered constant during modeling, changes in the reagents' quality, changes in process water quality, human factor, etc.
- The highest determination coefficients between actual and predicted values were obtained when modeling the copper recovery in the final concentrate, and the lowest when modeling the copper content in the final tailings. This can be applied to all models. The reason may lie in that the values of copper content in tailings vary in a relatively narrow range in relation to quality and recovery. Therefore, it may happen that the influences of completely different values of input parameters are integrated through very similar or the same copper contents in tailings, without this being taken into account during modeling. Such a situation could significantly affect the determination coefficient. Moreover, possible imperfections during tailings'

sampling (which are particularly linked to instabilities in the operation of the plant) should not be ignored, because the copper content in the samples is extremely low, and therefore proper sampling is of crucial importance for obtaining the precise chemical composition of the tailings.

- In accordance with the previous statement, the smallest standard deviations of the relative prediction error were obtained with the models that predict Cu recovery in the concentrate, and the largest with the models that predict Cu content in the tailings.
- By comparing the results of Mamdani and Takagi Sugeno fuzzy inference systems, it can be inferred that they demonstrate very similar predictive performance.
- Further research will be focused on the inclusion of other independent variables into the models, as well as on the modeling of individual parts of the flotation process depending on the availability (i.e., continual measurement) of process data to evaluate whether prediction results are improved or not.
- Sensitivity analysis of models, especially regarding the parameters considered constant, can be very helpful when it comes to improving the performances of the models and also represents the topic of future research.
- The application of more advanced software tools, as well as other soft computing methods, can also be effective in modeling such systems based on relatively large sets of input and output data.

Author Contributions: Conceptualization, I.J.; Methodology, I.J., F.N., D.K. and V.C.; Validation, F.N.; Investigation, I.J. and F.N.; Resources, I.J.; Data curation, D.K., V.C. and D.U.; Writing—original draft preparation, I.J. and F.N.; Writing—review and editing, D.K. and V.C.; Visualization, D.U.; Supervision, D.K., V.C. and D.U. All authors have read and agreed to the published version of the manuscript.

Funding: This work was financially supported by the Ministry of Education, Science and Technological Development of the Republic of Serbia, Grant No. 451-03-68/2022-14/200052.

Acknowledgments: We thank the Ministry of Education, Science and Technological Development of the Republic of Serbia.

Conflicts of Interest: The authors declare no conflict of interest.

Appendix A

During the observed period, in which measurement and data collection was carried out, the control and regulation system in the Veliki Krivelj flotation plant mainly relied on a manual mode of operation. This means that there is no fully automated and continuous monitoring of process parameters, but the role of controller is performed by experienced plant operators. In order to maintain the technological process in the stable state, the following technological parameters were measured and regulated:

- Particle size analysis of grinding products;
- Pulp density;
- Pulp level;
- Pulp pH value;
- Reagents' consumption.

Particle size distribution of the grinding product is one of the parameters that gives insight into the liberation degree of the mineral raw material. In order to determine the particle size distribution of the finally ground product, a sample of the hydrocyclone overflow is taken, and then sieved in the laboratory on the sieve with 0.074 mm aperture. The percentage mass content of the −0.074 mm class should be within a certain range (i.e., 58%–60%). If this content is lower than required, a correction is made at the feed to the rod mill by reducing the ore processing capacity or reducing the grinding pulp density within the prescribed range. If this content is higher than required, the ore processing capacity increases. When it comes to the overflow of the hydrocyclone in regrinding circuit, the regulation of the particle size distribution is conducted by reducing the amount of water (if the overflow is too coarse) or by increasing the amount of water (if the overflow is too fine) that is added to the regrinding mill.

In addition to the control of the finally ground product, if necessary, control of the grinding product of the rod mill, ball mill and regrinding mill is also carried out. Since the process is continuous and there is a need for quick reaction and process corrections, it is necessary to perform visual controls during the work. These controls are performed every hour and consist of rinsing a sample of the hydrocyclone overflow on the mining pan.

Pulp density has a significant impact on the grinding, classification and flotation process. The appropriate pulp density in the mills ensures the maximum capacity of the mill, while in the classifier (hydrocyclone), it affects the coarseness of the overflow. The optimal pulp density in the flotation cells provides adequate conditions for the optimization of the copper mineral flotation process.

Pulp density control is performed at various points in the process, such as for example, rod and ball mill discharges, hydrocyclone underflow and overflow, etc. Pulp density is determined by the glass pycnometer method in the laboratory. However, due to the timely reaction to changes in pulp density values in the process, it is necessary to perform a quick control of the same in the flotation plant. This procedure is performed by measuring the full "Denver" pycnometer on a "Denver" scale every 1–2 h.

Regulation of the pulp density is conducted by changing the amount of water added at certain points of the process or by changing the ore processing capacity. Water is added to the process manually, by means of a valve that corresponds to the given segment of the process in which the regulation is carried out.

The optimum value of the ***pulp level*** ensures the optimum time required for the hydrophobic mineral particles to adhere to the air bubbles. Regulation of the pulp level is done manually—by raising the plugs (when it is necessary to decrease the pulp level) or by lowering the plugs (when it is necessary to increase the pulp level), which are located in the boxes of the flotation machines.

The optimal ***pH value of the pulp*** creates optimal conditions for collector action on the surfaces of the mineral particles, as well as the selectivity of the flotation process. The pH value is measured with pH-meters whose electrodes are immersed in the hydrocyclone overflows. These values are read on the control panel display.

In addition, the pH values of all three cleaning tails are also determined. Samples are taken manually, and the pH is determined with manual pH-meters or the pulp titration method in the laboratory (this method is based on determining free lime in the clear part of the solution. The volumetric quantity of titrant (in this case HCl) needed to react with all $Ca(OH)_2$ in the solution (end point of the titration) is measured, according to the equation $Ca(OH)_2 + 2HCl = CaCl_2 + 2H_2O$).

To regulate the pH value of the pulp, lime is used, which is added to the process in the form of lime milk (6% aqueous solution of calcium oxide). When it comes to the grinding process, reducing or increasing the dose of milk of lime, depending on the desired pH value of the pulp, is conducted from the control panel, by adjusting the clearance of the pneumatic valves located at each dosing point. The regulation of lime milk consumption in each stage of cleaning is conducted manually.

Reagents are, as is known, one of the key elements for the success of the copper flotation process. Control of the reagent dosage (collector and frother) is performed by measuring the flow of reagents (which is read in the control room) and calculating their consumption per ton of processed ore.

Potassium ethyl xanthate (PEX) is used as collector and dosed into the process in the form of a 10% solution. PEX is added only to the roughing and scavenging circuit. The regulation of collector consumption is a complex task that depends on a number of factors in the plant. First of all, the information obtained from the mining pit about the ore mineralogical composition and copper content is considered. In addition, a visual analysis of the ground material, concentrate and tailings is performed by washing them on the mining pan every hour (the analysis of intermediate products is also performed in this way, but with a lower frequency of testing). The appearance of the flotation froth in terms of its color, bubble size, mineralization, etc., is also monitored.

Therefore, there are a number of factors that indicate whether it is necessary to increase or decrease the dose of the collector. Experienced plant operators recognize these indicators and take appropriate management action accordingly. Regulation of the collector consumption is conducted manually, using the valves located at each dosing point.

Dowfroth 250 is used as a frother and added to the process in liquid form, concentrated. Frother is only added in the agitation stage, just before the roughing. Its consumption primarily depends on the stability of the flotation froth (i.e., whether the froth "grows" or is prone to collapse), then on the presence of alumina in the feed, etc. The frother flow is regulated in one place, by means of a manual valve.

Besides the mentioned parameters, which are controlled and regulated directly in the "Veliki Krivelj" flotation plant, particle size distribution of the finally crushed ore is determined daily, which can indicate the efficiency of the crushing process and the eventual need to regulate the processing capacity in grinding. In addition, the copper content in the feed ore, concentrate and tailings is determined by chemical analysis. Samples for chemical analysis are taken every hour, and from the hourly increments, shift composite samples are formed. Based on results of chemical analyses, feed capacity and concentrate quantity, the recovery of copper in concentrate is calculated.

Data on the input parameters of the technological process, as well as data on the concentration products, are collected by daily monitoring and stored in the record documentation. Data are stored for each individual shift as a shift average.

For the purposes of model development, a two-and-a-half-year period of continuous plant production was chosen, with a total of 2553 shift data available to the authors. These data are arranged in a time-line order. However, considering that during the plant operation there is a stoppage due to various factors (such as device failures, the need for overhaul and lack of ore from the open pit), the analysis found that data are missing for certain shifts, either completely or partially. Moreover, due to errors that may occur during sampling, or due to the imperfection of chemical analyses, imperfection of measuring instruments, imperfection of calculations (human factor), etc., some shift data contained illogicalities in the sense of extremely low or extremely high values of process parameters. Such shift data were also eliminated from the primary dataset, so that the final and complete dataset contained 1910 shift data. The time-line sequence in the data arrangement is preserved.

Analyzing these final datasets, it was still possible to notice data scattering, as well as relatively large differences between minimum and maximum values for certain variables (e.g., recovery). Still, one of the modeling goals was to use a maximal quantity of industrial data and to, however is possible, follow the process continuity and the changes that occur over time. Therefore, these scatterings were considered normal during the industrial process, with the level of error in the process parameters determination up to 10%, according to the experiences of process and chemical engineers.

Appendix B

Appendix B.1. Input Datasets

Input datasets, as well as their distributions, are given in Figures A1–A10.

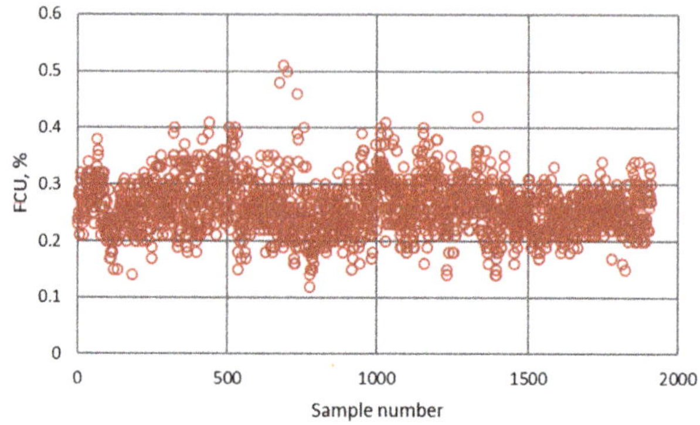

Figure A1. Copper content in feed (FCU)—input dataset.

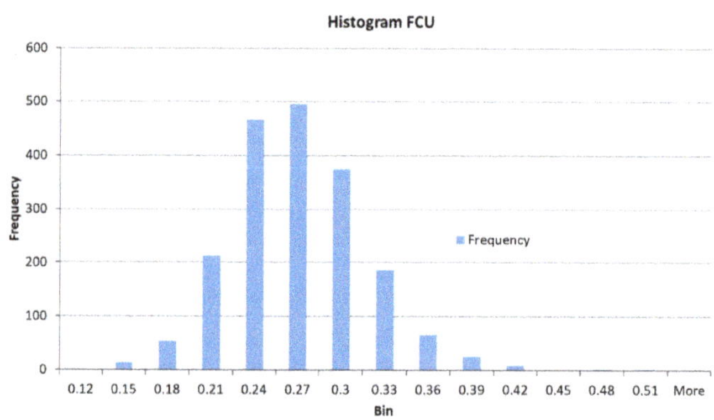

Figure A2. Copper content in feed (FCU)—dataset distribution.

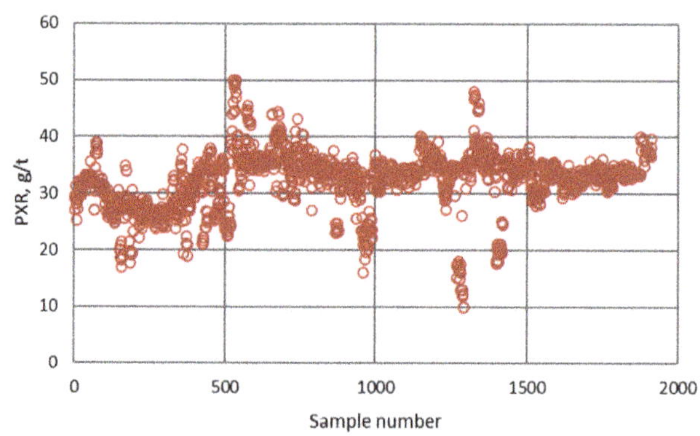

Figure A3. Collector consumption in rougher flotation stage (PXR)—input dataset.

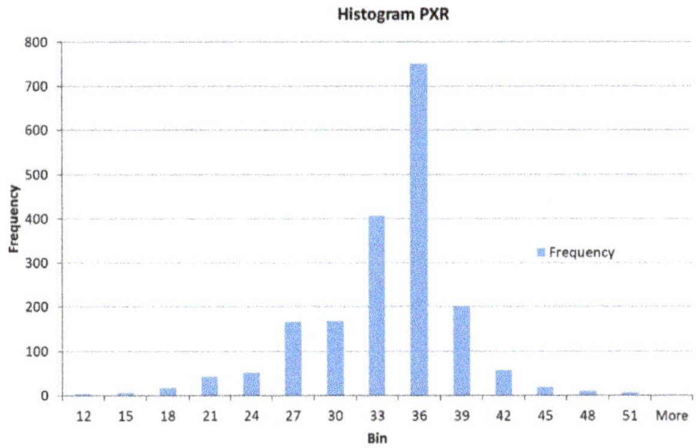

Figure A4. Collector consumption in rougher flotation stage (PXR)—dataset distribution.

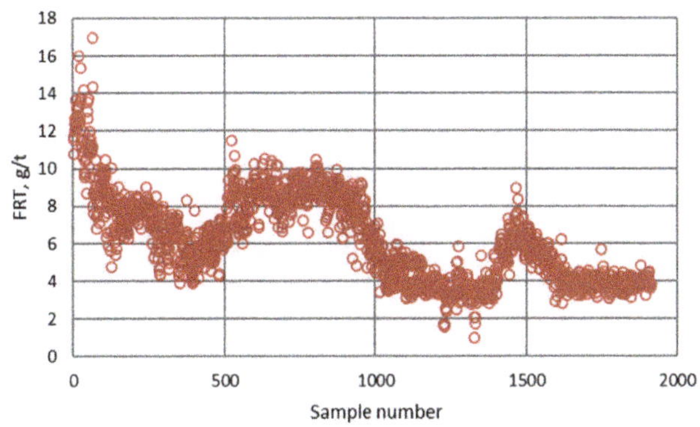

Figure A5. Frother consumption (FRT)—input dataset.

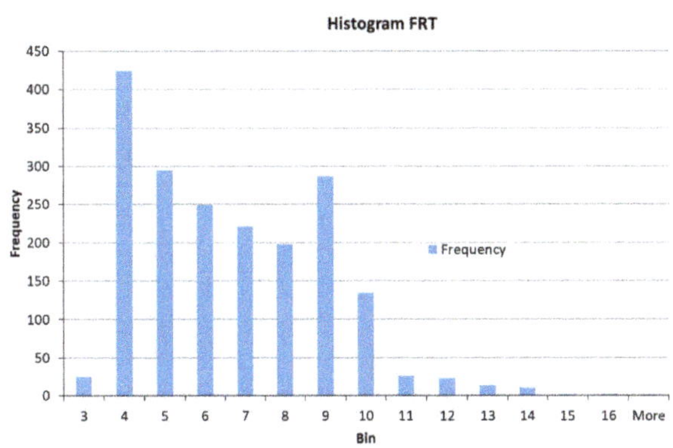

Figure A6. Frother consumption (FRT)—dataset distribution.

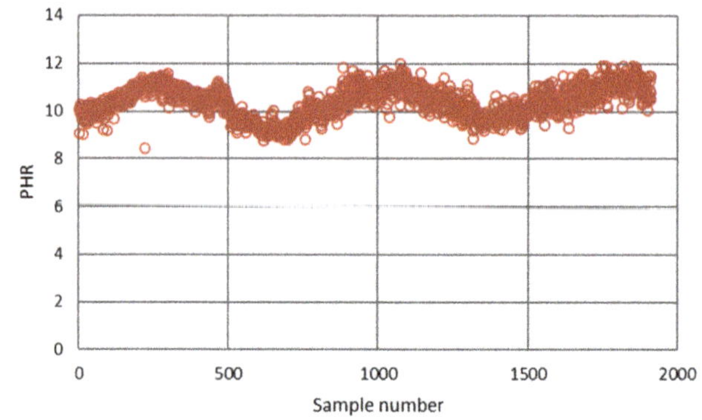

Figure A7. pH value of the pulp in rougher flotation stage (PHR)—input dataset.

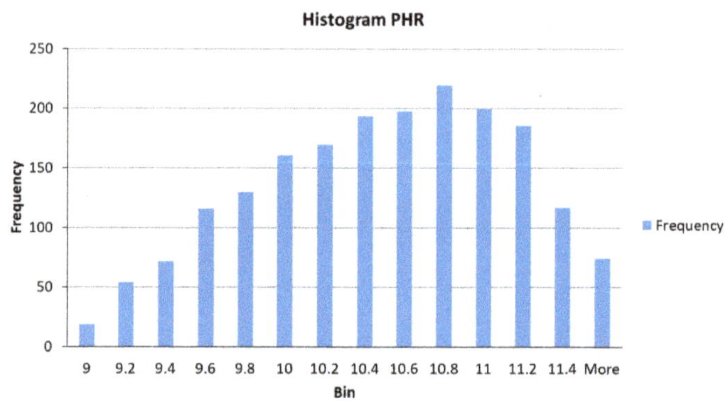

Figure A8. pH value of the pulp in rougher flotation stage (PHR)—dataset distribution.

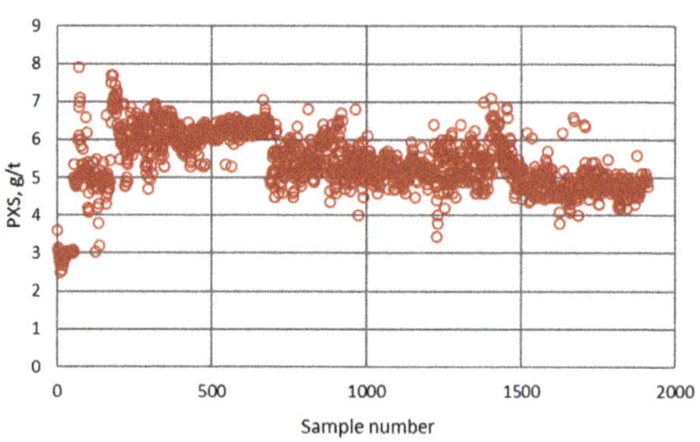

Figure A9. Collector consumption in scavenger flotation stage (PXS)—input dataset.

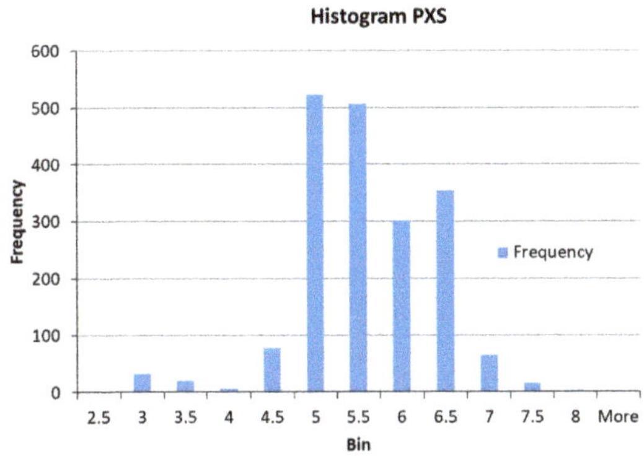

Figure A10. Collector consumption in scavenger flotation stage (PXS)—dataset distribution.

Appendix B.2. Output Datasets, Training, Testing and Validation Datasets

Distributions of output datasets are given in Figures A11–A13. Since all the data from the training testing and validation sets can be observed, the entire output datasets are not shown here.

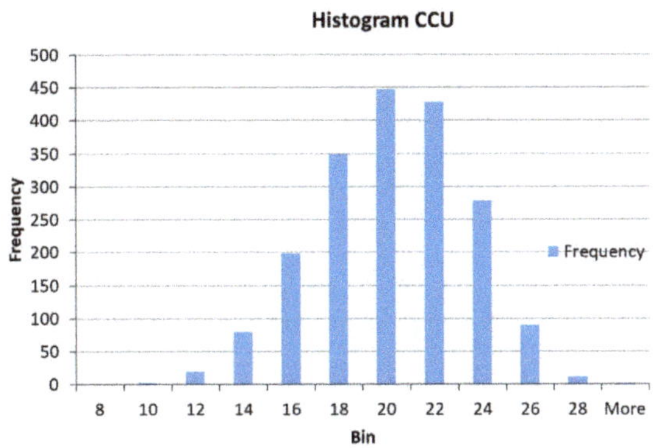

Figure A11. Copper content in final concentrate (CCU)—dataset distribution.

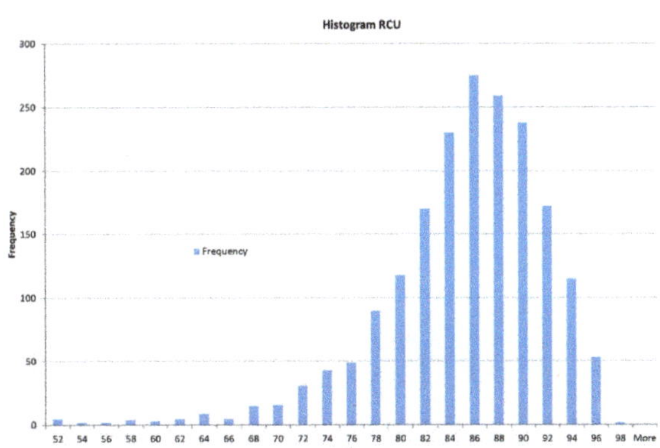

Figure A12. Copper recovery in final concentrate (RCU)—dataset distribution.

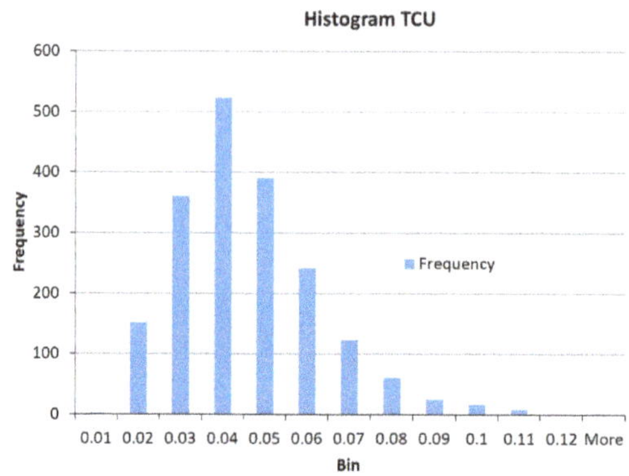

Figure A13. Copper content in final tailings (TCU)—dataset distribution.

Appendix B.2.1. Fuzzy Logic Model Based on Mamdani Inference System—PMM

Training and testing datasets of this model are selected on the following way:

- Data from every even shift (955 in total) are determined for training,
- Data from every odd shift (955 in total) are determined for testing.

Based on training results, the additional adjusting of fuzzy numbers within membership functions is performed. Fuzzy logic model based on the Takagi-Sugeno inference system (PSM) is obtained only by transformation of PMM to compare the results.

Training Datasets

Information about the PMM model training dataset is shown in Figures A14–A19 and in the Table A1.

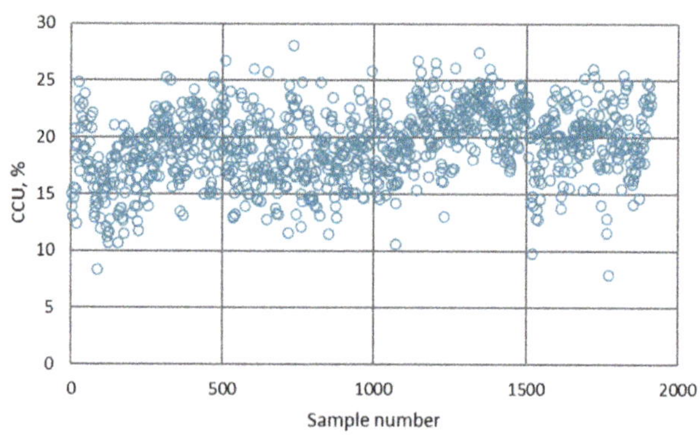

Figure A14. PMM model—training dataset for CCU output variable.

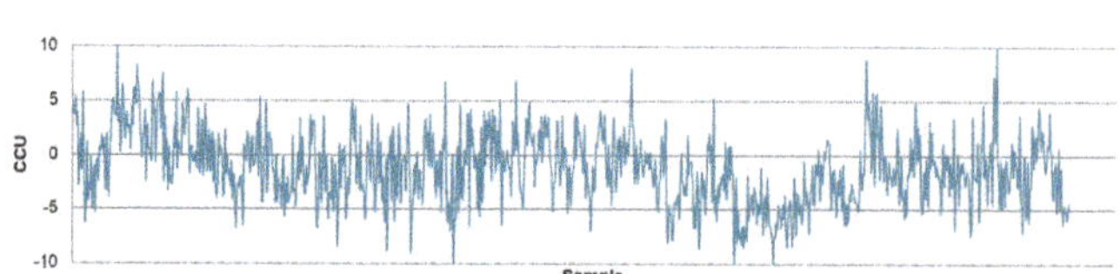

Figure A15. Prediction error of PMM model training dataset for CCU output variable.

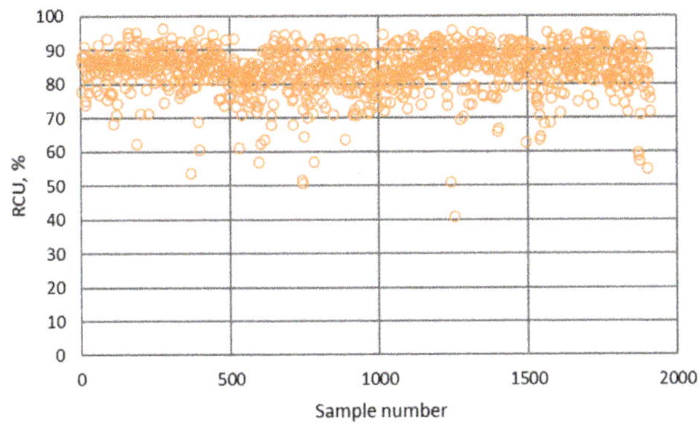

Figure A16. PMM model—training dataset for RCU output variable.

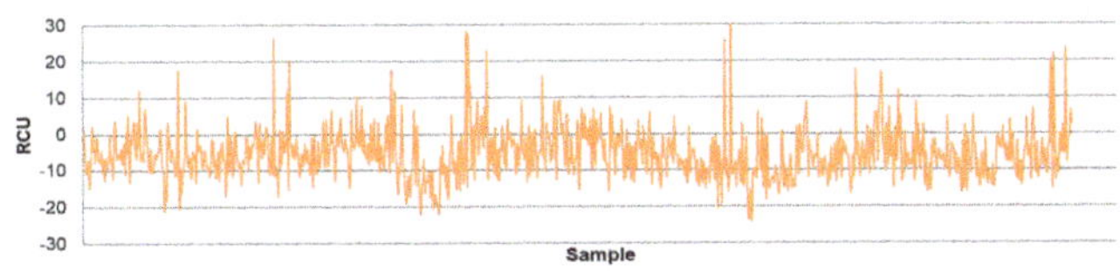

Figure A17. Prediction error of PMM model training dataset for RCU output variable.

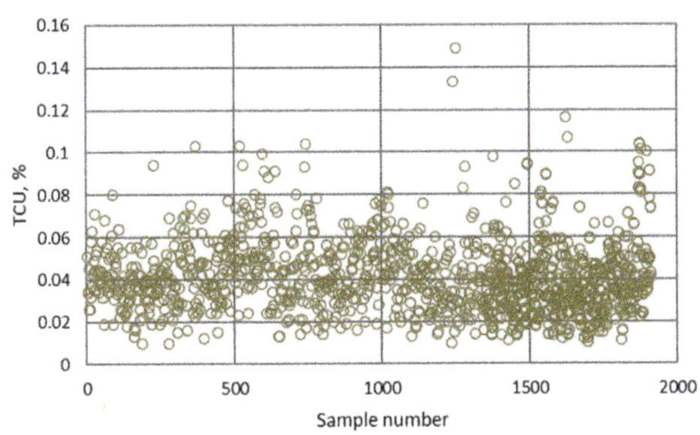

Figure A18. PMM model—training dataset for TCU output variable.

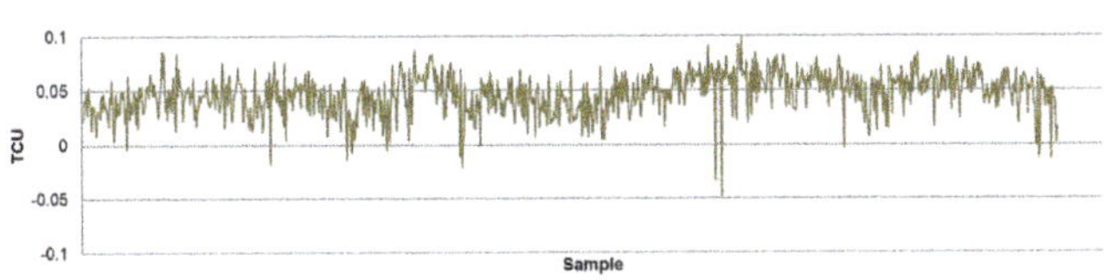

Figure A19. Prediction error of PMM model training dataset for TCU output variable.

Table A1. Regression statistics parameters of PMM model training dataset.

Statistical Parameters	Technological Indicator of the Flotation Process		
	CCU	RCU	TCU
R^2	0.971	0.992	0.851
RMSE	3.096	6.681	0.034

Testing Datasets

Information about the PMM model testing datasets is shown in Figures A20–A25 and in the Table A2.

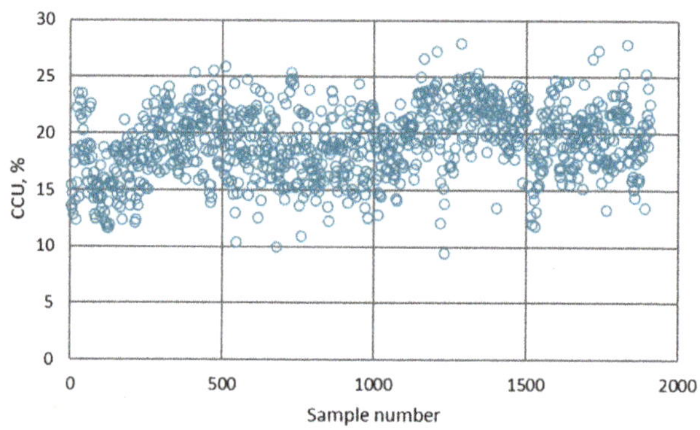

Figure A20. PMM model—testing dataset for CCU output variable.

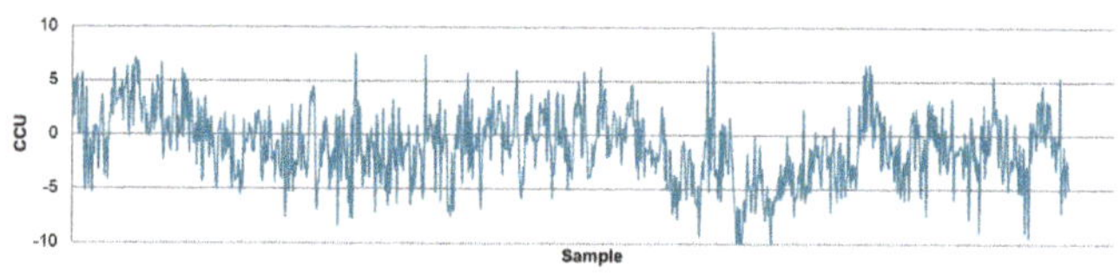

Figure A21. Prediction error of PMM model testing dataset for CCU output variable.

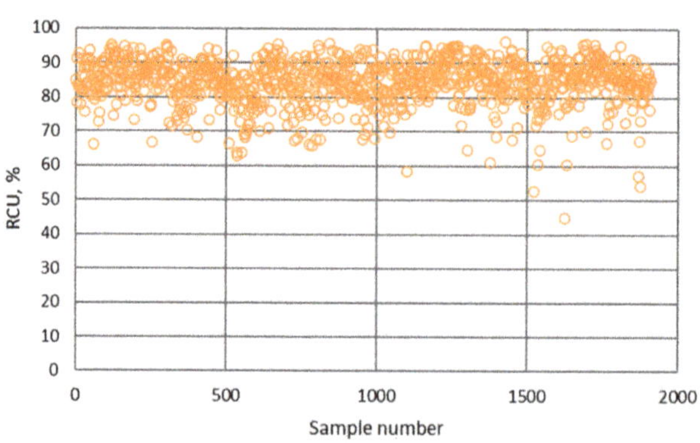

Figure A22. PMM model—testing dataset for RCU output variable.

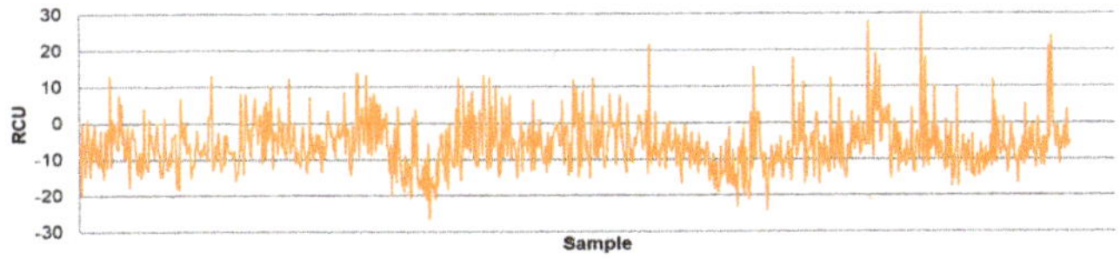

Figure A23. Prediction error of PMM model testing dataset for RCU output variable.

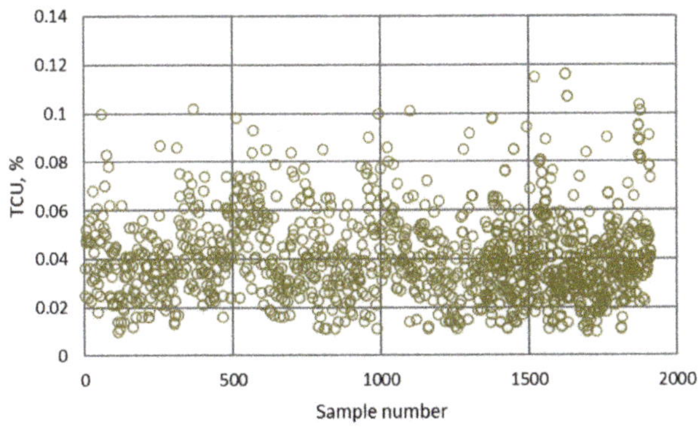

Figure A24. PMM model—testing dataset for TCU output variable.

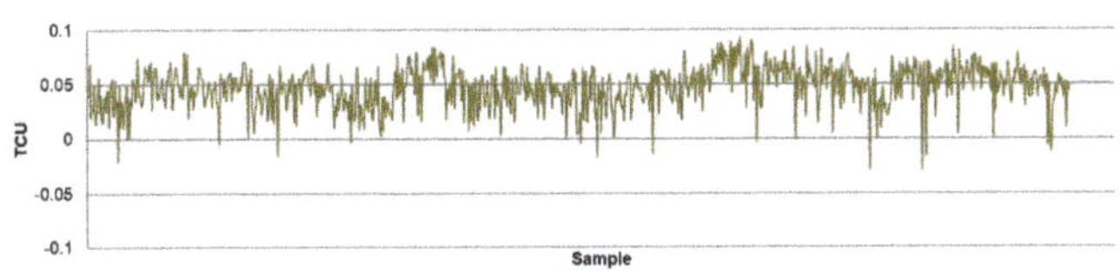

Figure A25. Prediction error of PMM model testing dataset for TCU output variable.

Table A2. Regression statistics parameters of PMM model testing dataset.

Statistical Parameters	Technological Indicator of the Flotation Process		
	CCU	RCU	TCU
R^2	0.972	0.993	0.840
RMSE	3.049	6.613	0.035

Appendix B.2.2. Artificial Neural Network-Based Model for CCU Prediction—NN1

Training, testing and validation data sets of this model are randomly chosen by the network in the following way:

- A total of 60% of the data (1336 in total) are determined for training;
- A total of 15% of the data (287 in total) are determined for testing;
- A total of 15% of the data (287 in total) are determined for validation.

Information about the NN1 model training, testing and validation datasets is shown in Figures A26–A31 and in the Table A3.

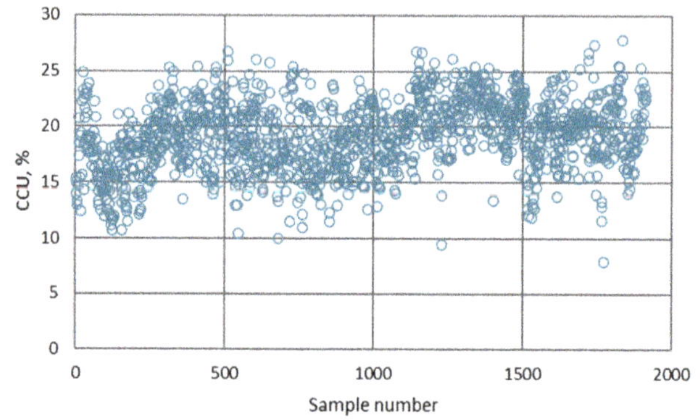

Figure A26. NN1 model—training dataset.

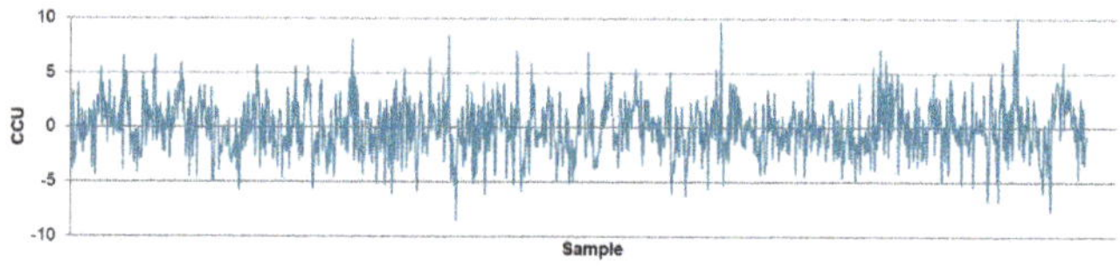

Figure A27. Prediction error of NN1 model training dataset.

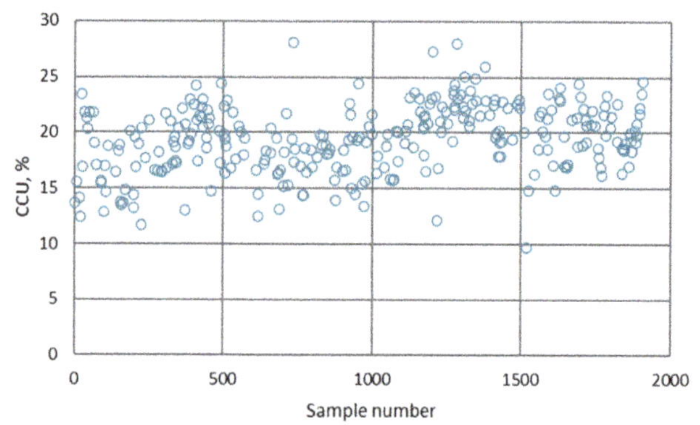

Figure A28. NN1 model—testing dataset.

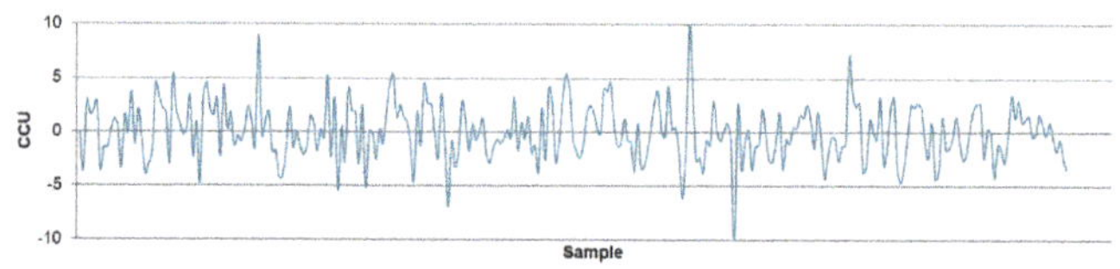

Figure A29. Prediction error of NN1 model testing dataset.

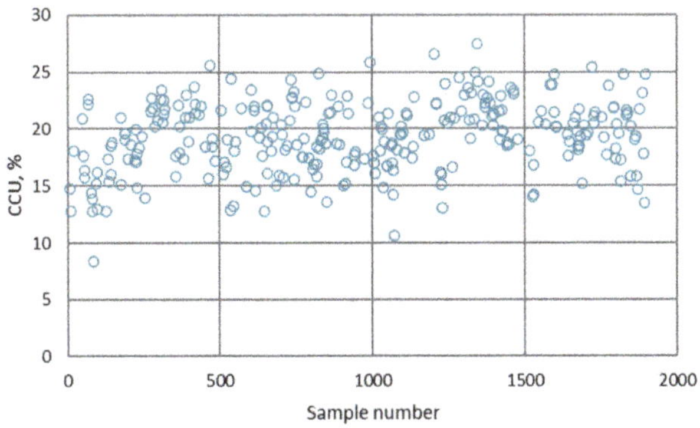

Figure A30. NN1 model—validation dataset.

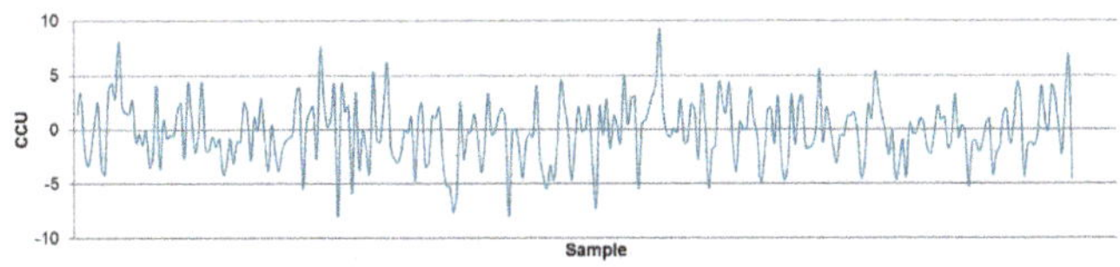

Figure A31. Prediction error of NN1 model validation dataset.

Table A3. Regression statistics parameters of NN1 model datasets.

Statistical Parameters	CCU		
	Training	Testing	Validation
R^2	0.983	0.981	0.978
RMSE	2.489	2.635	2.854

Appendix B.2.3. Artificial Neural Network-Based Model for RCU Prediction—NN2

Training, testing and validation data sets of this model are randomly chosen by the network in the following way (same as for NN1):

- A total of 60% of the data (1336 in total) are determined for training;
- A total of 15% of the data (287 in total) are determined for testing;
- A total of 15% of the data (287 in total) are determined for validation.

Information about the NN2 model training, testing and validation datasets is shown in Figures A32–A37 and in the Table A4.

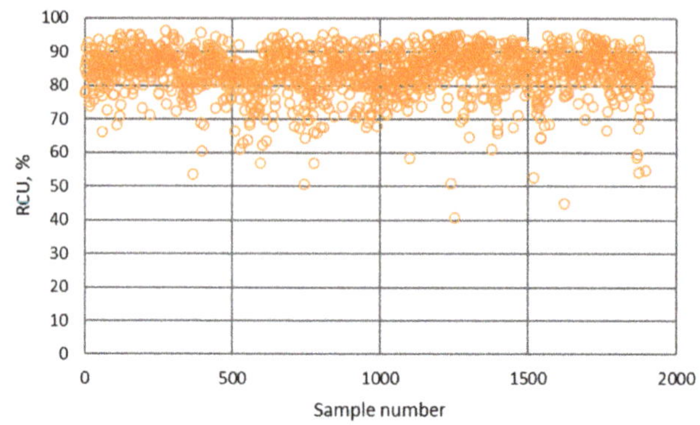

Figure A32. NN2 model—training dataset.

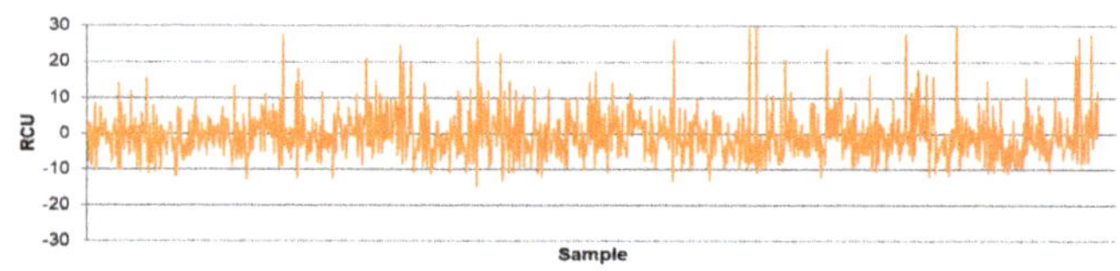

Figure A33. Prediction error of NN2 model training dataset.

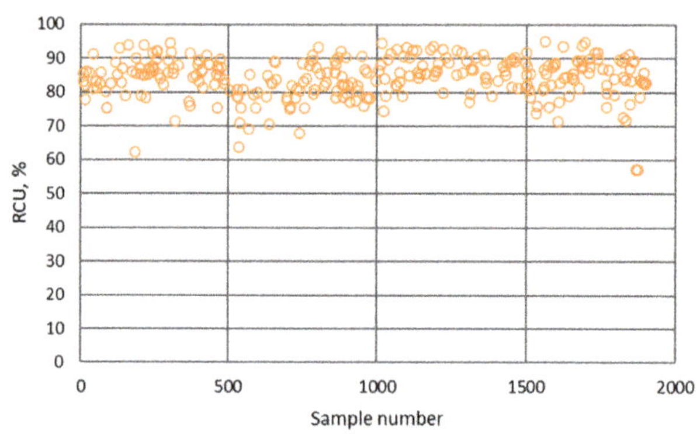

Figure A34. NN2 model—testing dataset.

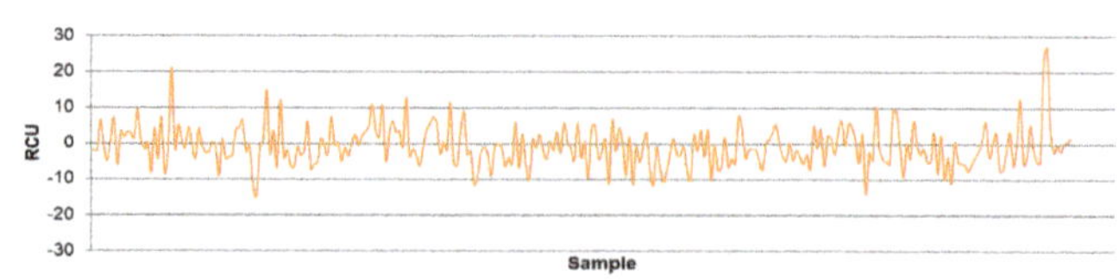

Figure A35. Prediction error of NN2 model testing dataset.

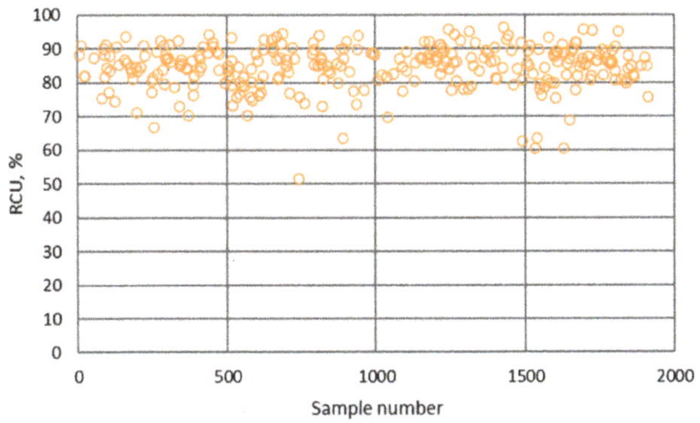

Figure A36. NN2 model—validation dataset.

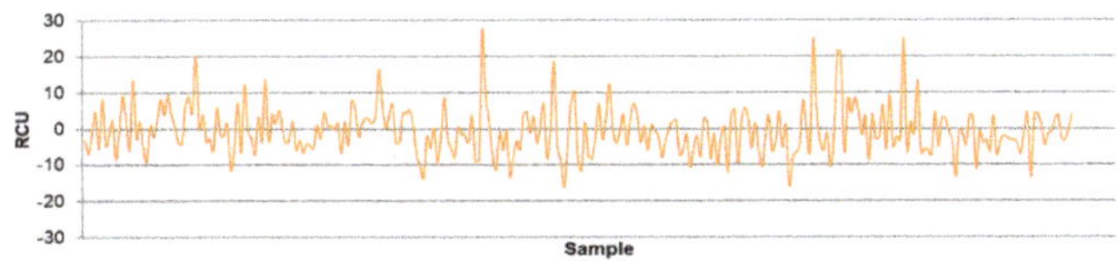

Figure A37. Prediction error of NN2 model validation dataset.

Table A4. Regression statistics parameters of NN1 model datasets.

Statistical Parameters	RCU		
	Training	Testing	Validation
R^2	0.994	0.995	0.993
RMSE	6.291	5.762	6.767

Appendix B.2.4. Artificial Neural Network-Based Model for TCU Prediction—NN3

Training, testing and validation data sets of this model are randomly chosen by the network in the following way (same as for NN1 and NN2):

- A total of 60% of the data (1336 in total) are determined for training;
- A total of 15% of the data (287 in total) are determined for testing;
- A total of 15% of the data (287 in total) are determined for validation.

Information about the NN3 model training, testing and validation datasets is shown in Figures A38–A43 and in the Table A5.

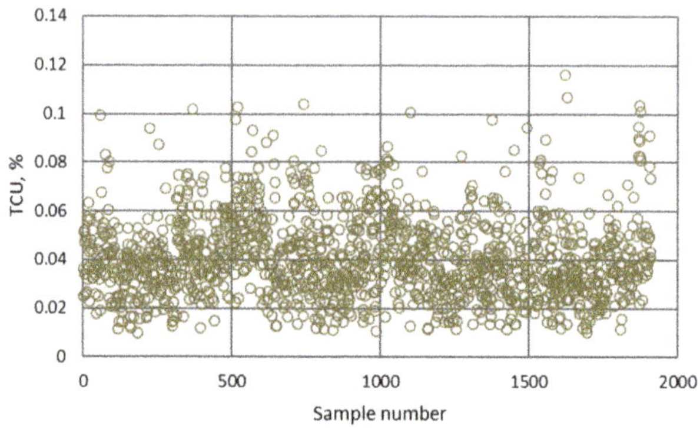

Figure A38. NN3 model—training dataset.

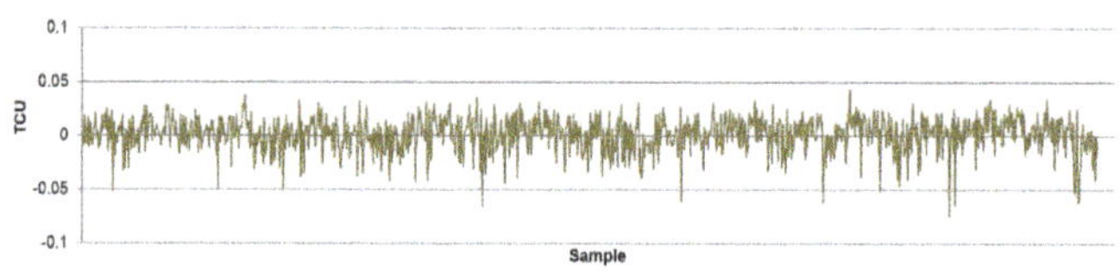

Figure A39. Prediction error of NN3 model training dataset.

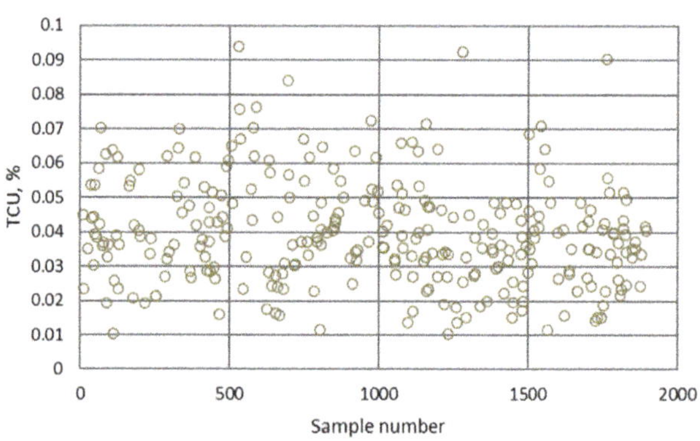

Figure A40. NN3 model—testing dataset.

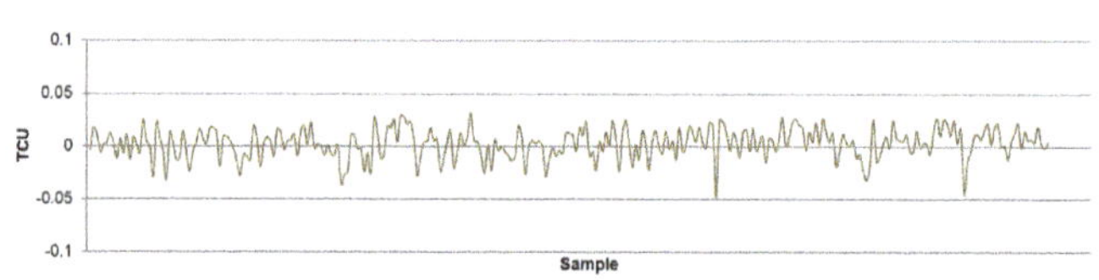

Figure A41. Prediction error of NN3 model testing dataset.

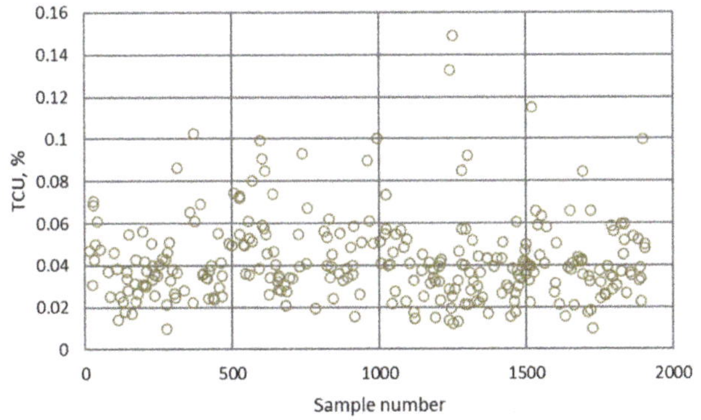

Figure A42. NN3 model—validation dataset.

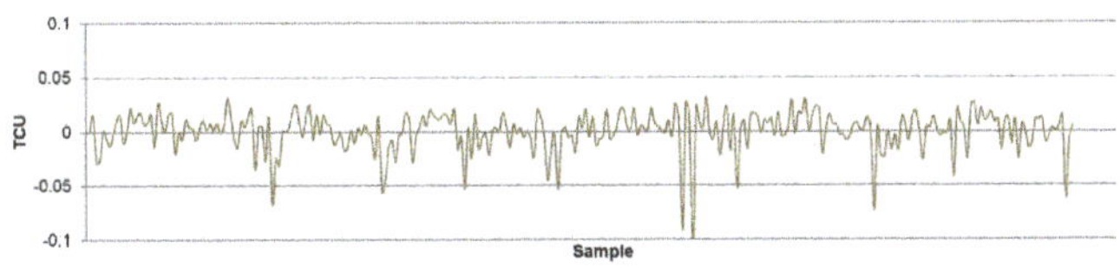

Figure A43. Prediction error of NN3 model validation dataset.

Table A5. Regression statistics parameters of NN3 model datasets.

Statistical Parameters	RCU		
	Training	Testing	Validation
R^2	0.873	0.884	0.829
RMSE	0.015	0.014	0.017

References

1. Jovanović, I.; Magdalinović, N.; Kržanović, D.; Rajković, R. Comparative analysis of AI models in the modeling of flotation process. In *Mining and Metallurgy Engineering Bor 3-4/2018*; Mining and Metallurgy Engineering Bor: Bor, Serbia, 2018; pp. 27–32, ISSN 2334-8836.
2. Nakhaei, F.; Sam, A.; Mosavi, M.R.; Zeidabadi, S. Prediction of copper grade at flotation column concentrate using artificial neural network. In Proceedings of the IEEE 10th International Conference on Signal Processing Proceedings, Beijing, China, 24–28 October 2010; pp. 1421–1424.
3. Quintanilla, P.; Neethling, S.J.; Brito-Parada, P.R. Modelling for froth flotation control: A review. *Miner. Eng.* **2021**, *162*, 106718. [CrossRef]
4. Jovanović, I. Model Inteligentnog Sistema Adaptivnog Upravljanja Procesom Prerade Rude. Ph.D. Thesis, Rudarsko-geološki Fakultet, Belgrade, Serbia, 2016; p. 217. (In Serbian)
5. Jovanović, I.; Miljanović, I.; Jovanović, T. Soft computing-based modeling of flotation processes—A review. *Miner. Eng.* **2015**, *84*, 34–63. [CrossRef]
6. Cisternas, L.A.; Lucay, F.A.; Botero, Y.L. Trends in Modeling, Design, and Optimization of Multiphase Systems in Minerals Processing. *Minerals* **2020**, *10*, 22. [CrossRef]
7. Dos Santos, N.A.; Savassi, O.; Peres, A.E.C.; Martins, A.H. Modelling flotation with a flexible approach—Integrating different models to the compartment model. *Miner. Eng.* **2014**, *66–68*, 68–76. [CrossRef]
8. Seppälä, P.; Sorsa, A.; Paavola, M.; Ruuska, J.; Remes, A.; Kumar, H.; Lamberg, P.; Leiviskä, K. Development and calibration of a dynamic flotation circuit model. *Miner. Eng.* **2016**, *96–97*, 168–176. [CrossRef]
9. Suazo, C.J.; Kracht, W.; Alruiz, O.M. Geometallurgical modelling of the Collahuasi flotation circuit. *Miner. Eng.* **2010**, *23*, 137–142. [CrossRef]
10. Al-Thyabat, S. On the optimization of froth flotation by the use of artificial neural network. *J. China Univ. Min. Technol.* **2008**, *18*, 418–426. [CrossRef]

11. Farghaly, M.; Serwa, A.; Ahmed, M. Optimizing the Egyptian Coal Flotation Using an Artificial Neural Network. *J. Min. World Express* **2012**, *1*, 27–33.
12. Jorjani, E.; Mesroghli, S.; Chelgani, S.C. Prediction of operational parameters effect on coal flotation using artificial neural network. *J. Univ. Sci. Technol. Beijing* **2008**, *15*, 528–533. [CrossRef]
13. Nakhaei, F.; Mosavi, M.R.; Sam, A.; Vaghei, Y. Recovery and grade accurate prediction of pilot plant flotation column concentrate: Neural network and statistical techniques. *Int. J. Miner. Process.* **2012**, *110–111*, 140–154. [CrossRef]
14. Nakhaei, F.; Irannajad, M. Comparison between neural networks and multiple regression methods in metallurgical performance modeling of flotation column. *Physicochem. Probl. Miner. Process.* **2013**, *49*, 255–266.
15. Nakhaei, F.; Irannajad, M. Forecasting grade and recovery of flotation column concentrate using radial basis function and layer recurrent neural networks. *AWERProcedia Inf. Technol. Comput. Sci.* **2013**, *4*, 454–473.
16. Jahedsaravani, A.; Marhaban, M.H.; Massinaei, M. Application of statistical and intelligent techniques for modeling of metallurgical performance of a batch flotation process. *Chem. Eng. Commun.* **2016**, *203*, 151–160. [CrossRef]
17. Massinaei, M.; Sedaghati, M.R.; Rezvani, R.; Mohammadzadeh, A.A. Using data mining to assess and model the metallurgical efficiency of a copper concentrator. *Chem. Eng. Commun.* **2014**, *201*, 1314–1326. [CrossRef]
18. Zhang, Y.; Chen, H.; Yang, B.; Fu, S.; Yu, J.; Wang, Z. Prediction of phosphate concentrate grade based on artificial neural network modeling. *Results Phys.* **2018**, *11*, 625–628. [CrossRef]
19. Abdulhussein, S.A.; Alwared, A.I. The use of Artificial Neural Network (ANN) for modeling of Cu (II) ion removal from aqueous solution by flotation and sorptive flotation process. *Environ. Technol. Innov.* **2019**, *13*, 353–363. [CrossRef]
20. Gholami, A.; Movahedifar, M.; Khoshdast, H.; Hassanzadeh, A. Hybrid Serving of DOE and RNN-Based Methods to Optimize and Simulate a Copper Flotation Circuit. *Minerals* **2022**, *12*, 857. [CrossRef]
21. Aldrich, C.; Marais, C.; Shean, B.J.; Cilliers, J.J. Online monitoring and control of froth flotation systems with machine vision: A review. *Int. J. Miner. Process.* **2010**, *96*, 1–13. [CrossRef]
22. Marais, C.; Aldrich, C. The estimation of platinum flotation grade from froth image features by using artificial neural networks. *J. South. Afr. Inst. Min. Metall.* **2011**, *111*, 81–85.
23. Marais, C.; Aldrich, C. Estimation of platinum flotation grades from froth image data. *Miner. Eng.* **2011**, *24*, 433–441. [CrossRef]
24. Jahedsaravani, A.; Marhaban, M.H.; Massinaei, M. Prediction of the metallurgical performances of a batch flotation system by image analysis and neural networks. *Miner. Eng.* **2014**, *69*, 137–145. [CrossRef]
25. Hosseini, M.R.; Haji Amin Shirazi, H.; Massinaei, M.; Mehrshad, N. Modeling the Relationship between Froth Bubble Size and Flotation Performance Using Image Analysis and Neural Networks. *Chem. Eng. Commun.* **2015**, *202*, 911–919. [CrossRef]
26. Wang, J.-S.; Han, S.; Shen, N.-N.; Li, S.-X. Features Extraction of Flotation Froth Images and BP Neural Network Soft-Sensor Model of Concentrate Grade Optimized by Shuffled Cuckoo Searching Algorithm. *Sci. World J.* **2014**, *2014*, 208094. [CrossRef]
27. Haiyang, Z.; Yali, K.; Guanghui, W.; Li, J. Soft Sensor Model for Coal Slurry Ash Content Based on Image grey characteristics. *Int. J. Coal Prep. Util.* **2014**, *34*, 24–37. [CrossRef]
28. Nakhaei, F.; Irannajad, M.; Mohammadnejad, S. Column flotation performance prediction: PCA, ANN and image analysis-based approaches. *Physicochem. Probl. Miner. Process.* **2019**, *55*, 1298–1310.
29. Nakhaei, F.; Irannajad, M.; Mohammadnejad, S. A comprehensive review of froth surface monitoring as an aid for grade and recovery prediction of flotation process. Part A: Structural features. *Energy Sources Part A Recovery Util. Environ. Eff.* **2019**. [CrossRef]
30. Zhang, H.; Tang, Z.; Xie, Y.; Gao, X.; Chen, Q.; Gui, W. Long short-term memory-based grade monitoring in froth flotation using a froth video sequence. *Miner. Eng.* **2021**, *160*, 106677. [CrossRef]
31. Fu, Y.; Aldrich, C. Flotation froth image recognition with convolutional neural networks. *Miner. Eng.* **2019**, *132*, 183–190. [CrossRef]
32. Pu, Y.; Szmigiel, A.; Chen, J.; Apel, D.B. FlotationNet: A hierarchical deep learning network for froth flotation recovery prediction. *Powder Technol.* **2020**, *375*, 317–326. [CrossRef]
33. Zarie, M.; Jahedsaravani, A.; Massinaei, M. Flotation froth image classification using convolutional neural networks. *Miner. Eng.* **2020**, *155*, 106443. [CrossRef]
34. Wen, Z.; Zhou, C.; Pan, J.; Nie, T.; Zhou, C.; Lu, Z. Deep learning-based ash content prediction of coal flotation concentrate using convolutional neural network. *Miner. Eng.* **2021**, *174*, 107251. [CrossRef]
35. Gao, X.; Tang, Z.; Xie, Y.; Zhang, H.; Gui, W. A layered working condition perception integrating handcrafted with deep features for froth flotation. *Miner. Eng.* **2021**, *170*, 107059. [CrossRef]
36. Jahan, A.; Edwards, K.L.; Bahraminasab, M. Multi-Attribute Decision-Making for Ranking of Candidate Materials. In *Multi-criteria Decision Analysis for Supporting the Selection of Engineering Materials in Product Design*, 2nd ed.; Jahan, A., Edwards, K.L., Bahraminasab, M., Eds.; Butterworth-Heinemann: Oxford, UK, 2016; pp. 81–126.
37. Naderloo, L.; Alimardani, R.; Omid, M.; Sarmadian, F.; Javadikia, P.; Torabi, M.Y.; Alimardani, F. Application of ANFIS to predict crop yield based on different energy inputs. *Measurement* **2012**, *45*, 1406–1413. [CrossRef]
38. Amiryousefi, M.R.; Mohebbi, M.; Khodaiyan, F.; Asadi, S. An empowered adaptive neuro-fuzzy inference system using self-organizing map clustering to predict mass transfer kinetics in deep-fat frying of ostrich meat plates. *Comput. Electron. Agric.* **2011**, *76*, 89–95. [CrossRef]

39. Sitorus, F.; Brito-Parada, P.R. Equipment selection in mineral processing—A sensitivity analysis approachfor a fuzzy multiple criteria decision making model. *Miner. Eng.* **2020**, *150*, 106261. [CrossRef]
40. Zhang, W.; Liu, D.; Du, Y.; Liu, R.; Wang, D.; Yu, L.; Wen, S. Mill Feed Control System and Algorithm Based on Python. *Minerals* **2022**, *12*, 804. [CrossRef]
41. Jovanović, I.; Nešković, J.; Petrović, S.; Milanović, D. A hybrid approach to modeling the flotation process from the "Veliki Krivelj" plant. In *Mining and Metallurgy Engineering Bor 1-2/2018*; Mining and Metallurgy Engineering Bor: Bor, Serbia, 2018; pp. 1–10. ISSN 2334-8836.
42. Lott, G.D.; Da Silva, M.T.; Cota, L.P.; Guimarães, F.G.; Euzébio, T.A.M. Fuzzy Decision Support System for the Calibration of Laboratory-Scale Mill Press Parameters. *IEEE Access* **2021**, *9*, 24901–24912. [CrossRef]
43. Carvalho, M.T.; Durão, F. Performance of a flotation column fuzzy controller. In *Computers and Computational Engineering in Control*; Mastorakis, N.E., Ed.; World Scientific and Engineering Society Press: Athens, Greece, 1999; pp. 220–225.
44. Carvalho, M.T.; Durão, F. Control of a flotation column using fuzzy logic inference. *Fuzzy Sets Syst.* **2002**, *125*, 121–133. [CrossRef]
45. Vieira, S.M.; Sousa, J.M.C.; Durão, F.O. Fuzzy modelling strategies applied to a column flotation process. *Miner. Eng.* **2005**, *18*, 725–729. [CrossRef]
46. Liao, Y.; Liu, J.; Wang, Y.; Cao, Y. Simulating a fuzzy level controller for flotation columns. *Min. Sci. Technol.* **2011**, *21*, 815–818. [CrossRef]
47. Jahedsaravani, A.; Mehrshad, N.; Massinaei, M. Fuzzy-based modeling and control of an industrial flotation column. *Chem. Eng. Commun.* **2014**, *201*, 896–908.
48. Zhou, K.; Zhou, X. Adaptive fuzzy local ternary pattern for mineral flotation froth image edge detection. *IFAC Pap.* **2018**, *51*, 235–240. [CrossRef]
49. Liang, Y.; He, D.; Su, X.; Wang, F. Fuzzy distributional robust optimization for flotation circuit configurations based on uncertainty theories. *Miner. Eng.* **2020**, *156*, 106433. [CrossRef]
50. Ali, S. Mathematical Models for the Efficiency of Flotation Process for North Waziristan Copper. Ph.D. Thesis, University of Education, Lahore, Pakistan, 2007, 2007; p. 167.
51. Yang, L.; Liu, T.; Ren, W.; Sun, W. Fuzzy Neural Network for Studying Coupling between Drilling Parameters. *ACS Omega* **2021**, *6*, 24351–24361. [CrossRef]
52. Zadeh, L.A. A new direction in AI: Toward a computational theory of perceptions. *AI Mag.* **2001**, *22*, 73.
53. Latha, B.; Senthilkumar, V. Analysis of thrust force in drilling glass fiber-reinforced plastic composites using fuzzy logic. *Mater. Manuf. Process.* **2009**, *24*, 509–516. [CrossRef]
54. Şahin, M.; Erol, R. Prediction of Attendance Demand in European Football Games: Comparison of ANFIS, Fuzzy Logic, and ANN. *Comput. Intell. Neurosci.* **2018**, *2018*, 5714872. [CrossRef]
55. Lerkkasemsan, N.; Achenie, L.E. Pyrolysis of biomass–fuzzy modeling. *Renew. Energy* **2014**, *66*, 747–758. [CrossRef]
56. Bhowmik, S.; Panua, R.; Ghosh, S.K.; Paul, A.; Debroy, D. Prediction of performance and exhaust emissions of diesel engine fuelled with adulterated diesel: An artificial neural network assisted fuzzy-based topology optimization. *Energy Environ.* **2018**, *29*, 1413–1437. [CrossRef]
57. Ghodrati, S.; Nakhaei, F.; VandGhorbany, O.; Hekmati, M. Modeling and optimization of chemical reagents to improve copper flotation performance using response surface methodology. *Energy Sources Part A Recovery Util. Environ. Eff.* **2020**, *42*, 1633–1648. [CrossRef]
58. Asghari, M.; Nakhaei, F.; VandGhorbany, O. Copper recovery improvement in an industrial flotation circuit: A case study of Sarcheshmeh copper mine. *Energy Sources Part A Recovery Util. Environ. Eff.* **2019**, *41*, 761–778. [CrossRef]
59. Ćalić, N. *Teorijski Osnovi Pripreme Mineralnih Sirovina*; Rudarsko-geološki fakultet Beograd—Faculty of Mining and Geology: Belgrade, Serbia, 1990; p. 496. (In Serbian)
60. Jassbi, J.; Alavi, S.H.; Serra, P.J.A.; Ribeiro, R.A. Transformation of a Mamdani FIS to First Order Sugeno FIS. In Proceedings of the IEEE International Conference on Fuzzy Systems, London, UK, 23–26 July 2007; pp. 5–10.
61. Nakhaei, F.; Sam, A.; Mosavi, M.R. Concentrate Grade Prediction in an Industrial Flotation Column using the Artificial Neural Network. *Arab. J. Sci. Eng.* **2013**, *38*, 1011–1023. [CrossRef]
62. Nakhaei, F.; Irannajad, M. Application and comparison of RNN, RBFNN and MNLR approaches on prediction of flotation column performance. *Int. J. Min. Sci. Technol.* **2015**, *25*, 983–990. [CrossRef]
63. Ying, X. An Overview of Overfitting and its Solutions. *J. Phys. Conf. Ser.* **2019**, *1168*, 022022. [CrossRef]
64. Available online: https://towardsdatascience.com/the-secret-neural-network-formula-70b41f0da767 (accessed on 20 October 2020).
65. Sarker, I.H. Deep Learning: A Comprehensive Overview on Techniques, Taxonomy, Applications and Research Directions. *SN Comput. Sci.* **2021**, *2*, 420. [CrossRef] [PubMed]
66. Nikolić, I.; Jovanović, I.; Mihajlović, I.; Miljanović, I. Analiza proizvodnje koncentrata bakra sistemskim pristupom (Analysis of copper concentrate production by systemic approach). *Bakar* **2016**, *40*, 33–50. (In Serbian)

minerals

MDPI

Article

Hybrid Serving of DOE and RNN-Based Methods to Optimize and Simulate a Copper Flotation Circuit

Alireza Gholami [1], Meysam Movahedifar [2], Hamid Khoshdast [3,*] and Ahmad Hassanzadeh [4,5,*]

[1] Department of Mineral Processing, Faculty of Engineering, Tarbiat Modares University, Tehran 14117-13116, Iran; ar.gholami2744@gmail.com
[2] Mineral Processing Division, Mining Engineering Department, Islamic Azad University, Sirjan, Iran; maysam.movahedi68@gmail.com
[3] Department of Mining Engineering, Higher Education Complex of Zarand, Zarand 77611-56391, Iran
[4] Department of Geoscience and Petroleum, Faculty of Engineering, Norwegian University of Science and Technology, 7031 Trondheim, Norway
[5] Maelgwyn Mineral Services Ltd., Ty Maelgwyn, 1A Gower Road, Cathays, Cardiff CF24 4PA, UK
* Correspondence: khoshdast@zarand.ac.ir (H.K.); ahmad.hassanzadeh@ntnu.no (A.H.)

Abstract: Prediction of metallurgical responses during the flotation process is extremely vital to increase the process efficiency using a proper modeling approach. In this study, two new variants of the recurrent neural network (RNN) method were used to predict the copper ore flotation indices, i.e., grade and recovery within different operating conditions. The model input parameters including pulp pH and solid content as well as frother and collector dosages were first analysed and then optimized using a two-step factorial approach. The statistical analysis showed a reliable correlation between operating parameters and copper grade and recovery with coefficients of 99.86% and 94.50%, respectively. The main effect plots indicated that pulp pH and solid content positively affect copper grade while increasing the frother and collector dosages negatively influenced the quality of the final concentrate. Despite the same effect from pulp pH, reverse effects from other variables were observed for copper recovery. Process optimization revealed that maximum copper recovery of 44.39% with a grade of 11.48% could be achieved under the optimal condition as pulp pH of 10, solid content of 20%, and frother and collector concentrations of 25 g/t and 9.9 g/t, respectively. Then, the predictive efficiency of long short-term memory (LSTM) and gated recurrent unit (GRU) networks with proper structure were evaluated using mean square error (*MSE*), root mean square error (*RMSE*), mean absolute percentage error (*MAPE*), and correlation coefficient (R^2). The simulation results showed that the LSTM network with higher R^2 of 0.963 and 0.934 for copper grade and recovery, respectively, was more effective than the GRU algorithm with the corresponding values of 0.956 and 0.919, respectively. The results show that the LSTM model could be useful in predicting the copper flotation behaviour in response to changes in the operating parameters.

Keywords: copper ore flotation; recurrent neural network; predictive geometallurgy; long short-term memory (LSTM); gated recurrent unit (GRU)

check for updates

Citation: Gholami, A.; Movahedifar, M.; Khoshdast, H.; Hassanzadeh, A. Hybrid Serving of DOE and RNN-Based Methods to Optimize and Simulate a Copper Flotation Circuit. *Minerals* **2022**, *12*, 857. https://doi.org/10.3390/min12070857

Academic Editor: Lev Filippov

Received: 25 May 2022
Accepted: 30 June 2022
Published: 5 July 2022

1. Introduction

Nowadays, simulation and modelling science are widely used as a reliable, fast, and low-cost solution to predict the behaviour of mineral processing units. Using simulations and modelling before or instead of experimental studies involving a large number of laboratory tests can significantly reduce executive and manpower costs. Furthermore, the flexibility of such approaches makes it possible to change or modify various stages of the process design without the need for a significant change in the initial redesign [1,2]. So far, several methods have been developed for modelling and simulation of mineral processing units, of which empirical and semi-empirical types and statistical models are the most common.

Despite the widespread use of these models, especially in simulation software development, each of the above approaches is associated with limitations that make their application in real operational environments challenging. For example, empirical and semi-empirical parametric models are developed based on a series of data sets from a specific process or multiple processes. Therefore, their use in new processes with different operating conditions leads to inefficiency and reduced accuracy in some cases. However, most models have tried to manage this limitation to some extent by considering the calibration coefficients [3,4]. The development of statistical models is easier than empirical models. However, the most important challenge of statistical models, in addition to being limited to the initial data of model development, is to include parameters that usually lack physical meaning. For example, variables that are the product of more than two major effects (the primary variable) and are abundantly observed in the design of experiment (DOE)-based models and cannot be interpreted in the model [5,6]. Besides, although mineral processing processes have relatively simple mechanical and physical aspects, they have very complex mechanisms, some of which still remain unknown. Metallurgical and operational factors as well as their interactions are major factors responsible for such complexities. As such, modelling and simulation of these processes have always posed a controversial topic.

In recent years, the use of expert system methods (ESM) such as artificial neural networks, and genetic algorithms have been proposed to simulate complicated separation techniques [7–9]. Intelligent algorithms process experimental data and transmit underlying knowledge in order to create a network structure. It is possible to apply ESM to implement intricate functions in a wide range of fields such as process automation, control and monitoring, medical diagnosis, and image analysis. Nowadays, we can solve difficult problems for humans and usual computers using efficiently trained intelligent algorithms [10–12]. Forecasting with ESMs has been one of the main uses of the algorithms, which have also shown excellent results. The good performance of ESMs has made them popular in different scientific fields, including mining and mineral processing. For instance, Jorjani et al. [13] used artificial neural networks to simulate the process of leaching rare earth elements from apatite concentrate on an industrial scale and showed that a reasonably accurate model could be developed using the improved ANN algorithm. In a later study, Milivojevic et al. [14] simulated the nickel ore leaching process to demonstrate that expert systems are more reliable than linear regression-based statistical models. In another research, Hoseinian et al. [15] used a hybrid neural-genetic algorithm to simulate copper recovery during a column leaching process of a copper ore sample on a pilot scale, and found that using an appropriate algorithm could yield reliable prediction results. In a recent study, using hybrid artificial neural networks and particle swarm optimization (PSO), Sobouti et al. [16] simulated lead recovery during the leaching of lead concentrate. The noteworthy aspect of the study was the wide range of operating parameters used in the simulation, such as temperature, liquid/solid ratio, stirring speed, fluoroboric acid concentration, and leaching time. They showed that an effective simulation of the process could be achieved using an optimized ANN-PSO algorithm. Artificial neural networks have also been successfully used to simulate mineral processing operations reported by Vyas et al. [17]. In their study, ANN was used to simulate and predict the spent catalyst bioleaching process with acceptable accuracy, and simulation results were presented in both numerical and graphical forms. According to Ghobadi et al. [18], the copper flotation circuit was modelled and optimized simultaneously using a genetic algorithm. They found that in comparison to conventional mathematical methods, an oriented genetic algorithm reduces the calculation time by 1/60 for a two-stage flotation system and can provide higher optimization accuracy. In another study, Gholami and Khoshdast [19] found that with a limited number of operating data, multiple metallurgical responses of the bioflotation process of coal can be accurately simulated using the ANN method. Their investigation into different algorithms for developing the ANN model demonstrated that the simulation accuracy is greatly influenced by the choice of the network algorithm. Recently, Gholami et al. [20] coupled historical data and deep learning techniques to predict the flotation behaviour of a copper mine in response to mineralogical and operational variables. They showed that mixed statistical/intelligent methods can be a promising approach to accurately simulate the flotation process with an accuracy of more than 95%.

The effective application of the recurrent neural network (RNN) method has also been reported by a few researchers. For example, radial basis function neural network, recurrent neural network, and multivariate nonlinear regression have been used to predict the metallurgical performance of the flotation column [21]. It was shown that the recurrent neural network forecasted the metallurgical performance of the flotation column better than the radial basis function neural network and multivariate nonlinear regression models. The same conclusions have also been recently reported by some other researchers [22,23] to estimate the mineral grade in the field of flotation. Inapakurthi et al. [24] proposed a method using RNN for simulation of industrial grinding circuit in the lead–zinc ore beneficiation process. They also demonstrated that the RNN model can successfully control a GC while tracking its set point without violating any constraints.

The main conclusion stemming from the above studies is that intelligent modelling methods along with limited amount of data of operational/process parameters can be used to simulate mineral processing operations successfully. Considering the results reported in the above studies, the simulation has a high degree of accuracy in most cases, which is highly desirable from an application perspective [25]. Hence, in this study, a mid-size copper mineral processing plant was targeted to intelligent simulation using long short-term memory (LSTM) and gated recurrent unit (GRU) networks as two well-known RNN methods. Among intelligent modelling methods for time series and sequential data predictions, RNNs are the commonly used ones. LSTM and GRU have a similar overall structure, but LSTM is more complex. Therefore, in addition to accurate prediction of output variables, a comparison of these two methods is also considered in this paper. The main parameters on the efficiency of the flotation process including pulp pH, solid content, and concentrations of frother and collector were first optimized by the DOE method and then, were considered as inputs in the model. Moreover, an accuracy analysis was performed to investigate the reliability of copper grade and recovery prediction.

2. Materials and Methods

2.1. A Brief Description of the Processing Plant

Studies were performed in the Takht Gonbad copper processing plant (Sirjan, Iran). With a practical capacity of 230 t/h, this plant beneficiates a copper sulfidic ore with an average copper grade of 0.45% to a concentrate with a grade of 22 ± 2% through several stages of roughing and cleaning. The copper grade in the final tailings of this plant is about 0.1%. According to Figure 1, the crushed ore is first broken down by four ball mills with capacities of 50 to 80 t/h to reduce the particle sizes to 80% finer than 100 μm. Comminuted particles are fed to the hydrocyclone unit, classifying them into two fine and coarse parts. The hydrocyclone overflow, consisting of particles 80% finer than 75 μm, is directed to a conditioning tank with a capacity of 50 m^3 to prepare the feed for the flotation circuit. At this stage, flotation reagents including sodium isopropyl xanthate (Z11) and sodium dithiophosphate (DTU) as copper mineral collectors, methyl isobutyl carbinol (MIBC) and a polypropylene glycol with molecular weight of 395.61 g/mol (A65) as frothers, lime as pH regulator, and NaHS as pyrite depressant (according to the mineralogical composition of the input feed) are added to the pulp. The concentration of reagents is adjusted according to the metallurgical conditions of the plant. The range of changes in the amount of consumption of these reagents is presented in Table 1. Prepared pulp with a solid content of about 25% (*w*/*w*) is first introduced into the rougher line with five 50 m^3 tank cells. The concentrate of this step is directed to an eight-cell row, including four cleaner and four recleaner cells to increase the copper grade. The final concentrate of this stage leaves the circuit as the final concentrate after several stages of recirculation in the cleaning circuit. The tailing is also returned to the beginning of the rougher circuit. Rougher tailing transfers to the scavenging circuit consisting of two 6-cell rows. The tailing of this unit is considered the final tailings of the plant, but the concentrate of the first two cells from both lines is returned to the cleaning circuit, and the concentrate of the last four cells is fed to the rougher circuit.

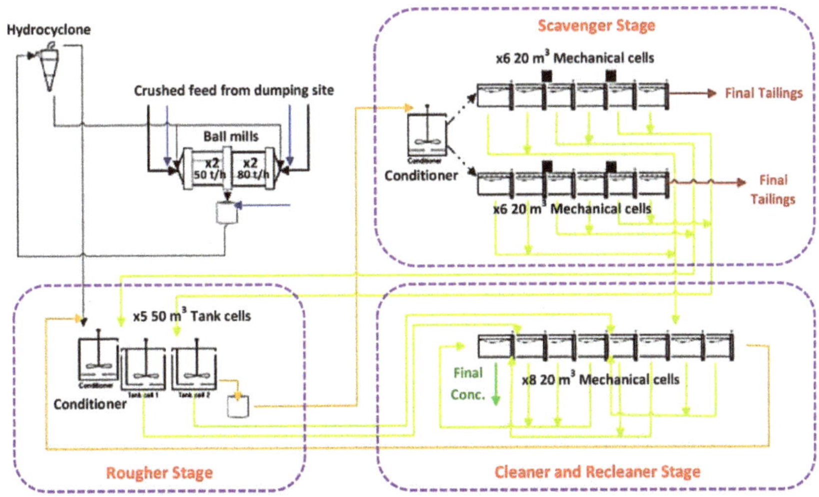

Figure 1. The processing circuit of Takht Gonbad copper plant (Feed —, Water —, Cells concentrate —, Cells tailings —, Final concentrate —, Final tailings —).

Table 1. Operating factors and their levels studied in the screening experimental design.

Factor	Variable	Unit	Low Actual	High Actual	Mid Level	Std. Dev.
A	pH	-	10	12	11	0.85
B	Solid content	(%)	20	30	25	4.26
C	MIBC conc.	(g/t)	5	35	20	12.79
D	A65 conc.	(g/t)	5	15	10	4.26
E	DTU conc.	(g/t)	5	35	20	12.79
F	Z11 conc.	(g/t)	5	25	15	8.53
G	NaHS conc.	(g/t)	0	10	5	4.26

2.2. Ore Sample and Reagents Used

A representative sample was taken from the hydrocyclone overflow as the feed of the flotation circuit (Figure 1) to determine the physical and mineralogical characteristics of the studied ore. Samples were collected using an automatic scoop sampler with an adjustable container size. Light microscopy (Axio Imager 2 Pol, Zeiss, Jena, Germany) was used to determine the mineralogical composition of the ore. The particle size distribution of the sample was measured using the standard dry sieve analysis method. All the applied reagents, including MIBC and A65 frothers, Z11 and DTU collectors, and NaHS depressant were sourced from the company's warehouse.

2.3. Screening of Operational Variables

Due to the variety of the operating parameters, first, the effectiveness of each operational variable was evaluated using a fractional factorial experimental design utilizing Design-Expert software (Demo version 7.0.0, from Stat-Ease Inc., Minneapolis, MN, USA). The levels of each variable were selected based on monitoring the process over six months (autumn and winter 2021). Table 1 lists the variables and their levels considered in the screening design. The experimental design used for screening studies and their practical results are presented in Table 2. These studies evaluated the grade and recovery of copper in the rougher stage concentrate as the process responses.

Table 2. The structure and practical results of the screening experimental design.

	Operating Factors							Responses	
Run	A: pH	B: Solid Content (%)	C: MIBC (g/t)	D: A65 (g/t)	E: DTU (g/t)	F: Z11 (g/t)	G: NaHS (g/t)	Cu Grade (%)	Cu Recovery (%)
1	10	20	35	15	5	5	10	4.90	71.53
2	12	20	35	5	35	5	0	4.90	86.87
3	11	25	20	10	20	15	5	3.60	53.88
4	10	30	5	5	35	5	10	2.96	64.92
5	12	30	35	15	35	25	10	4.30	44.00
6	10	20	5	15	35	25	0	4.70	58.28
7	12	30	5	15	5	5	0	4.40	58.17
8	12	20	5	5	5	25	10	5.20	55.82
9	11	25	20	10	20	15	5	2.60	62.42
10	12	30	5	15	5	5	0	4.80	59.59
11	12	20	35	5	35	5	0	5.10	68.90
12	10	20	35	15	5	5	10	5.17	56.97
13	12	30	35	15	35	25	10	4.60	50.91
14	10	30	35	5	5	25	0	3.57	52.75
15	11	25	20	10	20	15	5	3.15	58.14
16	11	25	20	10	20	15	5	3.10	58.83
17	12	20	5	5	5	25	10	4.90	58.38
18	10	20	5	15	35	25	0	4.50	68.48
19	11	25	20	10	20	15	5	2.44	41.96
20	10	30	35	5	5	25	0	3.61	53.32
21	11	25	20	10	20	15	5	3.54	62.53
22	10	30	5	5	35	5	10	2.78	59.05

2.4. Statistical Optimization Studies

Influential operating variables for optimization studies were selected according to the results of the screening studies as listed in Table 3. Then, a full factorial design was used to investigate the real impact of each parameter, their interactions, as well as process optimization. The optimization design with practical results is presented in Table 4. To evaluate the nonlinear effects of each parameter, the centre point was used with six replications. Moreover, each main experiment was replicated twice to achieve more reliable results and eliminate bias in the effects. Alike the screening tests, copper grade, and recovery at the rougher stage were considered as the process responses.

Table 3. Operating factors and their levels studied in the optimization experimental design.

Factor	Parameter	Unit	Low Level	High Level	Mid Level	Std. Dev.
A	pH		10	12	11	0.92
B	Solid Content	(%)	10	20	15	4.59
C	A65 Conc.	(g/t)	10	25	17.5	6.88
D	Z11 Conc.	(g/t)	0	20	10	9.18

Table 4. The structure and practical results of the optimization experimental design.

	Operating Factors				Responses	
Run	A: pH	B: Solid Content (%)	C: A65 (g/t)	D: Z11 (g/t)	Cu Grade (%)	Cu Recovery (%)
1	11	15	17.5	10	11.07	41.01
2	10	20	25	20	7.45	57.32
3	12	10	25	20	9.36	53.72
4	12	10	25	0	8.33	41.69
5	10	20	10	0	12.50	45.33

Table 4. *Cont.*

	Operating Factors				Responses	
Run	A: pH	B: Solid Content (%)	C: A65 (g/t)	D: Z11 (g/t)	Cu Grade (%)	Cu Recovery (%)
6	11	15	17.5	10	11.00	44.18
7	10	10	10	0	12.30	40.67
8	12	10	25	20	9.21	58.05
9	12	20	25	20	12.93	44.11
10	10	10	10	20	3.82	27.09
11	10	20	10	20	14.10	38.82
12	12	10	25	0	8.39	42.41
13	10	20	10	20	14.05	38.59
14	10	20	25	0	16.83	37.40
15	10	10	25	0	7.81	50.11
16	10	10	25	20	4.00	62.39
17	11	15	17.5	10	10.97	46.31
18	11	15	17.5	10	10.89	50.27
19	12	20	25	20	12.89	45.49
20	11	15	17.5	10	10.99	53.26
21	12	10	10	0	10.80	55.61
22	10	20	10	0	12.40	47.01
23	10	10	25	20	4.20	61.19
24	11	15	17.5	10	11.10	50.42
25	12	20	10	0	10.60	47.07
26	10	20	25	0	16.11	39.09
27	12	20	10	20	12.80	44.21
28	12	10	10	0	10.50	56.77
29	12	20	10	20	12.50	47.39
30	10	10	10	0	12.50	44.46
31	12	10	10	20	7.81	40.29
32	12	10	10	20	7.75	40.85
33	12	20	25	0	11.90	45.40
34	10	10	25	0	7.88	50.21
35	12	20	25	0	11.70	45.66
36	12	20	10	0	10.80	46.32
37	10	20	25	20	7.51	58.60
38	10	10	10	20	3.75	27.00

2.5. Flotation Experiments and Calculations

All the experiments were carried out in a standard D-12 Denver® flotation machine equipped with a 4 L cell. To perform each test in the screening (Table 2) and optimization (Table 4) experimental designs, after setting the operating conditions according to the run number in each design, the pulp mixture was agitated at impeller speed of 1000 rpm for 5 min in the flotation cell to ensure that all ore particles were well suspended. After conditioning, water was added to the cell to a specified level after conditioning. The pulp level was maintained constant during each test by constantly adding water as required. At the end of each experiment, the collected concentrates and tailings were weighed and dried in an oven at 60 °C for 24 h. Samples were then sent to the analysis laboratory to determine their copper content. The efficiency of the flotation process was evaluated in terms of final recovery and the grade of Cu using the following equation [26,27]:

$$R = \frac{C}{F} \times \frac{c}{f} \times 100, \quad (1)$$

where R (%) is recovery, F (kg) and C (kg) are the total mass of feed and concentrate, respectively, f (%) and c (%) are elemental grades (%) of feed and concentrate, respectively.

2.6. Recurrent Neural Network Simulation

2.6.1. RNN-Based Methods

(A) Recurrent neural network: Since the long short-term memory (LSTM) and gated recurrent unit (GRU) networks are improved variants of the recurrent neural network (RNN) and are used in state-of-the-art deep learning applications, the structure of RNN was discussed first. The RNN is a type of feedforward neural network that maintains internal memory and is able to remember information throughout time. This property makes it proper for processing time series and sequential data [28]. The network structure and circuit diagram of RNN are shown in Figure 2.

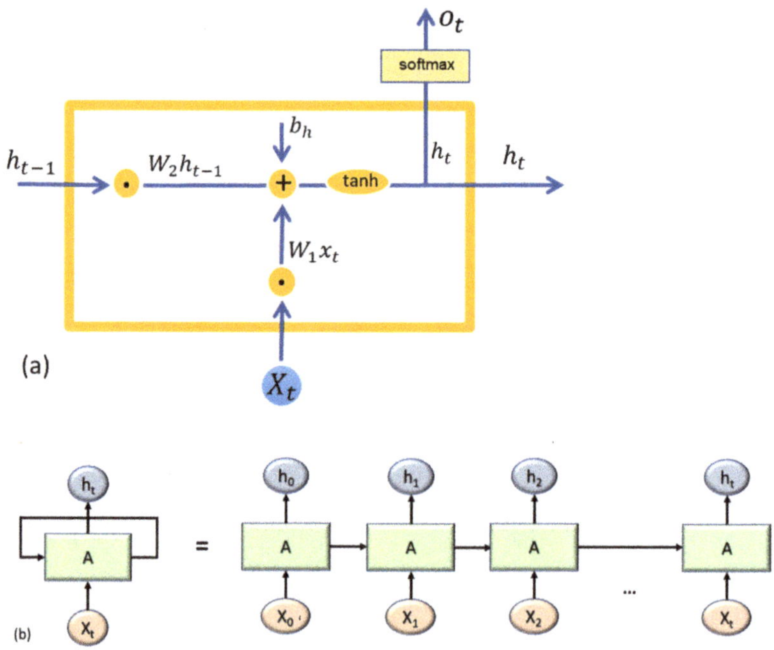

Figure 2. The network structure (**a**) and unrolled circuit diagram (**b**) of the recurrent neural network.

In the circuit diagram, RNN takes X_0 from the input sequence and then delivers h_0 as the output, which together with X_1 are the inputs of the next step. Similarly, h_1 along with X_2 are the inputs to the next step, and so on. In such a manner, RNN constantly remembers the information during the training. The current state formula is as follows:

$$h_t = f(h_{t-1}, x_t), \tag{2}$$

where x_t and h_t are the input and output sequence of a RNN unit, respectively. Equation (3) is also used to apply tanh as the activation function, which helps to regulate the values that flow through the network. Finally, y_t is the network output [29]:

$$h_t = \tanh(W_{hh}h_{t-1} + W_{xh}x_t), \tag{3}$$

$$y_t = W_{hy}h_t, \tag{4}$$

where W and h represent the weight and hidden vectors, respectively. W_{hh} is the weight in the previous hidden state, W_{hx} is the weight in the current input state, and W_{hy} represents the weight in the output state. tanh is an activation function that implements nonlinearity. Although RNN is theoretically designed to predict time series, it is difficult to predict long time series in practice because of the length of information, which can cause a vanishing gradient [30]. LSTM and GRU are the updated versions of RNN to overcome this challenge.

(B) Long short-term memory network (LSTM): LSTM is a special kind of RNN network that was introduced first by Hochreiter and Schmidhuber [31] and is able to tackle the problem of vanishing gradient and long length input processing of RNN. The LSTM network has internal mechanisms called gates. These gates (including input gates, forget gates, and output gates) control the data flow and also specify what data is important in the sequence and should be retained and what data should be removed. In this way, the network passes important information along the sequence chain to obtain the final output [32]. A structure diagram of LSTM is shown in Figure 3.

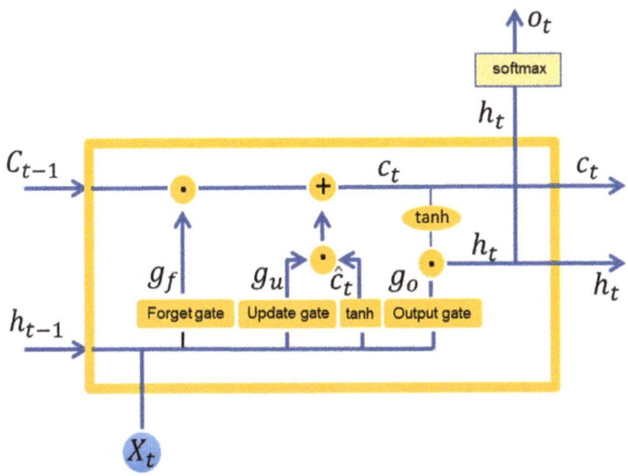

Figure 3. The structure of the long-short term memory neural network.

In Figure 3, x_t, h_t, and C_t are the input, the output, and the cell state at time t, respectively. g_f controls the information flow from the previous time step and is called the forget gate. This gate determines whether or not the information from the previous step is used. The update gate, which is indicated with g_u, is responsible for controlling the new information flow. This gate determines whether new information should be used at the current time step. g_o is the output gate and specifies how much of the previous information of time steps (previous and current) is transferred to the next time step. W, b, and σ are the weight of the model, the bias of the model and the activation function, respectively. The following equations represent the calculation process:

$$\hat{C}_t = \tanh(W_c \cdot [h_{t-1}, X_t] + b_c), \tag{5}$$

$$C_t = g_f \cdot C_{t-1} + g_u \cdot \hat{C}_t, \tag{6}$$

$$g_f = \sigma\left(W_f \cdot [h_{t-1}, X_t] + b_f\right), \tag{7}$$

$$g_u = \sigma(W_u \cdot [h_{t-1}, X_t] + b_u), \tag{8}$$

$$g_o = \sigma(W_o \cdot [h_{t-1}, X_t] + b_o), \tag{9}$$

$$h_t = g_o \cdot \tanh(C_t). \tag{10}$$

(C) Gated recurrent unit network (GRU): GRU was introduced by Chung et al. [33] to address the shortcomings of the traditional recurrent neural network and also to reduce the overload of LSTM architecture. The GRU has only two gates, including update and reset gates, compared with the LSTM. These gates are basically two vectors that are used to decide whether the information is transmitted to the output or not [34]. A structure diagram of GRU is shown in Figure 4 and relevant equations can be found below:

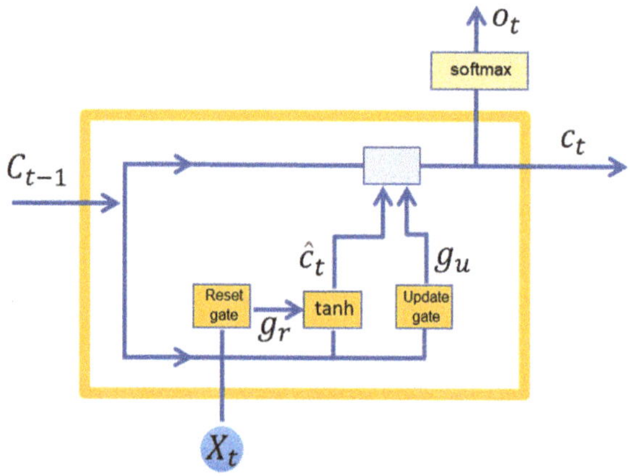

Figure 4. The structure of a gated recurrent unit network.

The reset gate specifies how much of the previous information is not needed (forgotten) and how much of the previous step information is used in the current step. The update gate specifies whether to use the previous state or current input (or a combination of both) at a time step. x_t is the input at time t, and σ is the activation function and h_t and h_{t-1} are the output at time t and $t-1$, respectively. The calculation process is expressed by the following equations:

$$\hat{C}_t = \tanh(W_1 \cdot [g_r \cdot C_{t-1}, X_t] + b_c), \tag{11}$$

$$g_u = \sigma(W_2 \cdot [C_{t-1}, X_t] + b_u), \tag{12}$$

$$g_r = \sigma(W_3 \cdot [C_{t-1}, X_t] + b_r), \tag{13}$$

$$C_t = g_u \cdot \hat{C}_t + (1 - g_u) \cdot C_{t-1}. \tag{14}$$

where g_r and g_u are the outputs of reset gate and update gate at time t.

2.6.2. Modelling Process

RNNs are considered the state-of-the-art algorithm for sequential data. LSTM and GRU are RNN-based algorithms that were developed with new designs to address the weaknesses of traditional RNNs. In this study, four inputs were included as pH, solid content (%, w/w), A65 concentration (g/t), and Z11 concentration (g/t); and the outputs were Cu grade (%) and its recovery (%). To analyse the methods and estimate outputs, four models were developed using the inputs. Table 4 shows the inputs and outputs as the operation factors and responses, respectively. Although controlling the flow of information is the same in both LSTM and GRU, LSTM wraps the hidden state into a memory unit, and GRU just passes the full hidden content without any control directly to the next cell. The models' parameters, including number of hidden layers, number of epochs, etc., were carefully examined and selected based on the complexity of the data and trial and error procedure. Indeed, as a result of increasing the number of parameters, overfitting occurred. An overfitted model has high accuracy on training data but low accuracy on test data. For the number of epochs, increasing this parameter had no significant effect on the accuracy of the model; it only increased the time it took to develop the model. The optimal amounts of parameters were selected according to the mentioned points. In addition, the GRU and LSTM algorithms are similar in general structure, and the most important difference is that there are fewer parameters in GRU, which accounts for its faster training time. One of the purposes of using these two algorithms in this project, in addition to effectively predicting the outputs, is to examine the results of these two models in order to select a more appropriate algorithm. LSTM has a significant advantage over the GRU algorithm in accordance with its more complex structure. Figure 5 shows the flowchart of the modelling procedure to estimate the Cu grade and recovery using RNNs.

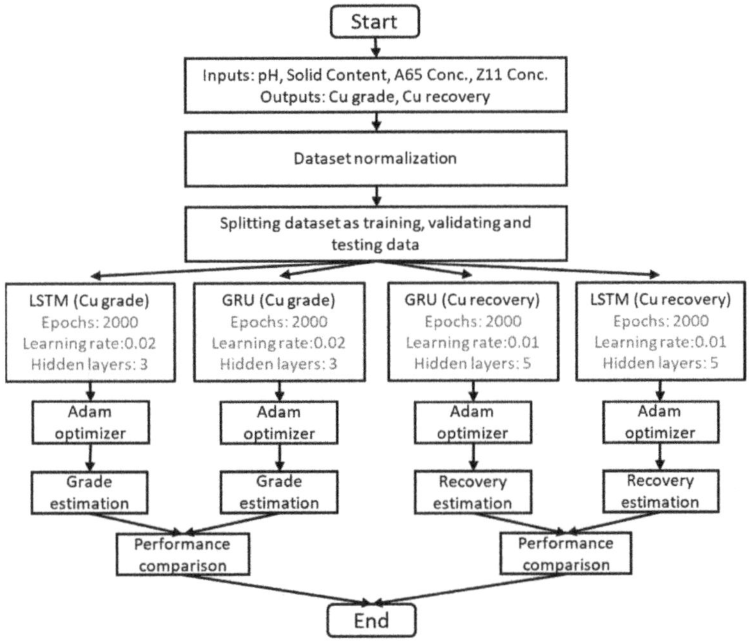

Figure 5. The flowchart used on applying the models in predicting the Cu grade and recovery.

Random forest (RF) and ANN with the Levenberg–Marquardt optimization algorithm (ANN-LMA) as the two other popular prediction methods were also implemented for comparison:

- Random forest is a powerful learning method for classification and regression problems by constructing a multitude of decision trees at training time. This non-parametric method uses ensemble learning to avoid overfitting. To find the optimum number of trees, different numbers were tested. Based on the results, 26 and 22 trees were found to have the most accurate prediction for Cu grade and Cu recovery, respectively, and increasing the number of trees did not have significant impact on the accuracy of results. The optimum depth of the tree was also found to be 4 for both Cu grade and recovery models, and increasing the maximum depth of trees resulted in overfitting;
- LMA is known as the preferred method for minimization in nonlinear least squares problems. LMA interpolates between the Gauss–Newton algorithm and the gradient descent method. Compared with Gauss–Newton, Levenberg–Marquardt is more robust and, in many cases, it will find a solution even if it starts very far from the final minimum. ANN-LMA was found to be superior in Ref. [19] against other metaheuristic algorithms. The proper structure for ANN-LMA was found to be 5-12-4-1 for the Cu recovery model and 5-10-5-1 for the Cu grade model based on trial and error.

The coding and modelling process was implemented using MATLAB software (MathWorks R2021b v9.11, MathWorks, Inc., Natick, MA, USA). Before modelling, normalization was applied to improve the networks training phase using Equation (15) [12]:

$$x_n = \left(\frac{x_i - x_{min}}{x_{max} - x_{min}} \right), \tag{15}$$

where X_n and X_i are normalized and actual values, respectively. X_{min} and X_{max} are the minimum and maximum values of each subset (inputs–outputs). Besides the modelling process, the Spearman correlation analysis, to find out the relationship between inputs and outputs, and sensitivity analysis to calculate the effectiveness of each input data on the outputs were also applied; the outcomes are presented in the Section 3.

3. Results and Discussion

3.1. Results of Ore Characterization

The mineralogical results showed that chalcopyrite and chalcocite are the predominant copper-bearing minerals in the ore. The most important gangue minerals were pyrite and clay minerals. The particle size distribution of the studied ore sample is shown in Figure 6. The feed to the flotation circuit contains 80% by weight of particles finer than 78 μm.

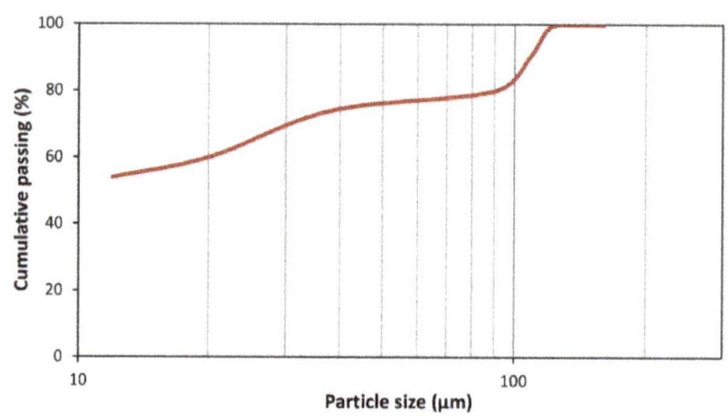

Figure 6. Particle size distribution of the copper ore fed to the flotation circuit.

3.2. Statistical Analysis of Screening DOE

To determine the most important parameters affecting metallurgical responses, Pareto charts were drawn based on the results of variance analysis. Pareto charts are quick tools used to assess the significance level and the type of impact (increasing or decreasing) of the parameters under study. In these charts, effects with a value greater than t-value are identified as the significant factors. The type of effect is also shown by the software and based on the relevant statistical calculations with different colours. According to Pareto charts of grade and copper recovery in Figure 7, it can be seen that the most significant operating parameters affecting the grade of concentrate are solid percentage (B) > pH (A) > and concentration of A65 frother (C). Generally, the copper grade decreases significantly with the increasing solid content mainly due to transporting gangue minerals (i.e., pyrite, silicates and clay minerals) into the concentrate through the false flotation process [35]. This result is in line with the outcomes presented by previous studies [20,36]. Copper grade is also significantly affected by pH and A65 concentration. Although the effects of DTU (E) and MIBC (C) concentrations are relatively high, these effects are not statistically significant.

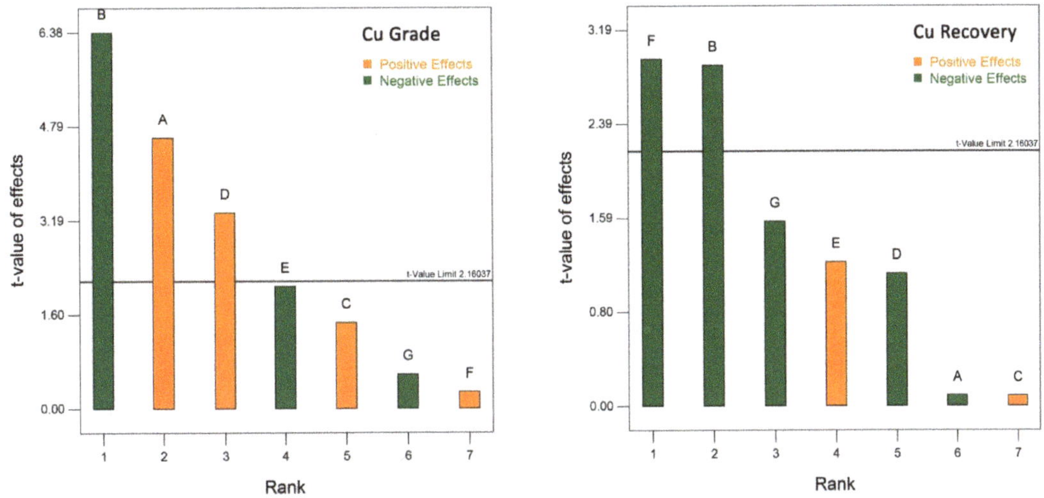

Figure 7. Pareto charts showing the significance and type of main effects for metallurgical responses.

Although the solid content (B) and the concentration of Z11 (F) are the most effective factors affecting copper recovery, their effects are both negative. It should be noted that since the interaction effects of the parameters in factorial designs cannot be analysed due to bias, providing any physical interpretation for the main effects observed in these diagrams cannot be reliable. For this reason, the results of this design have only been used to select the most effective operational variables for detailed studies. In addition, according to the type of effect, the amount of levels considered in the optimization plan was also modified. For example, according to Figure 7, the effect of solid content on both responses is negative; therefore, the levels of this parameter in the screening design (20–30) were reduced in the optimization design i.e., 10–20, and similarly, the amount of levels was increased for the frother concentration and decreased for the collector concentration. The pH levels were not changed due to the operational limitations.

3.3. Statistical Analysis of Optimization DOE

The first step in analysing the impact of operational variables on the process responses is developing a parametric model that can accurately predict the desired response in the operating space, i.e., within the low to high levels intended for the variables [37]. In the second step, after developing the initial model by the software, abnormal data were identified by examining the model parameters and the model was optimized by the user to achieve the best fitting results. The result of these measures for the data obtained in the flotation experiments was the development of nonlinear models for all process responses as below:

$$Cu\ Grade\ (\%) = 10.17 + 0.35A + 2.15B - 0.39C - 1.16D - 0.65AB + 0.46AC + 1.30AD + 0.24BC + 0.62BD - 0.17CD + 0.54ACD - 1.26BCD + 0.68ABCD, \tag{16}$$

$$Cu\ Recovery\ (\%) = 45.91 + 0.91A - 0.72B + 3.34C + 0.28D - 0.59AB - 3.41AC - 2.16BC + 0.91BD + 5.22CD + 1.71ABC - 0.97ABD - 1.53ACD - 1.78BCD - 1.90ABCD, \tag{17}$$

where the factors are described in a coded form. The validation parameters for the developed models are listed in Table 5. According to Fisher's F-test and marginal probability value (p model < 0.0001), which are shown in Table 5, all of the suggested prediction models are significant. For assessing the significance of a predictive model, the residuals normal probability plot is an effective tool [38]. According to the normal probability plots shown in Figure 8, all responses were relatively uniform, confirming the assumptions of normality and the independent nature of residuals during the statistical analyses. In addition, the high values of the adjusted correlation coefficients also indicate the significance of the prediction models. The Pred R^2 values were reasonably high, indicating that the model is able to explain variability in the prediction of new observations with adequate accuracy, which is in reasonable agreement with the Adj R^2 values [39]. Another statistical measure, called adeq precision, shows the signal-to-noise ratio, and any value over 4 is considered desirable [6]. In this investigation, the ratios were 134.99 and 20.59 for copper grade and recovery, respectively. These values show an adequate signal so that the models can be used to navigate the design space and predict appropriately.

Table 5. Validation parameters showing the significance of the models for metallurgical responses.

Model	F-Value	p-Value	R^2 (%)	Adj R^2 (%)	Pred R^2 (%)	Adeq Precision
Cu Grade	1251.23	<0.0001	99.86	99.78	99.57	134.99
Cu Recovery	24.07	<0.0001	94.50	90.58	88.62	20.59

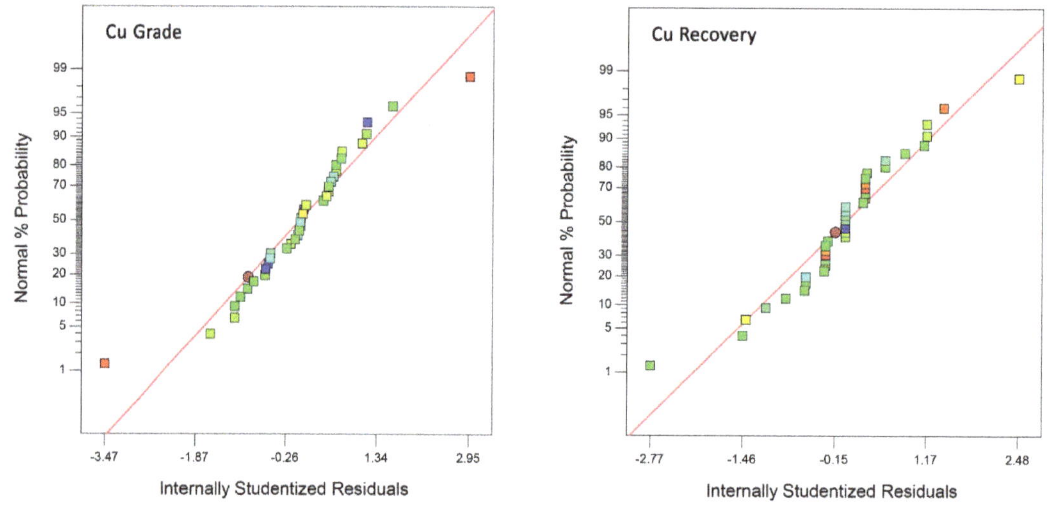

Figure 8. Normal plots of the residuals for the metallurgical responses of Cu grade (**left**) and recovery (**right**).

Model Equations (16) and (17) were used to assess the significance of operating variables on process responses. Tables 6 and 7 show the ANOVA results within a confidence interval of 95%. As shown in Tables 6 and 7, the effects of all operational variables on process responses are statistically meaningful due to p-values less than 0.05. As seen in the analysis of variance tables, no interaction was considered in the analyses. Moreover, except in some cases for the copper recovery analysis, the dual interaction effects are significant. The statistically meaningless interactions for copper recovery are those between pH and solid content and Z11 concentration as well as the interaction between solid content and Z11 concentration.

Table 6. Analysis of variance results for copper grade to the rougher concentrate.

Source	Sum of Squares	df	Mean Square	F-Value	p-Value (Prob > F)
Model	363.8967	13	27.99206	1251.234	<0.0001
A-pH	3.822613	1	3.822613	170.8693	<0.0001
B-Solid Content	147.3186	1	147.3186	6585.085	<0.0001
C-A65	4.8672	1	4.8672	217.562	<0.0001
D-Z11	43.29151	1	43.29151	1935.114	<0.0001
AB	13.4162	1	13.4162	599.699	<0.0001
AC	6.826513	1	6.826513	305.1425	<0.0001
AD	54.2882	1	54.2882	2426.662	<0.0001
BC	1.814513	1	1.814513	81.10801	<0.0001
BD	12.5	1	12.5	558.7452	<0.0001
CD	0.973012	1	0.973012	43.49328	<0.0001
ACD	9.46125	1	9.46125	422.9142	<0.0001
BCD	50.6018	1	50.6018	2261.881	<0.0001
ABCD	14.71531	1	14.71531	657.7688	<0.0001
Pure Error	0.485733	21	0.02313		
Cor Total	367.9095	37			

Table 7. Analysis of variance results for copper recovery to the rougher concentrate.

Source	Sum of Squares	df	Mean Square	F-Value	p-Value (Prob > F)
Model	2266.4	15	151.0933	24.07371	<0.0001
A-pH	31.7761	1	31.7761	10.309121	0.0043
B-Solid Content	35.5092	1	35.5092	12.092561	0.0031
C-A65	356.5785	1	356.5785	56.81369	<0.0001
D-Z11	44.2345	1	44.2345	14.291302	0.0021
AB	11.3288	1	11.3288	1.805019	0.1934
AC	372.5085	1	372.5085	59.35182	<0.0001
AD	16.4738	1	16.4738	2.624772	0.1201
BC	149.5585	1	149.5585	23.82917	<0.0001
BD	26.3538	1	26.3538	4.198954	0.0531
CD	873.411	1	873.411	139.1607	<0.0001
ABC	93.91351	1	93.91351	14.96325	0.0009
ABD	29.9538	1	29.9538	4.772542	0.0404
ACD	74.48101	1	74.48101	11.86707	0.0024
BCD	100.891	1	100.891	16.07498	0.0006
ABCD	115.596	1	115.596	18.41792	0.0003
Pure Error	131.8018	21	6.276278		
Cor Total	2408.51	37			

3.3.1. Interpretation of the Main Effects

The main effect plots can be used to analyse the effects of the operating variables on the process response. These plots are an effective tool for assessing the influence of each variable on the target response. The response variation is shown in these plots for the level of variables used in the experimental design. Design Expert software uses, by default, the developed prediction model (Equations (16) and (17)) to calculate the response values for parameters varying within their experimental levels. In contrast, other variables are maintained at their mid-levels. The main effects plots of different operating variables are shown in Figures 9 and 10 for copper grade and recovery, respectively.

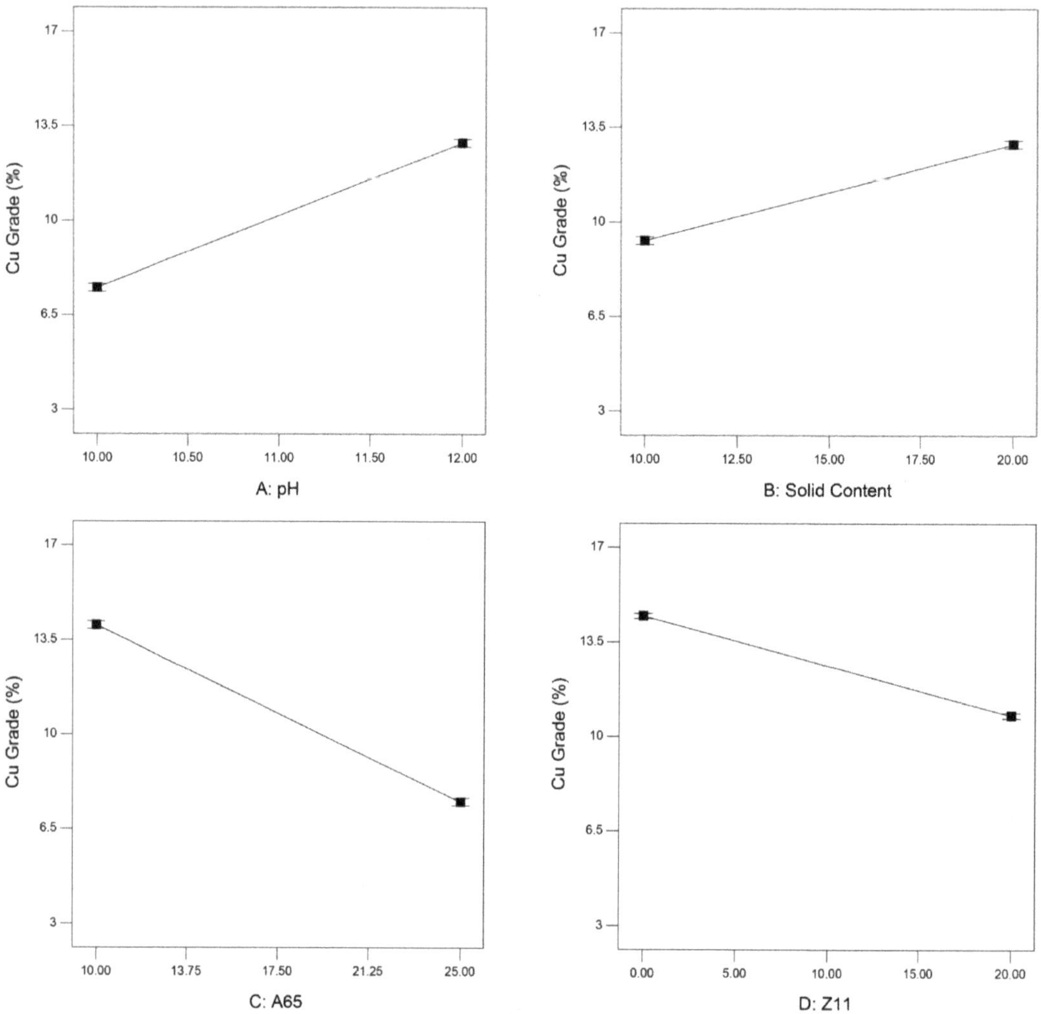

Figure 9. Main plots showing the effect of operating variables on copper grade.

According to Figure 9, the copper grade increases with the increasing pH. Obviously, as the pH increases, the pyrite depression rate also increases, and as a result, the copper grade in the concentrate improves. The effect of increasing the solid content on copper grade is also positive and increasing. This effect can be attributed to the improved particle distribution within the flotation cells, thereby increasing the probability of copper-containing particles colliding with the bubbles and transporting them to the froth and concentrate. However, as shown in Figure 10, copper recovery decreased with the increasing solid content, which can be ascribed to the interaction between the solid content and the frother concentration, as will be discussed later. Increasing the concentration of both frother and collector reduces the grade of copper in the concentrate. With an increasing frother concentration, due to increased bubble stability and increasing the rate of bubbly regime in the pulp zone, the entrainment rate will most likely increase due to the swarm phenomenon, and finally, the copper grade decreases as a consequence of the gangue particles transfer to the concentrate [40,41]. The effect of increasing the collector concentration is also due to the increase of hydrophobicity level and floatability of gangue minerals and, as a result, their transfer to the concentrate. Therefore, the copper grade in concentrate is expected to decrease. These phenomena, namely the increase in entrainment rate and the improvement of gangue floatability, will increase the solid transfer rate into the concentrate and, consequently, increase the recovery; this prediction is clearly seen in Figure 10.

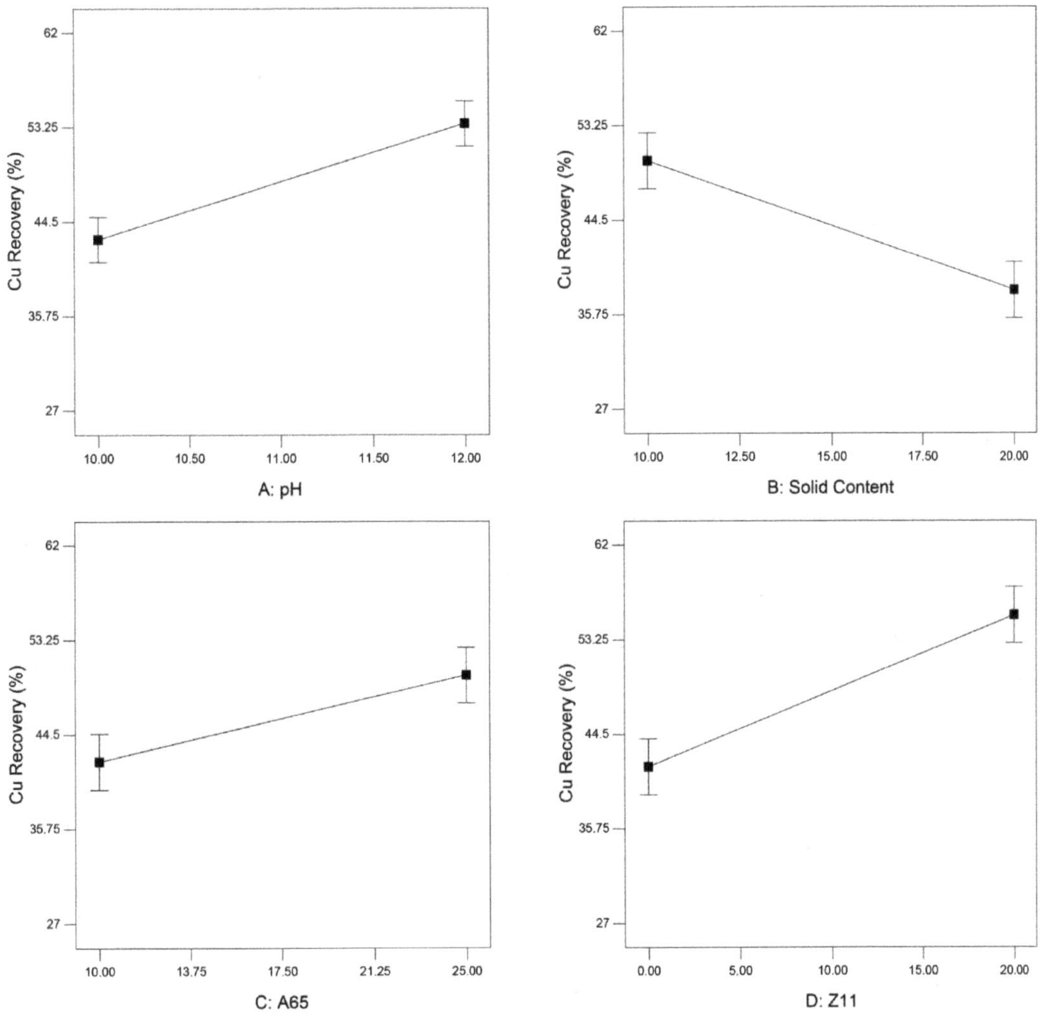

Figure 10. Main plots showing the effect of operating variables on copper recovery.

Increasing the pH, in addition to improving the grade of copper, has also increased the recovery (Figure 10). As the pH increases, the efficiency of the xanthate collectors (Z11 in this study) improves, resulting in increased particles floatability. For this reason, particle recovery is also expected to improve. The negative effect of solid content on the recovery can be evaluated according to the interaction of this factor with the frother concentration, which is presented in Section 3.3.2. Given that the main effect plot for each factor is plotted while other factors are kept constant at their mid-level, the individual effect of the solid content may lead to the misleading conclusion that increasing the solid content in any condition causes reduction, leading to recovery.

3.3.2. Evaluation of the Interactive Effects

The surface plots of the response of a process against other independent variables can provide valuable insights into not only the individual effect of operating factors, but the potential interaction between them as well [42,43]. Thus, the surface plots for the studied flotation experiments were presented. As shown in Figures 11 and 12, the surface response plots illustrate the nonlinear effects of most interactions among four variables. As mentioned earlier, the interaction between solid content and frother concentration (Figure 12) shows that maximum copper recovery can be obtained at a low level of frother concentration, whereas individual effects yielded contrary results. This conclusion can be directly attributed to the significant interaction between those variables. Due to this effect, it seems that the balance between the amount of particles in the system and the number of stable bubbles plays a significant role in improving the process efficiency. The same behaviours are observed for copper grade, as it may be concluded that the highest copper grade can be achieved at a high level for pH and solid content and a low level for reagents dosages when referring to the main effect plots. However, Figure 11 clearly shows different results. Similar results are observed in the case of copper recovery when evaluating the interactions given in Figure 12.

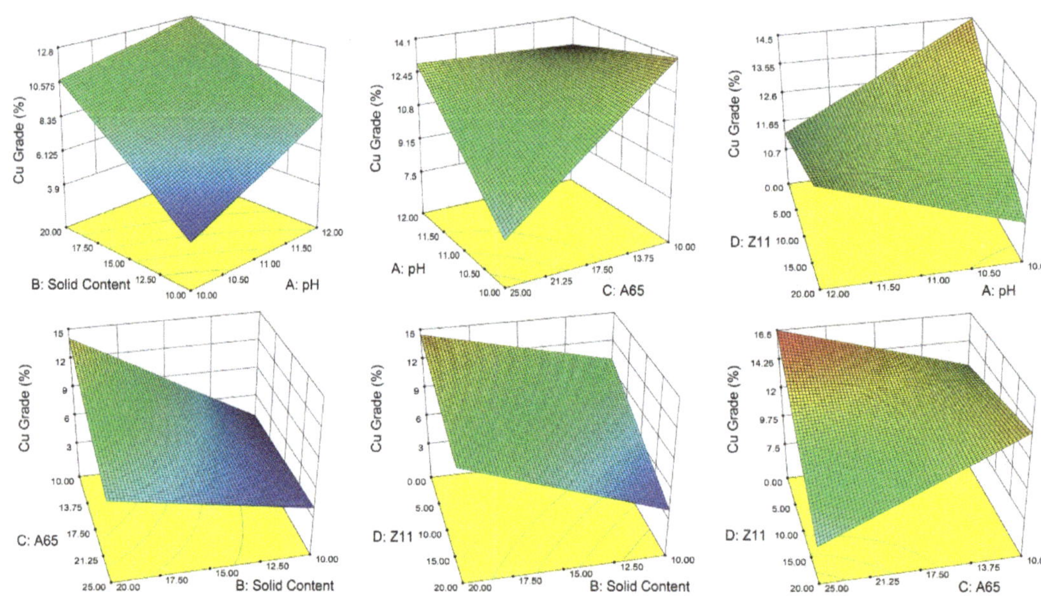

Figure 11. 3D surface plots showing the interaction effects between the operating variables on copper grade.

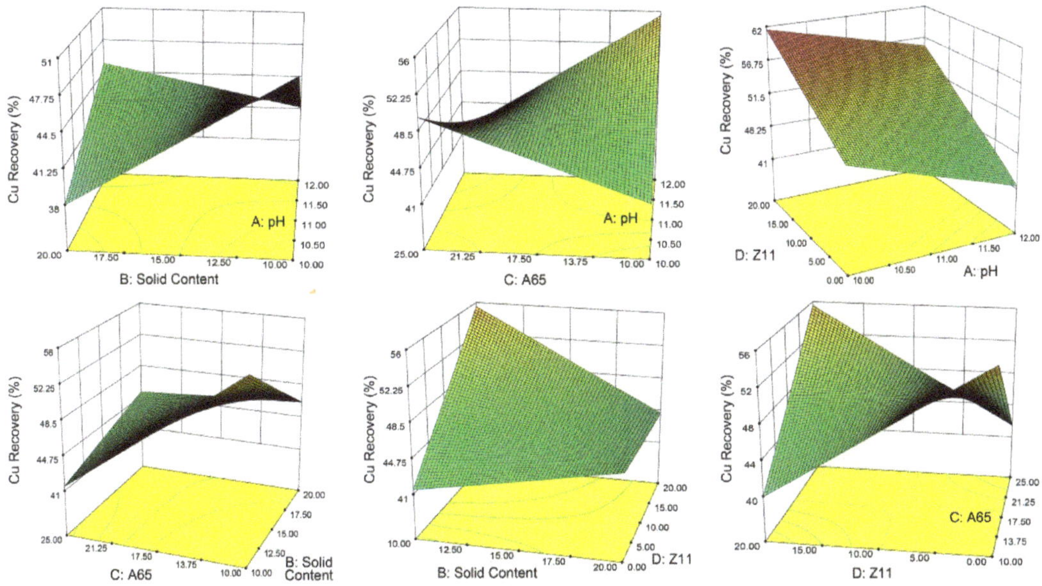

Figure 12. 3D surface plots showing the interaction effects between the operating variables on copper recovery.

3.3.3. Process Optimization and Verification Studies

Maximum Cu grade and recovery were separately defined as target responses to find suitable operating conditions for each case. The operating conditions and the predicted results suggested by the software are given in Table 8. To assess the accuracy of the predictions in Table 8, the plant was permitted to operate for 10 days under the suggested operating conditions and the metallurgical responses were monitored by regular sampling from each operating shift. Finally, an average value was reported as the final practical result, as given in Table 8. Compared to the grade values, the higher differences between the practical recovery results and predicted values are due to the lower accuracy of the prediction model developed by the software (Table 5).

Table 8. The predicted and practical results for copper grade and recovery under optimal conditions.

Goal (Max)	pH	Solid Content (%)	A65 Conc. (g/t)	Z11 Conc. (g/t)	Predicted Responses (%)		Practical Results (%)	
					G.	R.	G.	R.
Grade (G)	10.00	20.00	25.00	0.14	16.44	38.41	15.32	34.03
Recovery (R)	10.00	19.93	25.00	9.91	12.00	47.72	11.48	44.39

3.4. Simulation Results

3.4.1. Correlation Coefficient Analysis

Correlation coefficient is an indicator that is used to measure the dependence or relationship between two variables. The correlation coefficient between each variable and the outputs (Cu grade and Cu recovery) confirms the existence of correlation. In this study, the Spearman correlation coefficient was used, which is defined as Equation (18) [44]:

$$\rho = 1 - \frac{6\sum d_i^2}{n(n^2-1)}, \tag{18}$$

where ρ is Spearman's rank correlation coefficient, d_i is the difference between the two ranks of each variable, and n is the number of samples. The correlation coefficient between each variable and the outputs is presented in Figure 13 and Table 9. The coefficients indicate that the variables had a good correlation and could be used to estimate Cu grade and Cu recovery.

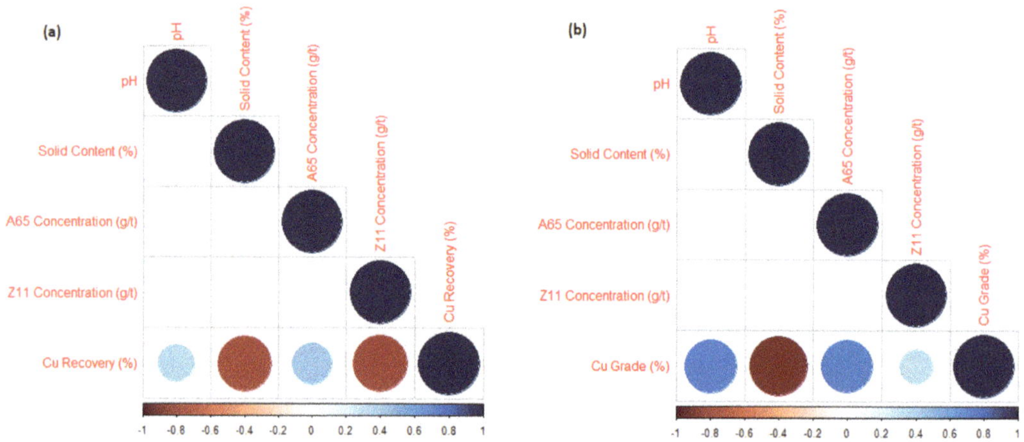

Figure 13. Spearman correlation assessments between input variables and outputs: (**a**) Cu recovery and (**b**) Cu grade.

Table 9. The correlation coefficient of the variables with Cu grade and Cu recovery.

Variable	pH	Solid Content (%)	A65 Conc. (g/t)	Z11 Conc. (g/t)
Cu grade (%)	0.7391	−0.8936	0.7234	0.3155
Cu recovery (%)	0.3653	−0.7880	0.4490	−0.8041

According to Table 9 and for this study, pH and A65 concentration had a good positive correlation with Cu grade; on the contrary, the solid content had a strong negative effect on Cu grade. It means that should the pH increase, the Cu grade decreases, as was also reported in Ref. [45]. For the Cu recovery, Z11 concentration and solid content had high negative correlation with this output, and pH and A65 concentration had positive correlation, although their correlation coefficient is lower than 0.5. It is important to note that in the correlation coefficient methods, the closer the coefficient is to 1 and −1, indicates a direct or inverse relationship between the two variables. However, a cause-and-effect relationship is not necessarily present, and the conclusion cannot be drawn on this basis.

3.4.2. Sensitivity Analysis

In order to determine the effect of each input on the amount of output, a sensitivity analysis was also applied. In this paper, the input parameters were pH, solid content, A65 concentration, and Z11 concentration, and output parameters were Cu recovery and Cu grade. To calculate the sensitivity analysis, the following equation was used [46]:

$$R_{ij} = \frac{\sum_{k=1}^{n}\left(x_{ik} \times x_{jk}\right)}{\sqrt{\sum_{k=1}^{n} x_{ik}^2 \sum_{k=1}^{n} x_{jk}^2}}, \quad (19)$$

where x_i and x_j are the input and output datasets, respectively. The effect of each input parameter on the outputs is shown in Figure 14.

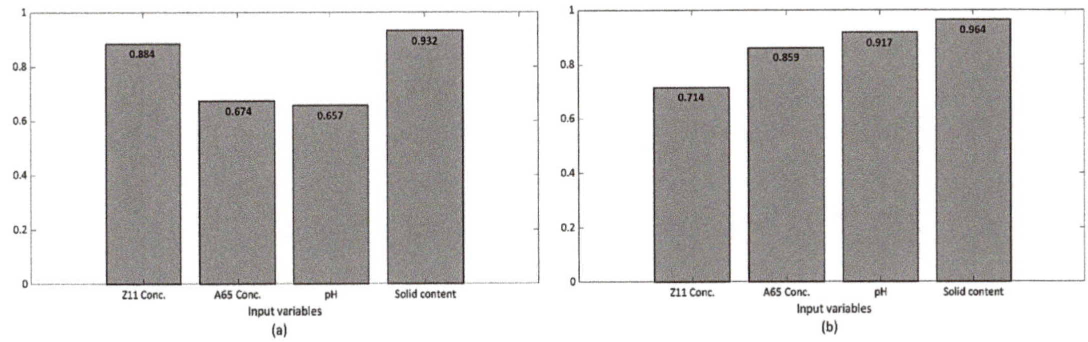

Figure 14. Sensitivity analysis to determine the effect of each input parameter on Cu recovery (**a**) and Cu grade (**b**).

It can be seen in the figure that the solid content influenced both Cu recovery and grade the most. The results also confirm the correlation results to a great extent. Using the results of different tests, the variables that had a greater impact on the outputs were selected, which ultimately led to simulation models with greater accuracy and less error.

3.4.3. Model Prediction Analysis

The dataset was divided into three categories, including training (65%), validating (15%), and testing (20%). The validation dataset helps to tune hyper-parameters during the training of the models to prevent the models from over-fitting in the testing phase. The test performance of the 'RNNs' models is shown in Figures 15 and 16.

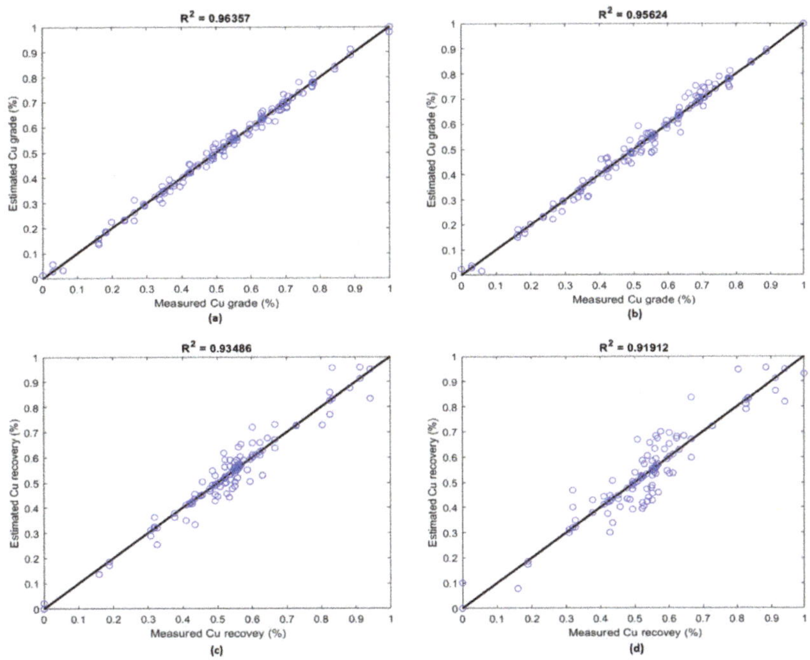

Figure 15. Correlation between the estimated and measured data by LSTM (**a**,**c**) and GRU (**b**,**d**) networks at the testing phase.

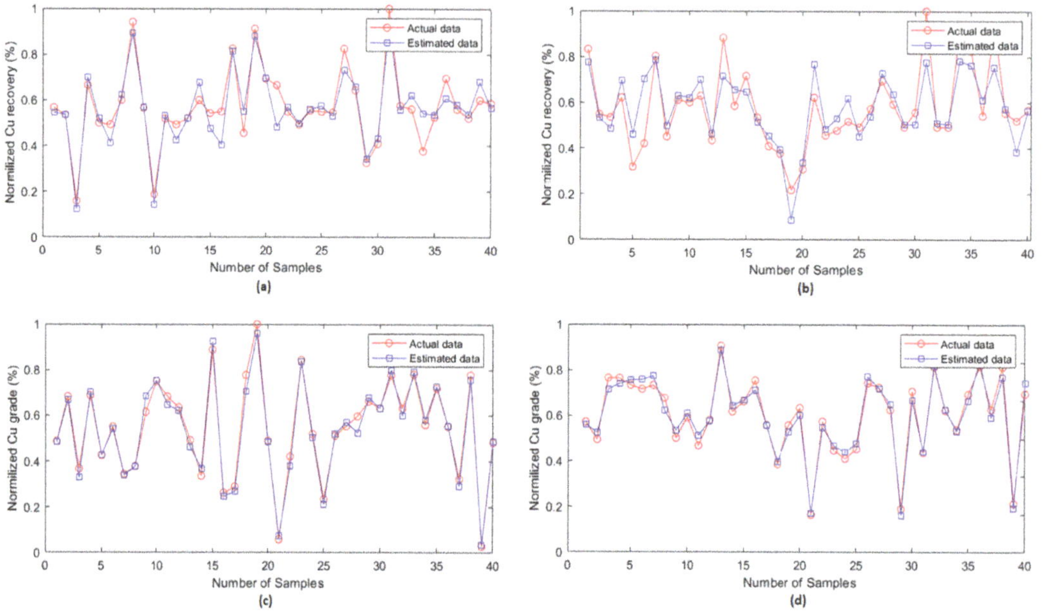

Figure 16. Comparison of estimated Cu grade and recovery vs. actual data by LSTM (**a**,**c**) and GRU (**b**,**d**) in the testing process.

Mean square error (*MSE*), root mean square error (*RMSE*), mean absolute percentage error (*MAPE*), and R^2 were computed using the following equations in order to evaluate the performance of each model [43,44]:

$$MSE = \frac{1}{n}\sum_{i=1}^{n}(y_i - \hat{y}_i)^2, \tag{20}$$

$$RMSE = \sqrt{\frac{1}{n}\sum_{i=1}^{n}(y_i - \hat{y}_i)^2}, \tag{21}$$

$$MAPE = \frac{1}{N}\sum_{i=1}^{N}\frac{|\hat{y}_i - y_i|}{\hat{y}_i}, \tag{22}$$

$$R^2 = \left(\frac{\sum_{i=1}^{N}(y_i - \bar{a})(\hat{y}_i - \bar{e})}{\sqrt{\sum_{i=1}^{N}(y_i - \bar{a})^2}\sqrt{\sum_{i=1}^{N}(\hat{y}_i - \bar{e})^2}}\right)^2, \tag{23}$$

where y is the actual data, $\hat{y}$ is the estimated data, $\bar{a}$ is the mean of actual data, $\bar{e}$ is the mean of the estimated data, and N is the number of sample sets.

The performance of the LSTM, GRU, RF, and ANN-LMA models in predicting Cu grade and its recovery is shown in Tables 10 and 11. Based on the results, the performance of the RNNs is better than the other models; however, in the estimation of Cu recovery, the RF and ANN-LMA results are so close to the GRU. The RNN results are also close to each other (LSTM provides better accuracy). Although RF has a better result than ANN-LMA due to its tree structure in Cu grade estimation, when there are many calculations, a large number of trees can make the algorithm too slow and ineffective for real-time predictions. Both GRU and LSTM could validly estimate Cu grade and recovery. The results demonstrate that LSTM and GRU are useful deep networks for predicting time series and sequential data.

Table 10. Performance of the models to estimate Cu grade.

Model	*MSE*	*RMSE*	*MAPE*	R^2
LSTM	5.5×10^{-3}	0.074	5.7×10^{-5}	0.963
GRU	8.7×10^{-3}	0.093	6.3×10^{-5}	0.956
RF	9.8×10^{-3}	0.098	7.4×10^{-5}	0.939
ANN-LMA	1.3×10^{-2}	0.114	8.6×10^{-5}	0.921

Table 11. Performance of the models to estimate Cu recovery.

Model	*MSE*	*RMSE*	*MAPE*	R^2
LSTM	0.017	0.132	7.8×10^{-5}	0.934
GRU	0.026	0.162	9.6×10^{-5}	0.919
RF	0.028	0.167	1.1×10^{-4}	0.915
ANN-LMA	0.029	0.170	1.3×10^{-4}	0.914

Although the accuracy of the RNN models is close, the number of parameters and time efficiency of the algorithms should be considered in practical applications [47,48]. The study showed that GRU performed faster with less CPU usage because it has fewer parameters for training. LSTM is more sophisticated and provides more parameters such as the number of weight matrices, number of bias vectors, learning rates, etc. This can be considered as a penalty for such models, as described in detail by Hassanzadeh et al. [49]. A model's learning rate controls how quickly it can adapt to a new problem and is a crucial parameter for efficient training [29]. Larger learning rates result in rapid changes and require fewer training epochs; however, smaller learning rates allow the model to learn a more optimal set of weights, but may take significantly longer to train. The structure of GRU is more straightforward. It has one gate less than LSTM, which reduces matrix multiplication, and it can save time. However, through empirical research and as reported by Cahuantzi et al. [50] and Yang et al. [51], the advantage of GRU is only relevant for small datasets and when you have little memory. In other scenarios, when dealing with more extensive sequences, LSTM is preferred. Generally, it can be concluded that if the dataset is small, the GRU algorithm is preferable, whereas LSTM is favourable and more accurate for larger datasets. Furthermore, parameter optimization for the purpose of studying the influence of different setting parameters on these networks could be investigated in future research. To find insights on the structure of the estimation models, the importance and impact of each input variable on the outputs should be specified. The model estimation error was calculated when input values were randomly shuffled in order to assess the features importance. It is expected that permutation will increase the model's estimation error if the model relies on an input value for estimation. Due to the superior performance of LSTM models for Cu grade and recovery, this process was applied to LSTM models and the results are shown in Figure 17. Features importance results are well aligned with results obtained through the Spearman correlation and sensitivity analysis. Besides, LSTM's success in developing models and extracting meaningful characteristics from training data can also be confirmed by the results of feature importance and statistical analysis (Section 3.2).

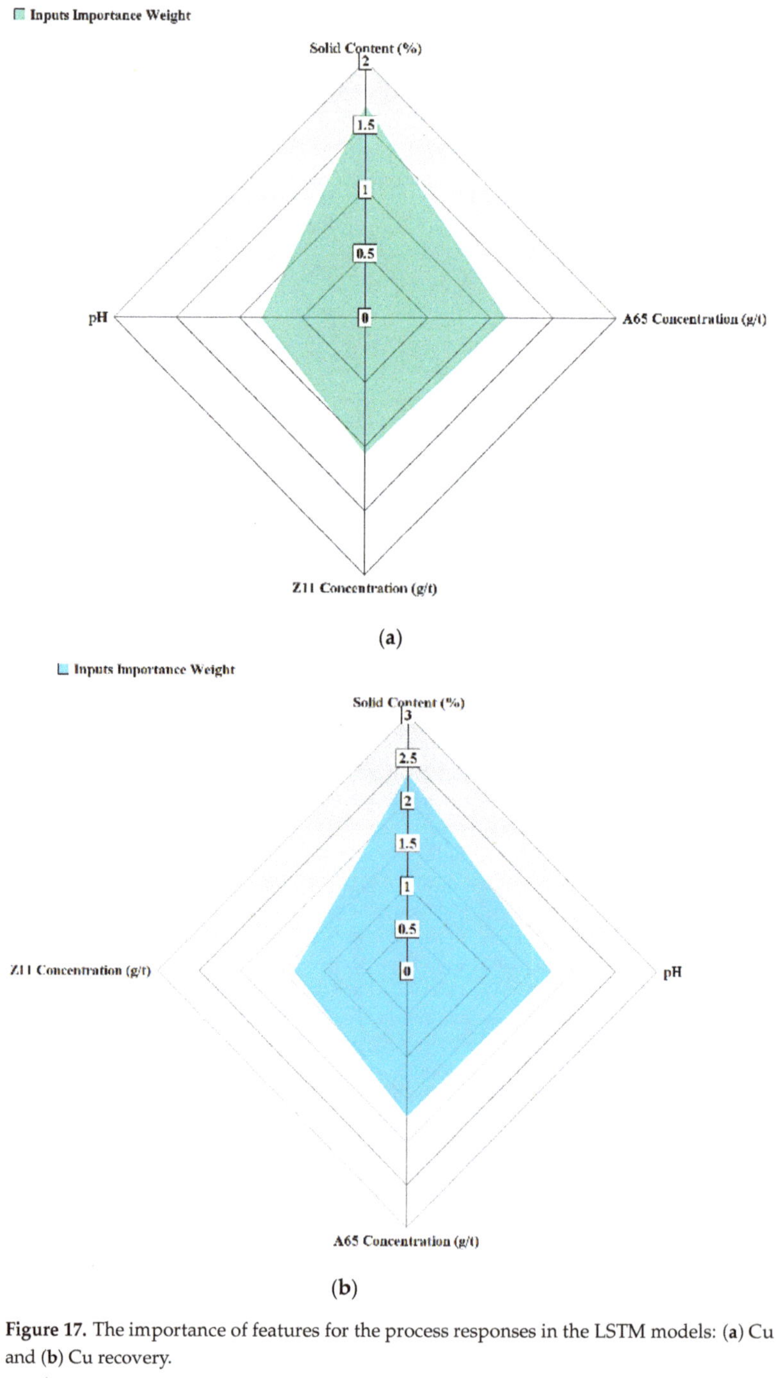

Figure 17. The importance of features for the process responses in the LSTM models: (**a**) Cu grade and (**b**) Cu recovery.

4. Conclusions

The metallurgical response of a copper processing plant was predicted using two efficient variants of the recurrent neural network (RNN) method based on the effective operating parameters, including pulp pH and solid content as well as the concentrations of the frother and collector. For this purpose, the process was first evaluated using a two-step screening/optimization DOE based on factorial designs. Statistical results based on analysis of variance showed that there is a reliable correlation between the metallurgical responses, i.e., copper grade and recovery in concentrate, and operating variables. ANOVA results indicated that all operating variables significantly affect the metallurgical responses, such that the copper grade increased by increasing the pulp pH and solid content and decreased as the dosage of frother and collector were increased. Contrary results were observed with respect to the copper recovery, with the exception that, like grade, copper recovery increased with the increase of pulp pH due to interaction effect with other factors.

Afterwards, the optimization process was performed based on the identified significant factors and by applying particular adjustment in terms of the technical aspects of the process. The effect of the studied variables was also interpreted based on the main and interaction effects to obtain a meaningful vision for the simulation step. The correlation coefficient, mean, root mean square, and mean absolute percentage errors as well as variance account for values for the training and testing datasets for the copper grade and recovery using the long short-term memory (LSTM), and gated recurrent unit (GRU) networks were compared. The results showed that the LSTM algorithm was more efficient than the GRU network and can be applied to predict the metallurgical responses during the flotation process.

Author Contributions: Conceptualization, A.G. and H.K.; methodology, M.M., A.G. and H.K.; software, A.G. and H.K.; validation, M.M. and H.K.; formal analysis, H.K.; investigation, H.K.; resources, M.M. and H.K.; writing—original draft preparation, A.G. and H.K.; writing—review and editing, A.G., A.H. and H.K.; visualization, A.G., A.H. and H.K.; supervision, H.K. All authors have read and agreed to the published version of the manuscript.

Funding: This research received no external funding.

Acknowledgments: The authors would like to acknowledge the managerial and process divisions of Takht Gonbad Copper Complex for their outstanding technical and scientific cooperation.

Conflicts of Interest: The authors declare no conflict of interest.

References

1. Adams, M.D. *Advances in Gold Ore Processing*; Elsevier: Amsterdam, The Netherlands, 2005.
2. King, R.P. *Modeling & Simulation of Mineral Processing Systems*; Butterworth-Heinmann: Boston, MA, USA, 2012.
3. Scharnhorst, A.; Börner, K.; van der Besselaar, P. *Models of Science Dynamics*; Springer: Berlin, Germany, 2012.
4. Kuhn, M.; Johnson, K. *Applied Predictive Modeling*; Springer: Berlin, Germany, 2018.
5. Herrell, F.E. *Regression Modeling Strategies*; Springer: Berlin, Germany, 2016.
6. Montgomery, D.C. *Design and Analysis of Experiments*, 10th ed.; Wiley: Hoboken, NJ, USA, 2020.
7. Asadi, E.; Isazadeh, M.; Samadianfard, S.; Ramli, M.F.; Mosavi, A.; Nabipour, N.; Shamshirband, S.; Hajnal, E.; Chau, K. Groundwater quality assessment for sustainable drinking and irrigation. *Sustainability* **2020**, *12*, 177. [CrossRef]
8. Shamshirband, S.; Hashemi, S.; Salimi, H.; Samadianfard, S.; Asadi, E.; Shadkani, S.; Kargar, K.; Mosavi, A.; Nabipour, N.; Chau, K. Predicting standardized streamflow index for hydrological drought using machine learning models. *Eng. Appl. Comput. Fluid Mech.* **2020**, *14*, 339–350. [CrossRef]
9. Apaydin, H.; Feizi, H.; Sattari, M.T.; Colak, M.S.; Shamshirband, S.; Chau, K. Comparative analysis of recurrent neural network architectures for reservoir inflow forecasting. *Water* **2020**, *12*, 1500. [CrossRef]
10. Hoseinian, F.S.; Rezai, B.; Kowsari, E.; Safari, M. A hybrid neural network/genetic algorithm to predict Zn (II) removal by ion flotation. *Sep. Sci. Technol.* **2020**, *55*, 1197–1206. [CrossRef]
11. Khoshdast, H.; Gholami, A.R.; Hassanzadeh, A.; Niedoba, T.; Surowiak, A. Advanced simulation of removing chromium from a synthetic wastewater by rhamnolipidic bioflotation using hybrid neural networks with metaheuristic algorithms. *Materials* **2021**, *14*, 2880. [CrossRef]
12. Gholami, A.R.; Khoshdast, H.; Hassanzadeh, A. Using hybrid neural networks/genetic and artificial bee colony algorithms to simulate the bio-treatment of dye-polluted wastewater using rhamnolipid biosurfactants. *J. Environ. Manag.* **2021**, *299*, 113666. [CrossRef]
13. Jorjani, E.; Bagherieh, A.; Mesroghli, S.; Chelgani, S.C. Prediction of yttrium, lanthanum, cerium, and neodymium leaching recovery from apatite concentrate using artificial neural networks. *J. Uni. Sci. Technol. Beijing Min. Metal. Mat.* **2008**, *15*, 367–374. [CrossRef]

14. Milivojevic, M.; Stopic, S.; Friedrich, B.; Stojanovic, B.; Drndarevic, D. Computer modeling of high-pressure leaching process of nickel laterite by design of experiments and neural networks. *Int. J. Miner. Metal. Mat.* **2012**, *19*, 584–594. [CrossRef]
15. Hoseinian, F.S.; Abdollahzade, A.; Mohamadi, S.S.; Hashemzadeh, M. Recovery prediction of copper oxide ore column leaching by hybrid neural genetic algorithm. *Trans. Nonfer. Metal Soc. China* **2017**, *27*, 686–693. [CrossRef]
16. Sobouti, A.; Hoseinian, F.S.; Rezai, B.; Jalili, S. The lead recovery prediction from lead concentrate by an artificial neural network and particle swarm optimization. *Geosys. Eng.* **2019**, *22*, 319–327. [CrossRef]
17. Vyas, S.; Das, S.; Ting, Y.-P. Predictive modeling and response analysis of spent catalyst bioleaching using artificial neural network. *Biores. Technol. Rep.* **2020**, *9*, 100389. [CrossRef]
18. Ghobadi, P.; Yahyaei, M.; Banisi, S. Optimization of the performance of flotation circuits using a genetic algorithm oriented by process-based rules. *Int. J. Miner. Process.* **2011**, *98*, 174–181. [CrossRef]
19. Gholami, A.R.; Khoshdast, H. Using artificial neural networks for the intelligent estimation of selectivity index and metallurgical responses of a sample coal bioflotation by rhamnolipid biosurfactants. *Energy Sources Part A Recovery Util. Environ. Eff.* **2020**, *2020*, 1857477. [CrossRef]
20. Gholami, A.R.; Asgari, K.; Khoshdast, H.; Hassanzadeh, A. A hybrid geometallurgical study using coupled Historical Data (HD) and Deep Learning (DL) techniques on a copper ore mine. *Physicochem. Prob. Miner. Process.* **2022**, *58*, 147841. [CrossRef]
21. Nakhaei, F.; Irannajad, M. Application and comparison of RNN, RBFNN and MNLR approaches on prediction of flotation column performance. *Int. J. Min. Sci. Technol.* **2015**, *25*, 983–990. [CrossRef]
22. Zhang, D.; Gao, X. Soft sensor of flotation froth grade classification based on hybrid deep neural network. *Int. J. Prod. Res.* **2021**, *59*, 4794–4810. [CrossRef]
23. Pu, Y.; Szmigiel, A.; Apel, D.B. Purities prediction in a manufacturing froth flotation plant: The deep learning techniques. *Neural Comput. Appl.* **2020**, *32*, 13639–13649. [CrossRef]
24. Inapakurthi, R.K.; Miriyala, S.S.; Mitra, K. Recurrent neural networks based modelling of industrial grinding operation. *Chem. Eng. Sci.* **2020**, *219*, 115585. [CrossRef]
25. Khoshdast, H.; Soflaeian, A.; Shojaei, V. Coupled fuzzy logic and experimental design application for simulation of a coal classifier in an industrial environment. *Physicochem. Prob. Miner. Process.* **2019**, *55*, 504–515. [CrossRef]
26. Khoshdast, H.; Sam, A.; Vali, H.; Akbari Noghabi, K. Effect of rhamnolipid biosurfactants on performance of coal and minerals flotation. *Int. Biodeter. Biodegrad.* **2011**, *65*, 1238–1243. [CrossRef]
27. Khoshdast, H.; Sam, A.; Manafi, Z. The use of rhamnolipid biosurfactants as a frothing agent and a sample copper ore response. *Miner. Eng.* **2012**, *26*, 41–49. [CrossRef]
28. Che, Z.; Purushotham, S.; Cho, K.; Sontag, D.; Liu, Y. Recurrent neural networks for multivariate time series with missing values. *Sci. Rep.* **2018**, *8*, 6085. [CrossRef]
29. Mandic, D.; Chambers, J. *Recurrent Neural Metworks for Prediction: Learning Algorithms, Architectures and Stability*; Wiley: Hoboken, NJ, USA, 2001.
30. Fei, H.; Tan, F. Bidirectional grid long short-term memory (bigridlstm): A method to address context-sensitivity and vanishing gradient. *Algorithms* **2018**, *11*, 172. [CrossRef]
31. Hochreiter, S.; Schmidhuber, J. Long short-term memory. *Neur. Comp.* **1997**, *9*, 1735–1780. [CrossRef]
32. Graves, A. Supervised sequence labelling. In *Supervised Sequence Labelling with Recurrent Neural Networks*; Graves, A., Ed.; Springer: Berlin, Germany, 2012; pp. 5–13.
33. Chung, J.; Gulcehre, C.; Cho, K.; Bengio, Y. Empirical evaluation of gated recurrent neural networks on sequence modeling. *arXiv* **2014**, arXiv:1412.3555.
34. Fu, R.; Zhang, Z.; Li, L. Using LSTM and GRU neural network methods for traffic flow prediction. In Proceedings of the 31st Youth Academic Annual Conference of Chinese Association of Automation (YAC), Wuhan, China, 11 November 2016; pp. 324–328. [CrossRef]
35. Agheli, S.; Hassanzadeh, A.; Hassas, B.V.; Hassanzadeh, M. Effect of pyrite content of feed and configuration of locked particles on rougher flotation of copper in low and high pyritic ore types. *Int. J. Min. Sci. Technol.* **2018**, *28*, 167–176. [CrossRef]
36. Azizi, A.; Hassanzadeh, A.; Fadaei, B. Investigating the first-order flotation kinetics models for Sarcheshmeh copper sulfide ore. *Int. J. Min. Sci. Technol.* **2015**, *25*, 849–854. [CrossRef]
37. Boveiri, R.; Shojaei, V.; Khoshdast, H. Efficient cadmium removal from aqueous solutions using a sample coal waste activated by rhamnolipid biosurfactant. *J. Environ. Manag.* **2019**, *231*, 1182–1192. [CrossRef]
38. Yetilmezsoy, K.; Demirel, S.; Vanderbei, R.J. Response surface modeling of Pb(II) removal from aqueous solution by Pistacia vera L.: Box–Behnken experimental design. *J. Hazard. Mater.* **2009**, *171*, 551–562. [CrossRef]
39. Shak, K.P.Y.; Wu, T.Y. Optimized use of alum together with unmodified *Cassia obtusifolia* seed gum as a coagulant aid in treatment of palm oil mill effluent under natural pH of wastewater. *Ind. Crop. Prod.* **2015**, *76*, 1169–1178. [CrossRef]
40. Zahab-Nazouri, A.; Shojaei, V.; Khoshdast, H.; Hassanzadeh, A. Hybrid CFD-experimental investigation into the effect of sparger orifice size on the metallurgical response of coal in a pilot-scale flotation column. *Int. J. Coal Prep.* **2022**, *42*, 349–368. [CrossRef]
41. Khoshdast, H.; Hassanzadeh, A.; Kowalczuk, P.B.; Farrokhpay, S. Characterization of flotation frothers–A review. *Miner. Process. Extrac. Metal. Rev.* **2022**, *43*, 2024822. [CrossRef]
42. Azizi, A.; Masdarian, M.; Hassanzadeh, A.; Bahri, Z.; Niedoba, T.; Surowiak, A. Parametric optimization in rougher flotation performance of a sulfidized mixed copper ore. *Minerals* **2020**, *10*, 660. [CrossRef]

43. Shami, R.B.; Shojaei, V.; Khoshdast, H. Removal of some cationic contaminants from aqueous solutions using sodium dodecyl sulfate-modified coal tailings. *Iran. J. Chem. Chem. Eng.* **2021**, *40*, 1105–1120. [CrossRef]
44. Dodge, Y. *The Concise Encyclopedia of Statistics*; Springer Science & Business Media: Berlin, Germany, 2008.
45. Asadi, M.; Soltani, F.; Mohammadi, M.R.; Khodadadi, D.A.; Abdollahy, M. A successful operational initiative in copper oxide flotation: Sequential sulphidisation-flotation technique. *Physicochem. Prob. Miner. Process.* **2019**, *55*, 356–369. [CrossRef]
46. Monjezi, M.; Hasanipanah, M.; Khandelwal, M. Evaluation and prediction of blast-induced ground vibration at Shur River Dam, Iran, by artificial neural network. *Neur. Comp. Appl.* **2013**, *22*, 1637–1643. [CrossRef]
47. Gulec, M.; Gulbandilar, E. Determination of the lower calorific and ash values of the lignite coal by using artificial neural networks and multiple regression analysis. *Physicochem. Prob. Miner. Process.* **2019**, *55*, 400–406. [CrossRef]
48. Krause, P.; Boyle, D.P.; Bäse, F. Comparison of different efficiency criteria for hydrological model assessment. *Adv. Geosci.* **2005**, *5*, 89–97. [CrossRef]
49. Hassanzadeh, A.; Hoang, D.H.; Brockmann, M. Assessment of flotation kinetics modeling using information criteria; case studies of elevated-pyritic copper sulfide and high-grade carbonaceous sedimentary apatite ores. *J. Dispers. Sci. Technol.* **2020**, *41*, 1083–1094. [CrossRef]
50. Cahuantzi, R.; Chen, X.; Güttel, S. A comparison of LSTM and GRU networks for learning symbolic sequences. *arXiv* **2021**, arXiv:2107.02248.
51. Yang, S.; Yu, X.; Zhou, Y. LSTM and GRU neural network performance comparison study: Taking yelp review dataset as an example. In Proceedings of the 2020 International Workshop on Electronic Communication and Artificial Intelligence (IWECAI), Shanghai, China, 12–14 June 2020; pp. 98–101. [CrossRef]

minerals

MDPI

Article

Control Structure Design Using Global Sensitivity Analysis for Mineral Processes under Uncertainties

Oscar Mamani-Quiñonez [1], Luis A. Cisternas [1], Teresa Lopez-Arenas [2] and Freddy A. Lucay [3,*]

[1] Departamento de Ingeniería Química y Procesos de Minerales, Universidad de Antofagasta, Antofagasta 1240000, Chile; oscar.maquiz@gmail.com (O.M.-Q.); luis.cisternas@uantof.cl (L.A.C.)
[2] Departamento de Procesos y Tecnología, Universidad Autónoma Metropolitana-Cuajimalpa, Ciudad de México 05348, Mexico; mtlopez@cua.uam.mx
[3] Escuela de Ingeniería Química, Pontificia Universidad Católica de Valparaíso, Valparaíso 2340000, Chile
* Correspondence: freddy.lucay@pucv.cl

Abstract: Multiple-input and multiple-output (MIMO) systems can be found in many industrial processes, including mining processes. In practice, these systems are difficult to control due to the interactions of their input variables and the inherent uncertainty of industrial processes. Depending on the interactions in the MIMO process, different control strategies can be implemented to achieve the desired performance. Among these strategies is the use of a decentralized structure that considers several subsystems and for which a SISO controller can be designed. In this study, a methodology based on global sensitivity analysis (GSA) to design decentralized control structures for industrial processes under uncertainty is presented. GSA has not yet been applied for this purpose in process control; it allows us to understand the dynamic behavior of systems under uncertainty in a broad value range, unlike approaches proposed in the literature. The proposed GSA is based on the Sobol method, which provides sensitivity indices used as interaction measures to establish the input–output pairing for MIMO systems. Two case studies based on a semi-autogenous grinding (SAG) mill and a solvent extraction (SX) plant are presented to demonstrate the applicability of the proposed methodology. The results indicate that the methodology allows the design of 2 × 2 and 3 × 3 decentralized control structures for the SAG mill and SX plant, respectively, which exhibit good performance compared to MPC. For example, for the SAG mill, the determined pairings were fresh ore flux/fraction of mill filling and power consumption/percentage of critical speed.

Keywords: global sensitivity analysis; uncertainty; control structure; SX process; SAG mill

Citation: Mamani-Quiñonez, O.; Cisternas, L.A.; Lopez-Arenas, T.; Lucay, F.A. Control Structure Design Using Global Sensitivity Analysis for Mineral Processes under Uncertainties. *Minerals* **2022**, *12*, 736. https://doi.org/10.3390/min12060736

Academic Editor: Saeed Chehreh Chelgani

Received: 6 May 2022
Accepted: 3 June 2022
Published: 8 June 2022

1. Introduction

Most industrial control systems are multiple-input and multiple-output (MIMO) systems, as the goal of multivariable control includes keeping multiple variables controlled at independent set points. For instance, mining plants, oil refineries, biorefineries, and, in general, chemical manufacturing plants contain MIMO processes. In these systems, each manipulated variable (input) can affect several controlled variables (outputs) causing interactions between them and consequently generating coupling in the system. In practice, such interactions result in difficulties in analyzing and controlling a given system. Furthermore, the parameters used to define the input and output variables may present uncertainty, vary with time, or be unknown [1]. For these reasons, the analysis of how to control MIMO systems is often more complex compared to single-input and single-output (SISO) systems.

Depending on the interactions in the MIMO process, different control strategies can be applied to achieve the desired performance: decentralized, centralized, or decoupled control. The centralized structure considers the design of a complete multivariate controller to control n output variables using n manipulated variables, yielding that n^2 number of controllers prevail. However, these control systems are complex and lack integrity [2].

The decentralized structure considers several subsystems for which a SISO controller is designed. Thus, only n controllers prevail for each n output variable, since it uses single-loop or diagonal controllers [3]. The decoupling structure uses separate elements, known as decouplers, or simply controllers, to compensate for the strong interactions present in the system [4]. Decoupling can be divided into static and dynamic decoupling according to the characteristics of time, or can be classified into total and approximate decoupling according to the degree of decoupling.

The decentralized control technique is still widely used in many industrial control systems due to its simple implementation, proficient maintenance, simple tuning, and robust performance even under model mismatches and uncertainties [5]. The key issue when designing a decentralized control system is the control structure design (CSD), that is, the selection of inputs and outputs and how they are paired [6]. The available literature proposes several mathematical measures to quantify the degree of interaction between input–output pairs. Probably the most widely used measure is relative gain array (RGA), which was proposed by Bristol in 1966 and requires only the steady-state gain of the plant model. This information can be obtained by step test methods. The simplicity of RGA is the main reason for its popularity [7]. RGA has been studied and used by several authors to propose new interaction measures. For example, Niederlinski [8] proposed the use of an index based on the gain matrix to provide direct information on the ability of a decentralized control to stabilize a 2×2 MIMO system. A variation of RGA was reported by Zhu [9], known as relative interaction array (RIA). This is based on the concept of viewing the interaction as an unmodeled term for a particular pairing. A dynamic extension of RGA was proposed by Kinnaert [10] which can be applied to analyze plants at any frequency. Mc Avoy et al. [11] also proposed a dynamic extension of RGA which assumes the availability of a dynamic process model that is used to design an optimal proportional output controller.

RGA provides limited knowledge; specifically, it does not indicate when to use multivariable controllers or how to carry out CSD. Therefore, some authors have proposed alternative approaches. Salgado and Conley [12] considered observability and controllability Gramians in so-called participation matrices (PMs). Using a similar approach, Wittenmark and Salgado [13] introduced the Hankel interaction index matrix (HIIA). These Gramian-based interaction measures help to overcome most of the disadvantages of RGA. Specifically, these measures seem to provide suggestions for designing controller structures. Hanzon [14] showed that the PM is closely related to the direct Nyquist array, which was introduced by Rosenbrock in 1970. Birk and Medvedev [15] proposed an alternative to HIIA. They used the $\mathcal{H}_2$ and $\mathcal{H}_\infty$ norms as the basis for new interaction measures. Meanwhile, Halvarsson et al. [16] proposed a different approach to obtain interaction measures based on linear quadratic Gaussian (LQG) control. Moreover, many MIMO systems present uncertainty, creating a set of possible systems for which the interaction measures may differ. Consequently, the control structure design (CSD) may differ between models. For example, Jain and Babu [17] analyzed the sensitivity of RGA to model uncertainty. Specifically, they studied how the process dynamics can affect CSD decisions proposed by RGA in systems under uncertainty.

As outlined above, there are currently many rigorous methods of CSD based on process control theory. On the other hand, there is a large gap between research and industrial application, which means process control engineers in industry today still use a strongly empirical approach to CSD, basing their decisions on practical knowledge or principles of common sense and experience [6]. For instance, in the milling process, input–output pairing has been established based on practical knowledge or trial and error under uncertainty [18–20], or not [21–23]. In solvent extraction (SX) [24], froth flotation [25], and melting furnaces [26], input–output pairing was settled using classical RGA. These works verify the need to have one methodology to help establish CDS under uncertainty for devices implemented in mineral processing.

Within this context, global sensitivity analysis (GSA) is proposed in this work as an alternative to decide on the CSD. Sensitivity analysis (SA) is a commonly used method of identifying the important input variables that determine the behavior of a model under uncertain conditions. SA can be performed locally or globally, and according to Saltelli [27], the latter is more robust and reliable even for nonlinear models. There are several methods of performing GSA and among them, due to their versatility and efficiency, those based on variance decomposition stand out [28]. Because these methods involve high computational cost, Homma and Saltelli [29] introduced the concept of total sensitivity indices to overcome this disadvantage. These indices indicate the average effect of a given input variable on a specific output of the model, taking into account all possible interactions with the other input variables of the system. A significant advantage of this method is that it can be used in both steady-state and dynamic systems [27].

In this work, GSA is proposed to determine the CSD for nonlinear MIMO systems under time-varying and uncertain conditions. GSA has not yet been applied to this purpose in process control; it allows us to understand the dynamic behavior of systems under uncertainty in a broad value range, unlike approaches proposed in the related literature. The first order and total sensitivity indices provided by GSA are used as measures of input–output interactions for the CSD. The methodology is illustrated with two case studies from the mining industry: a semi-autogenous grinding (SAG) mill and an SX plant. The open-loop modeling and simulation of these processes have been studied previously [30,31], so in this work, the analysis of CSD based on GSA is presented first. Subsequently, for the purpose of comparing control performance levels, several controllers were designed based on different control structures (reported structures vs. CSD proposed in this work) and control strategies (proportional–integral (PI) control and model predictive control (MPC)). The results of the GSA in both cases studies were obtained using the Sobol–Jansen method, which allowed quantification of the interactions of variables over time as well as observation of the changes in the output variables due to the uncertainty of input variables. The information generated by GSA allowed a reduction of the CSD of the SAG process from a 3×3 to a 2×2 MIMO system, and that of the SX process from a 4×5 to a 3×3 MIMO system. Finally, these control structures were implemented for the control strategies described above and better closed-loop performance was obtained when reduced CSDs were implemented.

2. Materials and Methods

2.1. Uncertainty Analysis (UA)

Mathematical models are fundamental tools in decision-making and are developed considering assumptions and sometimes little-known information, introducing uncertainty in the modeling. Uncertainty can be classified as either stochastic or epistemic [32]. The former is also known as variability, inherent uncertainty, irreducible uncertainty, or uncertainty due to chance and is related to variations inherent in a given system, usually as a result of the random nature of model inputs. The latter is also known as reducible uncertainty, subjective uncertainty, or uncertainty due to a lack of knowledge. This uncertainty type, as a source of non-deterministic behavior, derives from a lack of knowledge of the system or the environment. Uncertainty in numerical models has many origins: input data, model simplification, algorithm structure, calibration process, calibration and validation data, and equifinality. In this context, UA corresponds to determining the uncertainty in the output variables as a result of the uncertainty in the input variables. UA can be addressed using probability theory, imprecise probability, probability bound analysis, evidence theory, or possibility theory [33]. In this work, UA is applied using probability theory, with a procedure that includes four steps: first, the uncertain input variables are described using the probability distribution function (PDF); second, a sample is generated from the PDF using random sampling, such as the Monte Carlo method; third, the values of the model output variables are determined for each element of the sample; fourth, the behavior of the model output variables is characterized by graphs, descriptive statistics, and statistical tests.

2.2. Sensitivity Analysis (SA)

According to Saltelli [34], SA can be defined as an examination of how the uncertainty in the output of a model can be apportioned among different sources of uncertainty in the model's input variables. This analysis can be done locally or globally. The latter quantifies the importance of model inputs and their interactions with respect to model outputs. GSA provides an overall view of the influence of inputs on outputs, as opposed to the local view based on partial derivatives, which has the disadvantage of depending on the choice of the evaluation point. The general objectives of GSA are as follows [35]: to identify significant and insignificant variables in a given model, aiming to reduce its dimension; to improve the understanding of model behavior, specifically highlighting interactions between input variables and finding combinations of input variables that result in high or low values for the model output. GSA considers six steps [36]: (1) determine the objective function, (2) select the input variables of the model, (3) assign a range and type of PDF to the input variables, (4) apply a sampling design to generate samples, (5) assess the model for the generated samples, and (6) implement the results of step 5 to perform GSA and determine the importance of the input variables on the model outputs. The related literature reveals that there are several methods of performing GSA and those based on variance decomposition are used more often due to their versatility and efficiency [28]. In this category, approaches based on the method of Sobol can be found. The latter considers a squared-integrable function f on $\Omega^m = \{x/0 \le x_j \le 1,\ j = 1,2,\ldots,m\}$ that is represented in terms of increasing dimensions [27]:

$$f = f_0 + \sum_j f_j + \sum_j \sum_{k>j} f_{jk} + \ldots + f_{1,2,\ldots,m} \quad (1)$$

where $f_j = f_j(x_j)$, $f_{jk} = f_{jk}(x_j, x_k)$, and so on; whereas $f_0 = E(Y)$, $f_j = E(Y/x_j)$, $f_{jk} = E(Y/x_j, x_k) - f_j - f_k - E(Y)$, and so on. Here, $Y = f(x_1, x_2, \ldots, x_m)$ and E represents the mathematical expectation. Note that these last expressions have the following properties: $V_j = V(f_j(x_j)) = V(E(Y/x_j))$, $V_{jk} = V\left(f_{jk}(x_j, x_k)\right) = V(E(Y/x_j, x_k)) - V(E(Y/x_j)) - V(E(Y/x_k))$, and so on. Here, V represents the variance. The square integration of Equation (1) on Ω^m allows us to obtain the so-called ANOVA-HDMR decomposition or its normalized equivalent:

$$V(Y) = \sum_j V_j + \sum_j \sum_{k>j} V_{jk} + \ldots + V_{1,2,\ldots,m} \quad (2)$$

$$1 = \sum_j \frac{V_j}{V(Y)} + \sum_j \sum_{k>j} \frac{V_{jk}}{V(Y)} + \ldots + \frac{V_{1,2,\ldots,m}}{V(Y)} \quad (3)$$

In Equation (3), $j = 1,2,\ldots,m$, $V(Y)$ represents the model variance, V_j represents the first order effect for each input variable x_j, and V_{jk} to $V_{1,2,\ldots,m}$ represent the interactions of the m input variables. The calculation of Equation (3) has a high computational cost that can be overcome by calculating total sensitivity indices [29]. These indices allow us to determine the average effect of a given input variable, considering all possible interactions of the respective variable with all other input variables. In this work, the Sobol–Jansen method was used, which allows calculation of the first order sensitivity index (S_j) and the total sensitivity index (S_j^T) for input variable x_j of the mathematical model. The Sobol–Jansen method has been used to analyze flotation circuits [37,38], heap leaching [31], grinding [30], and the lithium supply chain [39]. In addition, it exhibits high performance when analyzing chemical processes [40]. This method considers 5 steps [28]: first, choose an integer N; second, generate a matrix of size $(N, 2r)$ of quasi-random numbers from the sampling of input variables of their respective PDF (r represents the number of input variables); third, divide the matrix into 2 submatrices, A and B, of size (N, k); fourth, form matrix D_j from the columns of matrix A, except the jth column, which is taken from matrix B, and similarly, form matrix C_j from the columns of matrix B, except the jth column, which is taken from matrix A; fifth, assess the model output in matrices A, B, C_j, and D_j, obtaining $Y_A = f(A)$, $Y_B = f(B)$, $Y_{C_i} = f(C_i)$, and $Y_{D_i} = f(D_i)$, and subsequently use the following equations:

$$S_j = \frac{V(E(Y/x_j))}{V(Y)} = \frac{V(Y) - \frac{1}{2N}\sum_{i=1}^{N}\left(Y_B^{(i)} - Y_{D_j}^{(i)}\right)^2}{V(Y)}, j = 1, 2, \ldots, m \tag{4}$$

and

$$S_j^T = 1 - \frac{V(E(Y/x_j))}{V(Y)} = \frac{\frac{1}{2N}\sum_{i=1}^{N}\left(Y_A^{(i)} - Y_{D_j}^{(i)}\right)^2}{V(Y)}, j = 1, 2, \ldots, m \tag{5}$$

Other expressions to estimate the sensitivity indices can be found elsewhere [28,36,40]; these use one or another matrix defined earlier, e.g., the Sobol–Jansen method uses matrices A, B, and D. The interpretation of the indices is straightforward: the higher the sensitivity index of an input variable, the greater its influence on the model output. The first order index allows us to determine the most important input variable, while the total sensitivity index allows us to identify the input variables that do not influence the model outputs. In this sense, if input variable x_j of the model does not interact with the other input variables, the sensitivity indices satisfy $S_j \approx S_j^T$, otherwise $S_j < S_j^T$. If $S_j^T \approx 0$, input variable x_j does not influence the model output and can be fixed at its nominal operating value and consequently the dimension of the mathematical model can be reduced [27]. Note that, ideally, UA precedes SA, as before uncertainty can be apportioned, it needs to be estimated [41].

2.3. Solving the Model in MATLAB–Simulink

The models were implemented as a Mask subsystem in MATLAB–Simulink^{TM}. Simulink^{TM} (R2020a-Academic Version) is a programming system that uses blocks, i.e., graphical programming, to solve differential equations. In this work, such equations were solved using the ode4 solver based on the fourth order Runge–Kutta formula. In addition, Simulink^{TM} allows users to program their own blocks across functions. This feature and the possibility to use specific toolboxes, such as PID control and MPC, provide a powerful platform for the development of prototypes.

2.4. Methodology for Control Structure Design (CSD)

A computational method of developing control structures is proposed and presented in Figure 1. In the first step, the multivariable system is modeled using mathematical and computational tools, such as differential equations and MATLAB software (R2020a-Academic Version), respectively. In the second step, the process variables are classified as manipulated, controlled, supervised, or disturbed. Furthermore, the variables manipulated under uncertainty are characterized by distribution functions after determining the type of uncertainty. In the third step, GSA is carried out using methods based on variance decomposition, such as the Sobol–Jansen method, after carrying out UA. Here, the GSUA toolbox [42] is implemented, and the sample size for each uncertain manipulated variable is defined as one thousand. According to [43], this value allows us to obtain robust results from UA and GSA. Subsequently, the input–output pairing is selected according to the total sensitivity indices provided by GSA. Here, it was important to analyze the behavior of total sensitivity indices over the simulation time to establish such pairing, which allowed us to obtain a decentralized structure whose SISO subsystems can be controlled using PID or PI controls. In the fourth step, the designed control structure is evaluated through simulations and experiments. Specifically, the control structure is subject to different set points and compared with other approaches proposed in the related literature. If the control structure provides satisfactory results, it is considered robust; otherwise, we return to the second step. The latter considers changing the nominal operating conditions or parameters used to define the distribution functions.

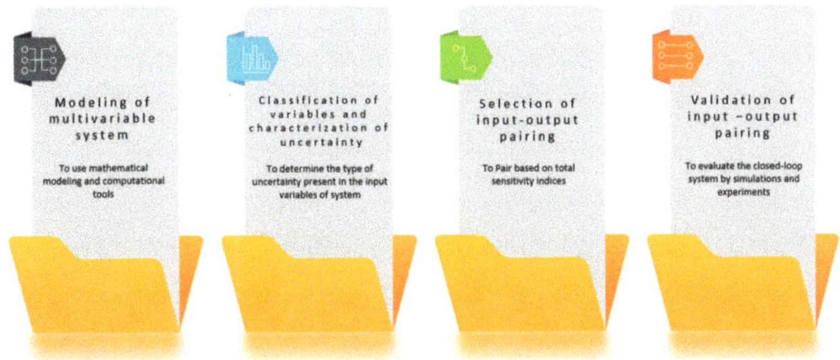

Figure 1. Methodology used for CSD using GSA and UA.

3. Results

The methodology proposed for CSD is illustrated considering grinding and SX processes to demonstrate the methodology's capacity to address systems with different degrees of freedom.

3.1. Semi-Autogenous Grinding (SAG)

Step 1. Modeling

From an energy point of view, mineral milling is decisive in the evaluation of operating cost, representing 50–80% of the total operating cost of a mineral concentrator plant. Various modeling trends are proposed in the literature based on the principles that govern the grinding phenomenon; this is how the models based on population models stand out, which were used for this case study. The milling model presented by Austin et al. [44,45] was considered. The SAG mill model is generally divided into two zones, the grinding chamber and the sorting zone, as shown in Figure 2. The F particles entering the mill are introduced into the grinding chamber. The product obtained, P^*, faces the classification zone, where, according to a classification probability c_i, the particles can return to the crushing chamber or become part of product P of the SAG mill.

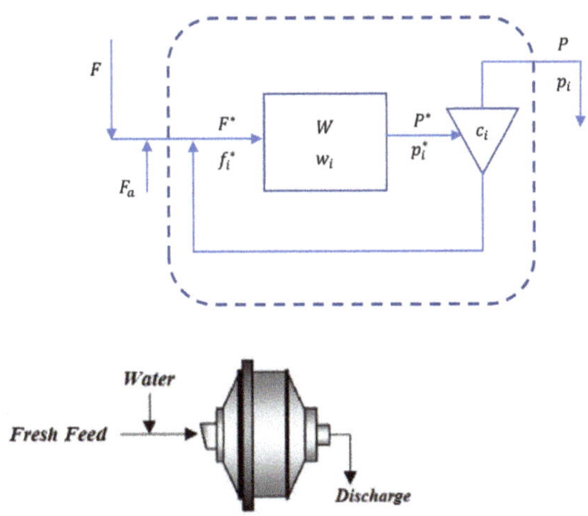

Figure 2. Schematic representation of a SAG mill, adapted from [30].

It is commonly assumed that the SAG mill behaves like a perfectly mixed reactor, with a mass retained (W) in the volume (V) of the mill and first order kinetics. Some authors indicate that second order kinetics might be able to model the breakage of coarse particles better than first order kinetics [46,47]. However, computational experiments were carried out under conditions where first order kinetics provided reliable estimations [48,49]. According to Austin et al. [44], F is the fresh ore flux fed to mill and the ith size fraction in F is $K_i w_i W$. In this expression, w_i is the weight fraction of retained mass in the mill and K_i is the specific breakage rate of the ith size fraction. When a fraction of size i breaks, a fraction b_{ij} of the broken material is sent to size j. The dynamic mass balance in each size i is:

$$\begin{gathered}\frac{d[w_i(t)\cdot W]}{dt} = F_i - P_i + W \sum_{\substack{j=1 \\ i>j}}^{i-1} b_{ij} K_j w_j - K_i w_i W, \\ w_i(0) = w_{i0},\ i = 1,\ 2,\ \ldots, n; n \geq i \geq j \geq 1\end{gathered} \tag{6}$$

where n is the number of species present in the fresh feed, F_i is the fraction of ore flux (F) fed to the mill, P_i is the fraction of flux discharged (P), and c_i is the classification efficiency of the internal grid, which affects the mass flow recirculated internally ($C^* = \sum\limits_i c_i w_i / \sum\limits_i w_i(1-c_i)$). The complete model equations and parameter values can be found in Appendix A and [30]. In the appendix, the reader can see the expressions used to model the cumulative breakage distribution function and its implementation to determine b_{ij}, as well as expressions used to estimate the classification efficiency of the internal grid mill, which is required to calculate recirculated mass flow. In [30], the reader can find comminution-specific energy, mill power consumption, and the fraction of mill filling expressions, among other equations.

The design of the grinding process by Magne et al. [49] is considered in this work. Here, a SAG mill was implemented to process copper sulfide ore 1.83 m in diameter (D) and 0.61 m in length (L). The operating conditions of the SAG mill were as follows: ore flux fed (F) at 3.45 t/h with granulometry of 12% for 4″, 8% between 4″ and 2″, and 80% below 2″; mill volume occupied by the discharge mill (J_b) equal to 8.5% by volume; percentage of solids in the discharge mill (Y_d) equal to 74%; operating speed equal to 72% of critical speed (φ_c); flow of water fed (F_a) equal to 1.2 m^3/h; a classification grill with an opening of $\frac{1}{2}$″. The SAG mill model was simulated using Simulink, obtaining fraction of mill filling (J), power consumption (M_p), and retained mass (W) equal to 0.22, 9.8 kW, and 0.41 t/h, respectively.

Step 2. UA and GSA

For the MIMO system (see Figure 3a), the main output variables are J, M_p, and W, while the possible manipulated input variables are F, F_a, φ_c, F_1, F_2, and F_3. The manipulated input variables were described using distribution functions that allow the effect of uncertainty in the system to be included. In this context, the SAG mill feed fractions exhibit stochastic uncertainty due to geological uncertainty, while the other SAG mill input variables exhibit epistemic uncertainty due to insufficient measurements, as reported in [49]. Considering that the particle size distribution of the feed to the mill can be represented by a normal distribution [50] and that the particle fragmentation exhibits a fractal nature [51], the fractions in the SAG mill feed are described using the normal distribution. According to the principle of indifference, a uniform distribution should be implemented to describe epistemic uncertainty in the absence of information [52]. Then, manipulated input variables were described as follows: $F \sim U[3.24, 3.65]$ t/h, $F_a \sim U[1.02, 1.39]$ m^3/h, $\varphi_c \sim U[0.7, 0.74]$, $F_1 \sim N[12, 0.70]$%, $F_2 \sim N[8, 0.80]$%, and $F_3 \sim N[80, 0.73]$%.

The MIMO system for the SAG mill is shown in Figure 3a, and Figure 3b–d show the UA results considering a sample of 6000 instances of operation. Here, it can be observed that J, M_p, and W present values around the responses obtained using nominal operating conditions (red line) when the SAG mill was simulated under operational uncertainty, which is consistent with previously reported results [30,49]. Thus, the grinding model provides robust estimates under uncertainty; subsequently, the uncertainty must be apportioned. The GSA results using the Sobol–Jansen method are shown in Figure 4.

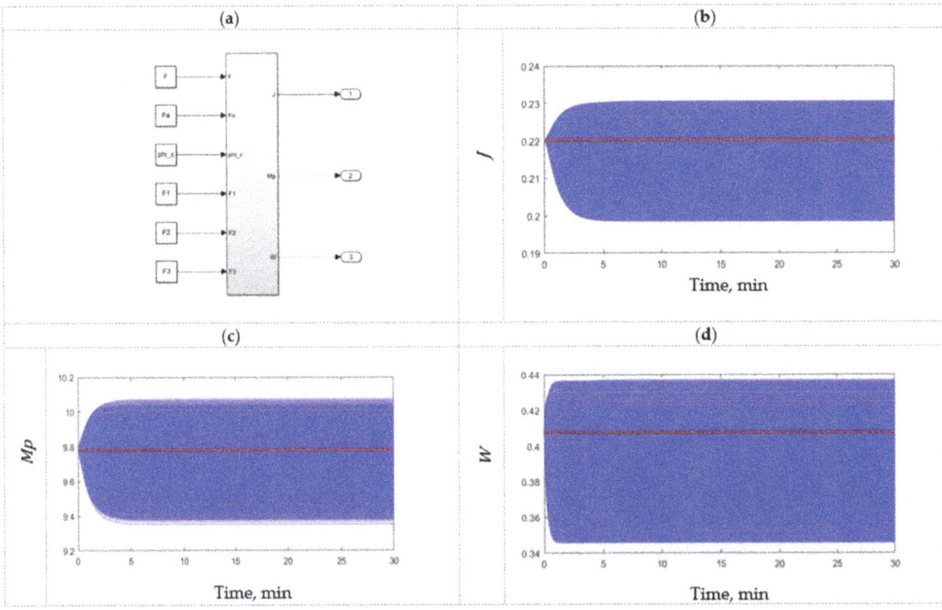

Figure 3. (**a**) SAG mill as MIMO system, and UA results using (**b**) mill filling, (**c**) power consumption, and (**d**) mass retained as output variables.

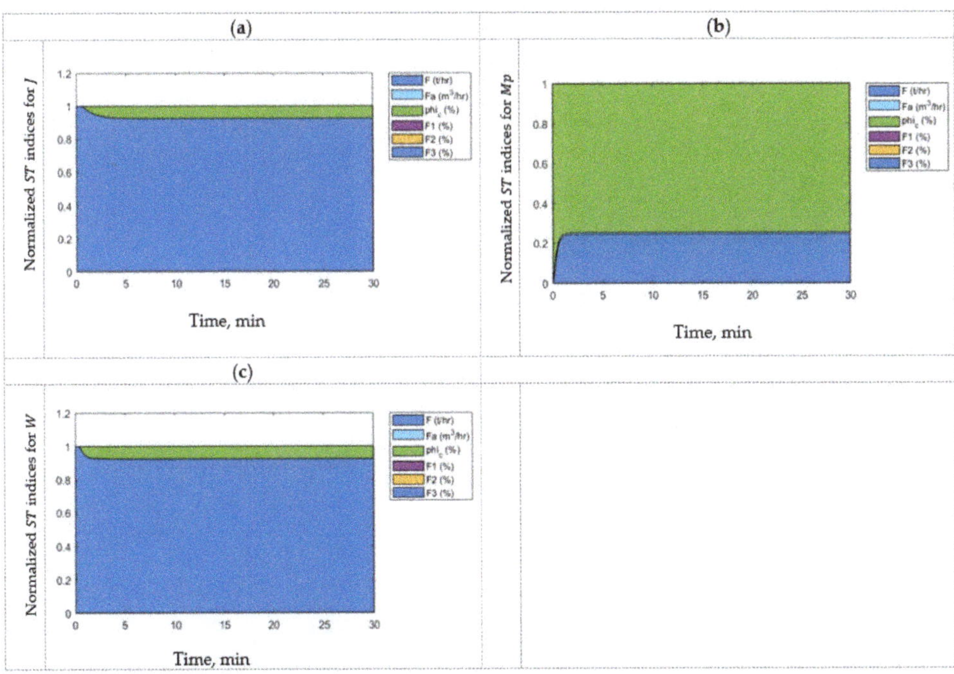

Figure 4. GSA results using (**a**) mill filling, (**b**) power consumption, and (**c**) mass retained as output variables.

Figure 4 shows the normalized total sensitivity indices of the input variables for each output variable of the SAG mill. Here, the first-order indices are not shown because the total sensitivity indices provide more relevant information for the purpose of this work. According to these results, it can be concluded that all output variables $(J,\ M_p,\ W)$ strongly depend on F and φ_c, and there is a negligible effect of F_a, F_1, F_2, and F_3. Then, the input–output pairing is selected using this information. These results suggest that the manipulated input variables should be F and φ_c, while the disturbances should be F_a, F_1, F_2, and F_3.

Step 3. CSD

Now, the process variables can be classified as manipulated, controlled, supervised, or disturbed, allowing the SAG mill model to be expressed in a standard control notation (Figure 5) as follows:

$$\dot{x} = f(x, u, d,\ p),\ x(0) = x_0\ ;\ \dim(x) = 3\ ,\ \dim(u) = n_u \tag{7}$$

$$y = h(x)\ ;\ dim(y) = n_y \tag{8}$$

where the vector-valued function of time f, defined by the right-hand side of Equation (6), depends on vectors of states (x), manipulated input variables (u), disturbances (d), and model parameters (p). For the SAG mill, the states (x) correspond to the weight fraction of the mill retention (w_i), while the outputs (y) in Equation (8) are the variables to be regulated at desired values (controlled variables).

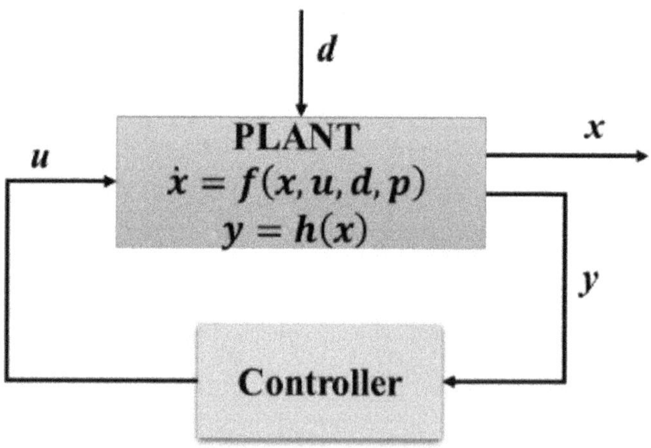

Figure 5. General control system.

For the MIMO system of the SAG mill (see Figure 3a), and for comparison purposes, two structures are defined for the CSD:

(a) A traditional structure (3 × 3) previously reported for the SAG mill [53], with $n_u = n_y = 3$:

$$u = \begin{bmatrix} F \\ F_a \\ \varphi_c \end{bmatrix},\ y = \begin{bmatrix} J \\ M_p \\ W \end{bmatrix},\ d = \begin{bmatrix} F_1 \\ F_2 \\ F_3 \end{bmatrix},\ x = \begin{bmatrix} w_1 \\ w_2 \\ w_3 \end{bmatrix} \tag{9}$$

(b) A reduced structure (2 × 2) obtained in step 2 (considering W as a supervised variable, which is kept in a range depending on grinding design capacity [54]), with $n_u = n_y = 2$:

$$u = \begin{bmatrix} F \\ \varphi_c \end{bmatrix}, \; y = \begin{bmatrix} J \\ M_p \end{bmatrix}, \; d = \begin{bmatrix} F_1 \\ F_2 \\ F_3 \\ F_a \end{bmatrix}, \; x = \begin{bmatrix} w_1 \\ w_2 \\ w_3 \end{bmatrix} \quad (10)$$

In addition to the two control structures designed using total sensitivity indices, two other control strategies are considered to evaluate performance and robustness: a conventional proportional–integral–derivative (PID) controller and an advanced MPC. So, in the next step, three control schemes are evaluated and analyzed, as shown in Figure 6: 2 × 2 PID control, 2 × 2 MPC, and 3 × 3 MPC.

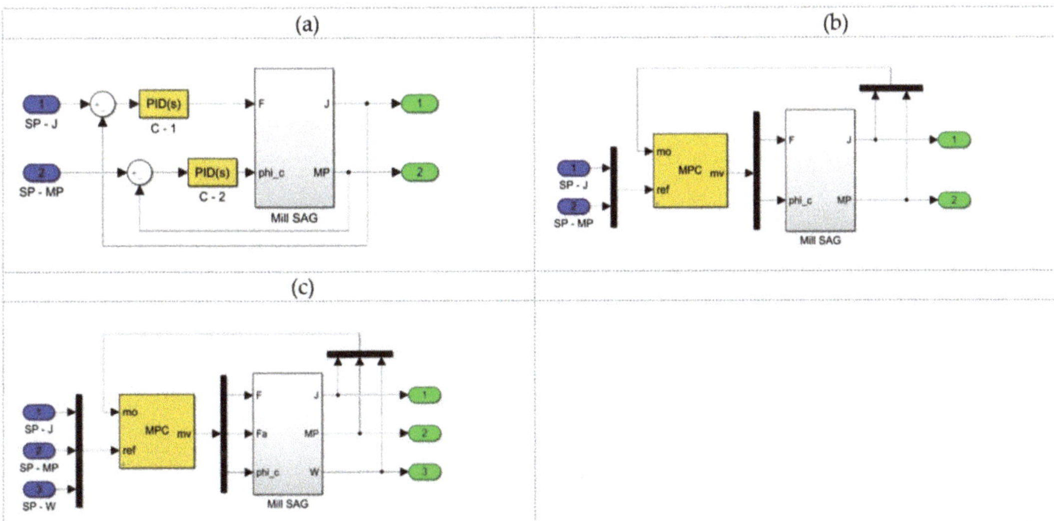

Figure 6. Control systems for SAG mill: (**a**) 2 × 2 PID control, (**b**) 2 × 2 MPC, (**c**) 3 × 3 MPC.

Step 4. Closed-loop validation

Subsequently, the designed control systems were implemented in MATLAB and compared with their corresponding open-loop dynamics, as shown in Figure 7. It is important to note that the tuning parameters for the PID and MPC controllers were determined using the automatic tuning tool included in Simulink. It can be seen in Figure 7a,b that the SAG mill load and energy consumption responses for a step change in the set point are satisfactory. To quantify the quality of these results, the integral absolute error (IAE) was calculated using the formula $IAE = \int |e(t)| dt$, where $e(t)$ is the difference between the set point and the controller response [55]. Table 1 shows a summary of IAE values obtained for the three controllers.

In the case of J, Table 1 shows that the 2 × 2 PID, 3 × 3 MPC, and 2 × 2 MPC controllers provide IAE values of 0.001, 0.020, and 0.075, respectively. Thus, the PID control designed using the pairing J/F proposed by the CSD methodology provides high performance compared to the other controllers. In the case of M_p, Table 1 shows that the 3 × 3 MPC, 2 × 2 MPC, and 2 × 2 PID controllers provide IAE values of 0.111, 0.299, and 0.630, respectively. Therefore, the PID control designed using the pairing M_p/ϕ_c proposed by the CSD methodology demonstrates sufficient performance compared to the others. In the first case, the high performance might be based on the strong influence of the feed flux on mill load, regardless of time and uncertainty (approximately 0.95; see Figure 4a). In the second case, the percentage of critical velocity and feed flux influence power consumption 0.75/0.25 regardless of time and uncertainty, this feed flux effect could explain the sufficient performance of the 2 × 2 PID controller. This could also explain the offset observed in the behavior of manipulated variables (Figure 7d).

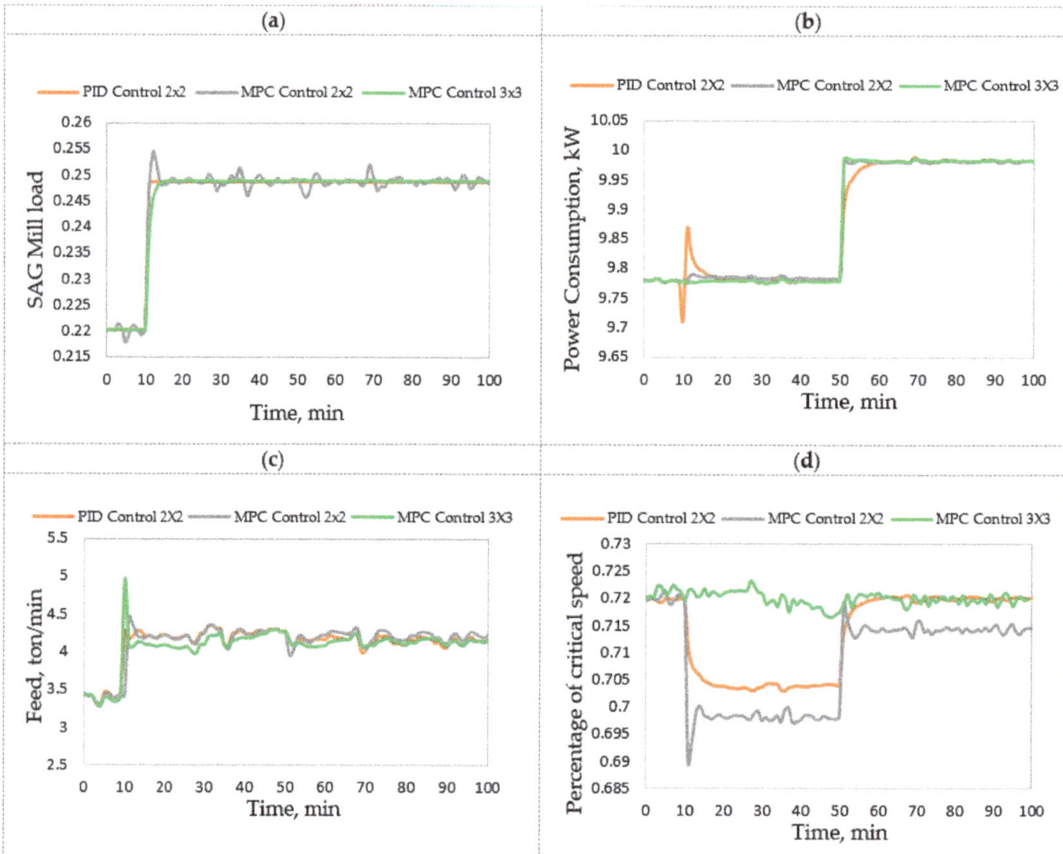

Figure 7. Dynamic responses of controlled variables: (**a**) SAG mill load (J); (**b**) power consumption (M_p) and manipulated variables; (**c**) feed (F); (**d**) percentage of critical speed (φ_c).

Table 1. Analysis and interpretation of results obtained by controllers, SAG mill.

Control / IAE	Controlled Variables		Manipulated Variables	
	J	M_p	F	ϕ_c
High performance	PID 2 × 2 0.001	MPC 3 × 3 0.111	PID 2 × 2 6.437	PID 2 × 2 0.793
Good performance	MPC 3 × 3 0.020	MPC 2 × 2 0.299	MPC 3 × 3 6.081	MPC 3 × 3 0.240
Sufficient performance	MPC 2 × 2 0.075	PID 2 × 2 0.630	MPC 2 × 2 9.594	MPC 2 × 2 0.995
Reference	Open-loop 0.255	11.536		

3.2. Solvent Extraction (SX) Process

Step 1. Modeling

The SX process includes an extraction and a re-extraction system (Figure 8), which can be located in different configurations, both in parallel and in series. The model proposed by Komulainen et al. [56] consists of four units, three for extraction and one for re-extraction. The flow input to the process comprises pregnant leach solutions (PLS) with mineral to be recovered, F_{1a} and F_{2a}; solution with a copper concentration, c_{0a}; poor electrolyte flow, F_{1e}, with a concentration of c_{0e}; the flow of the loaded organic (LO) solution, F_{LO}. The LO solution is recycled in the process, but the flow can be managed through the organic storage tank, c_{3o}. The results of the process are rich copper concentration, c_{1e}, and refined copper concentrations, c_{1a} and c_{3a}.

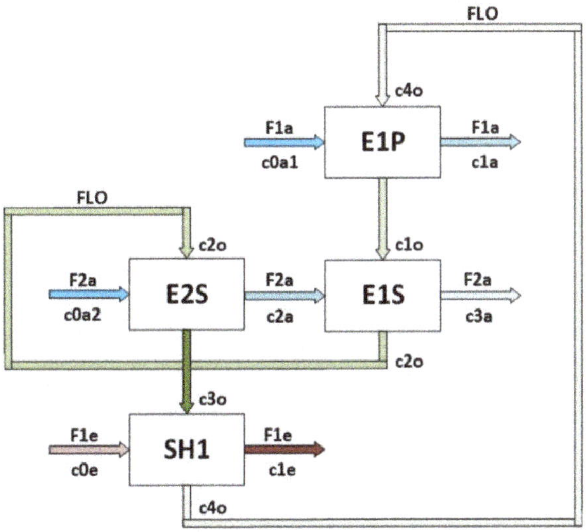

Figure 8. Schematic representation of SX process, adapted from [24].

In the extraction, the copper is transferred from the aqueous to the organic phase. Each of the three extraction units is modeled using dynamic mass balances for organic (c_{io}), aqueous (c_{ia}), and electrolyte (c_{ie}) phases:

$$\frac{dc_{i,phase}(t)}{dt} = \frac{F_{i,phase}(t)}{V_{mix,i}(t)}\left[c_{i-1,phase}(t-t_0) - c_{i,phase}(t)\right] + K_i\left[c_{i,phase}(t) - c^*_{i,phase}(t)\right]$$
$$c_{i,phase}(0) = c_{i,phase,0} \tag{11}$$
$$i = 1, \ldots, 4\ for\ phase = o;\ i = 1,2,3\ for\ phase = a; i = 1\ for\ phase = e$$

where c_i represents concentrations, F_i represents flow rates, V_{mix} represents the mixing volumes, K_i represents the mass transfer coefficients, and the settler, always following the mixer, is described by a pure time delay, t_i. The complete model equations and parameter values can be found in Appendix B and Komulainen et al. [56]. Here, the SX process considers the following operating conditions: organic flow (F_{LO}) of 17.83 m^3/min, electrolyte flow (F_{1e}) of 6.26 m^3/min, aqueous flow (F_{1a}) of 16.88 m^3/min with concentration (c_{oa}) of 1.53 g/L, and aqueous flow (F_{2a}) of 16.88 m^3/min with concentration (c_{2a}) of 3.37 g/L. The SX process model was simulated using Simulink (see Figure 9b–f), obtaining c_{1e}, c_{1a}, c_{3a}, c_{3o}, and c_{4o} values of 51.58, 0.254, 1.38, 6.62, 1.01 and 3.532 g/L, respectively.

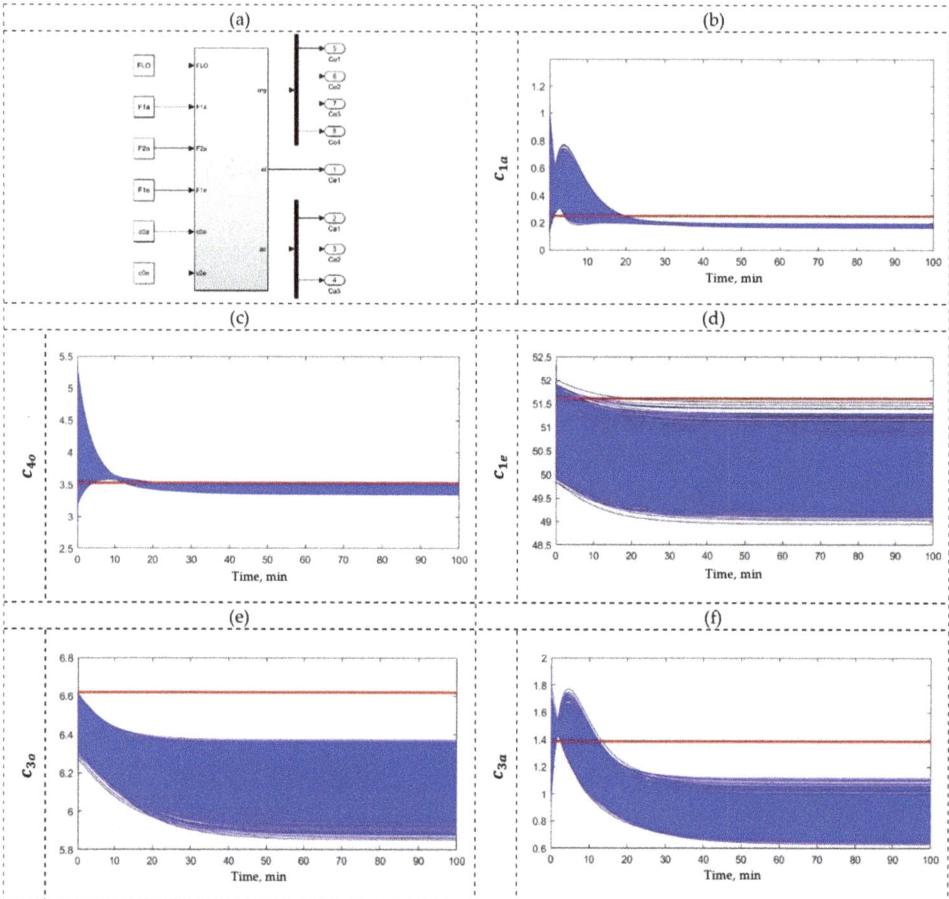

Figure 9. (**a**) SX plant as a MIMO system, and UA results using (**b**) c_{1a}, (**c**) c_{4o}, (**d**) c_{1e}, (**e**) c_{3o}, and (**f**) c_{3a} as output variables.

Step 2. UA and GSA

For the MIMO system (see Figure 9a), the main output variables are c_{1e}, c_{1a}, c_{3a}, c_{3o}, and c_{4o}, and the possible manipulated input variables are F_{LO}, F_{1a}, F_{2a}, F_{1e}, c_{0a1}, c_{0a2} , and c_{0e}. The manipulated input variables of the SX process exhibit epistemic uncertainty due to insufficient information collected from the related literature. Again, according to the principle of indifference, a uniform distribution must be implemented to describe the epistemic uncertainty in the absence of information. In this way, manipulated input variables were described as follows: $FLO \sim U[16.04, 19.6]$ m^3/h, $F_{1a} \sim U[15.19, 18.56]$ m^3/h, $F_{2a} \sim U[14.19, 18.56]$ m^3/h, $F_{1e} \sim U[5.63, 6.88]$ m^3/h, $c_{0a1} \sim N[1.53, 0.5]$ %, and $c_{0a2} \sim N[3.37, 0.5]$%.

Figure 9b–f shows the SX plant responses when subjected to UA, considering a sample of 7000 instances of operation. Here, it can be observed that c_{1a} and c_{4o} present values similar to those obtained using nominal operating conditions despite the uncertainty, while c_{1e}, c_{3o}, and c_{3a} exhibit lower values than the responses obtained using nominal operating conditions (red line), which indicates the influence of operating uncertainty on SX plant responses. The estimates provided by the SX model under uncertainty are consistent with the related literature [24,56]. The next step is to study the uncertainty apportion using the Sobol–Jansen method.

Figure 10 shows the normalized total sensitivity indices of the input variables for each output variable of the SX plant. According to this figure, F_{LO}, F_{1a}, and F_{1e} have an influence on c_{1o}, c_{1a}, and c_{1e}, respectively, and they should be selected as manipulated input variables, while c_{0a1}, c_{0a2} , and c_{0e} should be selected as disturbances.

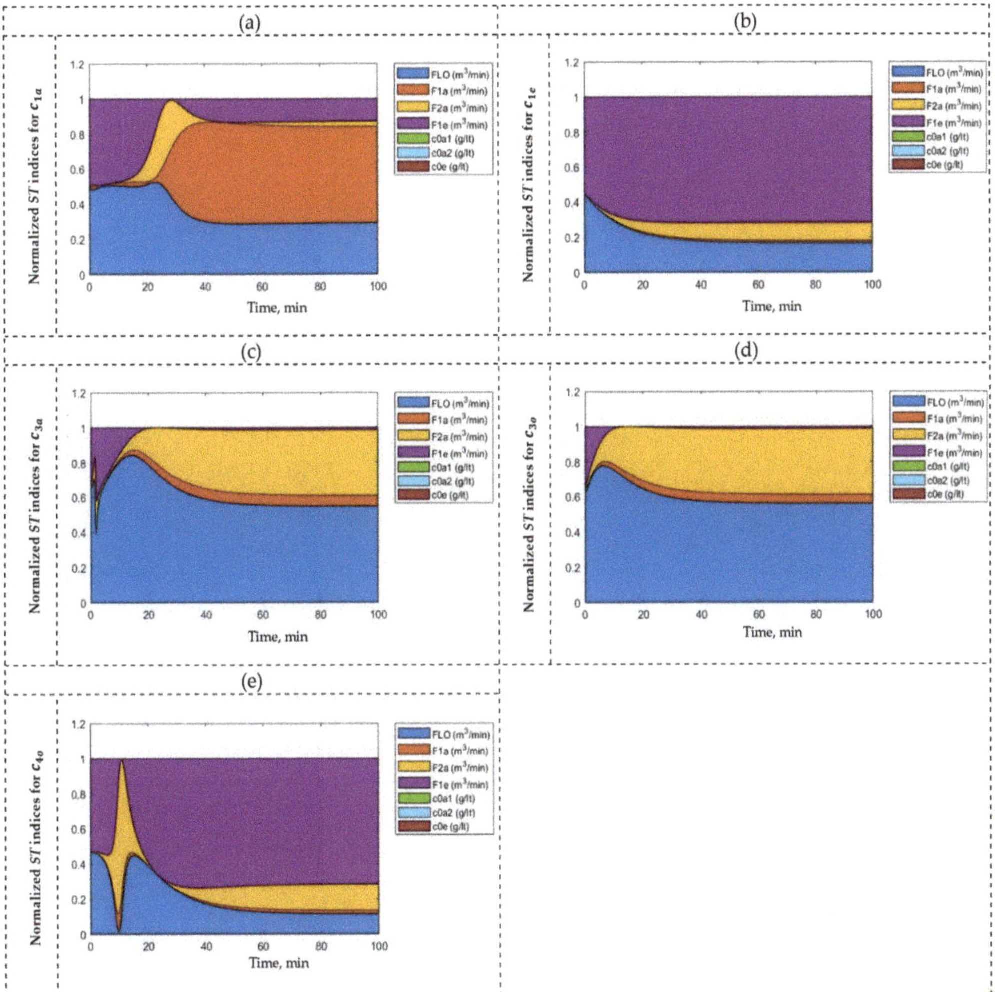

Figure 10. GSA using (**a**) c_{1a}, (**b**) c_{1e}, (**c**) c_{3a}, (**d**) c_{3o}, and (**e**) c_{4o} as output variables.

Step 3. CSD

In this step, the variables of the process are classified as manipulated, controlled, supervised, or disturbed, allowing the SX model to be expressed in a standard control notation (Figure 3) as follows:

$$\begin{gathered}\dot{x} = f(x,u,d,\ p)\ ,\ \ x(0) = x_0\ ;\ \dim(x) = 8,\ \dim(u) = n_u \\ y = h(x)\ ;\dim(y) = n_y\end{gathered} \tag{12}$$

where the vector-valued function of time, f, is defined by the right-hand side of Equation (11). For the SX process, the states (x) correspond to concentrations c_{10}, c_{20}, c_{30}, c_{40}, c_{1a}, c_{2a}, c_{3a}, and c_{1e}, while the manipulated input variables (u) are selected according to the output variables (y) to be regulated at desired values.

For the MIMO system of the SX process (see Figure 11a), and again for comparison purposes, the two structures are defined as follows:

(a) A traditional structure (4 × 5) reported previously [24], with $n_u = 4$ and $n_y = 5$:

$$u = \begin{bmatrix} F_{LO} \\ F_{1a} \\ F_{2a} \\ F_{1e} \end{bmatrix}, y = \begin{bmatrix} c_{1e} \\ c_{1a} \\ c_{3a} \\ c_{3o} \\ c_{4o} \end{bmatrix}, d = \begin{bmatrix} c_{Oa1} \\ c_{Oa2} \\ c_{Oe} \end{bmatrix} \tag{13}$$

(b) A reduced structure (3 × 3) obtained in step 2, with $n_u = n_y = 3$:

$$u = \begin{bmatrix} F_{LO} \\ F_{1a} \\ F_{1e} \end{bmatrix}, y = \begin{bmatrix} c_{1e} \\ c_{1a} \\ c_{3o} \end{bmatrix}, d = \begin{bmatrix} c_{Oa1} \\ c_{Oa2} \\ c_{Oe} \end{bmatrix} \tag{14}$$

These two CSDs are proposed using total sensitivity indices, and again, two control strategies are considered to evaluate performance and robustness: a conventional controller (PID) and an advanced controller (MPC). In the next step, three control systems for the SX plant are evaluated and analyzed, as shown in Figure 11: 3 × 3 PID, 3 × 3 MPC, and 4 × 5 MPC.

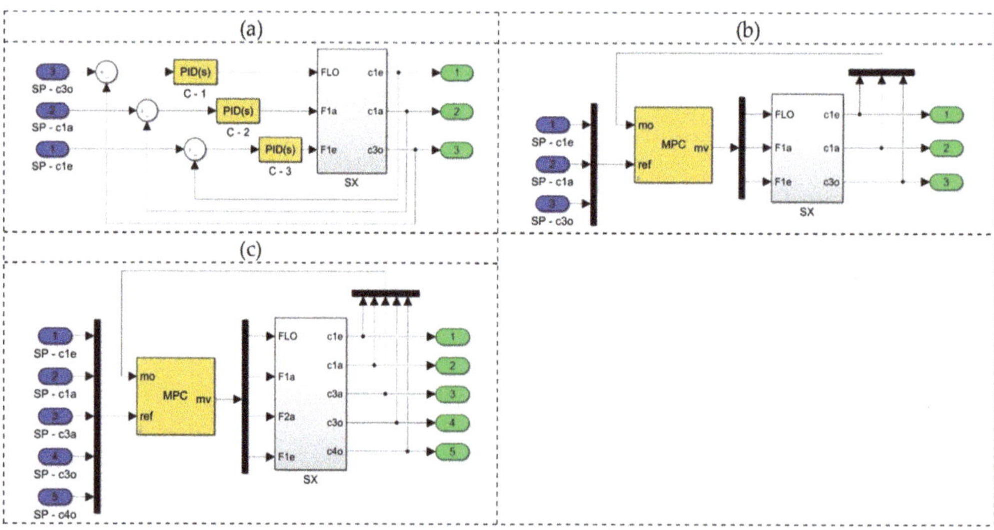

Figure 11. Control systems for SX plant: (**a**) 3 × 3 PID, (**b**) 3 × 3 MPC, (**c**) 4 × 5 MPC.

Step 4. Closed-loop validation

The designed control systems were implemented in MATLAB and compared with their corresponding open-loop dynamics, as shown in Figure 11. Again, the parameters of the PID and MPC controllers were determined via the automatic tuning tool included in Simulink. In Figure 12, it can be seen that the dynamic responses of all variables for the 3 × 3 PID and 3 × 3 MPC are satisfactory, while the performance of the 4 × 5 MPC controller is poor. To quantify the quality of these results, IAE was calculated, and the values are given in Table 2.

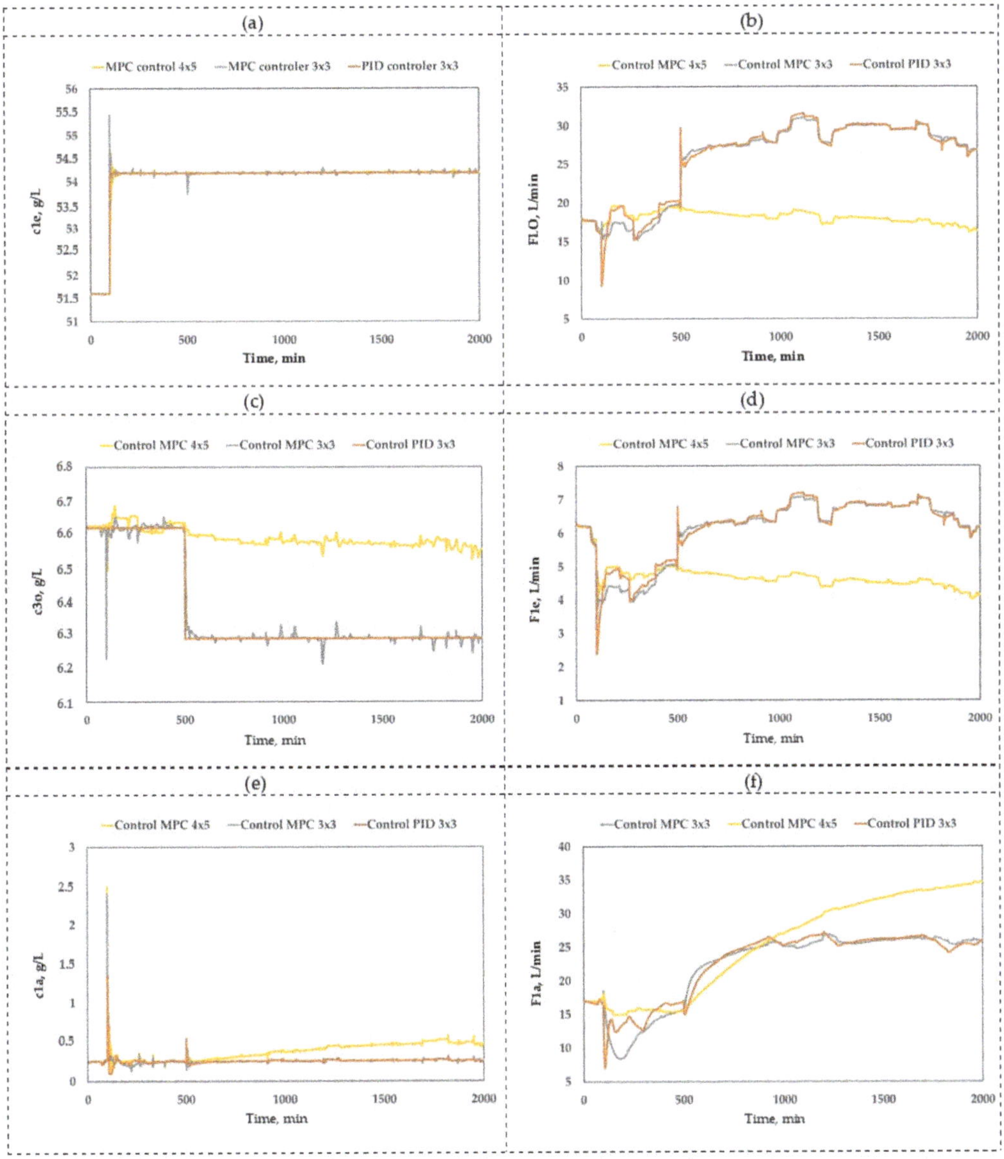

Figure 12. Dynamic response of controlled variables (**a**) c_{1e}, (**c**) c_{3o}, and (**e**) c_{1a} and manipulated variables (**b**) F_{LO}, (**d**) F_{1e}, and (**f**) F_{1a}.

In the case of c_{1e}, Table 2 shows that the 3 × 3 PID, 3 × 3 MPC, and 4 × 5 MPC controllers provide IAE values of 0.008, 0.096, and 0.112, respectively, for a step change in the set point. In other words, the PID controller designed using total sensitivity indices exhibits high performance compared to the other controllers, which is related to the strong influence of F_{1e} on c_{1e}. In the case of c_{1a}, Table 2 shows that the 3 × 3 PID, 3 × 3 MPC, and

4×5 MPC controllers provide IAE values of 0.072, 0.090, and 2.234, respectively. Again, the controller designed using total sensitivity indices shows high performance compared to the other controllers, even though FL_o and F_{1e} have a side effect on c_{1a}. In the case of c_{3o}, Table 2 shows that the 3×3 PID, 3×3 MPC, and 4×5 MPC controllers provide IAE values of 0.001, 0.060, and 2.636, respectively. Here, the controller designed using total sensitivity indices exhibits high performance compared to MPC despite the secondary influence of F_{2a} on c_{3o}. The good and sufficient performance of MPC could be explained by the cross-influence of some input variables on the SX plant responses detected in the GSA (Figure 10).

Table 2. Analysis and interpretation of results obtained by controllers, SX plant.

Control / IAE(10^{-3})	Controlled Variables			Manipulated Variables		
	c_{1e}	c_{1a}	c_{3o}	F_{1e}	F_{1a}	FL_o
High performance	PID 3×3 0.008	PID 3×3 0.072	PID 3×3 0.001	MPC 3×3 2.300	MPC 3×3 9.088	MPC 3×3 9.365
Good performance	MPC 3×3 0.096	MPC 3×3 0.090	MPC 3×3 0.060	PID 3×3 2.545	PID 3×3 10.230	PID 3×3 10.660
Sufficient performance	MPC 4×5 0.112	MPC 4×5 2.234	MPC 4×5 2.636	MPC 4×5 21.598	MPC 4×5 85.689	MPC 4×5 110.112
Reference	2.180	Open-loop 0.099	0.605			

4. Conclusions

A methodology was presented to design decentralized control structures. This methodology considers the use of GSA based on the Sobol–Jansen method to establish the control structure design (input–output pairing) for MIMO systems operating under uncertainty conditions. These control structures are made using total sensitivity indices provided by the Sobol–Jasen method, and their behavior depends on the dynamics of the studied process and the magnitude of the uncertainty. In this sense, the Sobol–Jansen method provides graphical results that help in understanding the dynamic behavior of systems under uncertainty. The methodology was illustrated using a SAG mill and an SX plant operating under uncertainty. For the SAG mill, the methodology allowed us to design a 2×2 decentralized control structure whose pairings J/F and M_p/ϕ_c exhibited high and sufficient performance, respectively, compared to MPC. For the SX plant, the methodology allowed us to design a 3×3 decentralized control structure whose pairings c_{1o}/F_{LO}, c_{1a}/F_{1a}, and F_{1e}/c_{1e} exhibited high performance compared to MPC. The proposed methodology for the design of the control structure using GSA was illustrated with mineral processes, and it can be applied to any other process that operates under uncertainty; therefore, it could provide satisfactory results for a wide range of operating conditions.

Author Contributions: Conceptualization, F.A.L. and L.A.C.; methodology, F.A.L. and O.M.-Q.; software, F.A.L. and O.M.-Q.; formal analysis, F.A.L., O.M.-Q., L.A.C. and T.L.-A.; investigation, O.M.-Q. and F.A.L.; data curation, O.M.-Q. and F.A.L.; writing—Original draft preparation, F.A.L. and O.M.-Q.; writing—Review and editing, O.M.-Q., F.A.L., L.A.C. and T.L.-A.; funding acquisition, L.A.C. All authors have read and agreed to the published version of the manuscript.

Funding: The authors are grateful for the support of Agencia Nacional de Investigación y Desarrollo de Chile (ANID) through Anillo-Grant No. ACT210027 and Fondecyt 1211498.

Data Availability Statement: Not applicable.

Acknowledgments: This publication was supported by ANID, Anillo-Grant ACT210027, and Fondecyt 1211498.

Conflicts of Interest: The authors declare no conflict of interest.

Nomenclature

For SAG model:

A	ore impact breakage parameter
a	parameter of specific breakage rate model
a_S	parameter of specific breakage rate model
B_{ij}	cumulative breakage distribution function
B	values of input variables that provide desired behavior of output milling model
B^-	values of input variables that provide unwanted behavior of output milling model
b_{ij}	ore impact breakage parameter
n	breakage distribution function
C^*	number of species present in fresh feed
c_i	mass flow recirculated internally by grill
c_f	classification efficiency of internal grid mill
c_f	solid weight percentage in mill charge
D	mill diameter
E_{cs}	comminution specific energy (kWh/t)
F	fresh ore flux fed to mill, t/h
f_i	fraction of fresh ore flux fed to mill
J	fraction of mill filling
J_b	percentage of mill volume occupied by steel balls
K_i	specific breakage rate
L	mill length
M	parameter of classification efficiency model
M_p	mill power consumption
m	total number of input variables in model Y
R	total number of simulations
S_j	first-order sensitivity index for input variable x_j
S_j^T	total sensitivity index for input variable x_j
V	mill volume
R	total number of simulations
$V(Y)$	variance of model Y
W	mass retained in mill
W_a	water in mill charge
w_a	ratio between ore mass and water mass retained inside mill
w_i	weight fraction of retained mass in mill
Y_d	percentage of solids in discharge mill
x_0	parameter of specific breakage rate model
x_i	particle size of species present in fresh feed
x_{50}	parameter of classification efficiency model
Z	parameter of classification efficiency model
α	characteristic parameter of material
α_1	parameter of specific breakage rate model
α_s	parameter of specific breakage rate model
β	fraction of fines produced in a single fracture event
β_1	parameter of classification efficiency model
γ	parameter of cumulative breakage distribution function
μ	parameter of specific breakage rate model
Ψ	parameter of specific breakage rate model
Λ	parameter of cumulative breakage distribution function
$\varnothing_j$	percentage of critical speed
$\varnothing_c$	parameter of classification efficiency model

For SX model:

F_{1a}	flow inputs to process, pregnant leach solution (PLS)
F_{2a}	flow inputs to process, pregnant leach solutions (PLS)
c_{0a}	solution with a copper concentration
F_{1e}	poor electrolyte flow

Appendix A. SAG Model

The mathematical expression used to model the breakage distribution function b_{ij} is based on B_{ij}, the cumulative breakage distribution function, given by:

$$B_{ij} = \phi_j \left(\frac{x_{i-1}}{x_j} \right)^{\gamma} + (1 - \phi_j) \left(\frac{x_{i-1}}{x_j} \right)^{\beta} \tag{A1}$$

where ϕ_j and γ are parameters with values ranging from 2.5 to 5 and 0.5 to 1.5, respectively, and β represents the fraction of fines produced in a single fracture event. Considering that B_{ij} is a cumulative distribution, this can be expressed as:

$$B_{i,j} = b_{n,j} + b_{n-1,j} + \ldots + b_{i,j} = \sum_{k=i}^{n} b_{kj} \tag{A2}$$

so that

$$B_{i,j} - B_{i+1,j} = b_{ij} \tag{A3}$$

The classification efficiency of the internal grid mill, c_i, is calculated with the following expression:

$$c_i = \psi \beta (x_i M)^{(\beta-1)} exp\left(-\psi (x_i M)^{\beta} \right) + \frac{1}{1 + \left(\frac{x_{50}}{x_i} \right)^{Z}} \tag{A4}$$

where ψ, Z, M, x_{50}, and β are parameters of the model. In the case of specific breakage rate K_i, related theory suggests that this parameter varies with particle size; the typical form of specific breakage rate has three regions. Magne et al. [48,49] proposed the following equation to estimate the specific breakage rate in the three regions:

$$K_i = a \left(\frac{x_i}{x_0} \right)^{\alpha} \frac{1}{1 + \left(\frac{x_i}{\mu} \right)^{\Lambda}} + a_s \left(\frac{x_i}{x_0} \right)^{\alpha_s} \tag{A5}$$

where α, μ, Λ, a, a_s, x_0, and α_s are parameters of the model.

Appendix B. SX Model

The mass balances of the SX process are given below.

Organic–aqueous balance in E1P:

$$\frac{dc_1^{org}(t)}{dt} = \frac{F_1^{org}(t)}{V_{mix,1}(t)} \cdot \left[c_4^{org}(t - t_0) - c_1^{org}(t) \right] + K_1 \left[c_1^{org}(t) - c_1^{org^*}(t) \right] \tag{A6}$$

$$\frac{dc_1^{aq}(t)}{dt} = \frac{F_1^{aq}(t)}{V_{mix,1}(t)} \cdot \left[c_0^{aq}(t) - c_1^{aq}(t) \right] - K_1 \left[c_1^{org}(t) - c_1^{org^*}(t) \right] \tag{A7}$$

Organic–aqueous balance in E1S:

$$\frac{dc_2^{org}(t)}{dt} = \frac{F_2^{org}(t)}{V_{mix,2}(t)} \cdot \left[c_1^{org}(t - t_1) - c_2^{org}(t) \right] + K_2 \left[c_2^{org}(t) - c_2^{org^*}(t) \right] \tag{A8}$$

$$\frac{dc_3^{aq}(t)}{dt} = \frac{F_2^{aq}(t)}{V_{mix,2}(t)} \cdot \left[c_2^{aq}(t) - c_3^{aq}(t) \right] - K_2 \left[c_2^{org}(t) - c_2^{org^*}(t) \right] \tag{A9}$$

Organic–aqueous balance in E2S:

$$\frac{dc_3^{org}(t)}{dt} = \frac{F_3^{org}(t)}{V_{mix,3}(t)} \cdot \left[c_2^{org}(t - t_2) - c_3^{org}(t) \right] + K_3 \left[c_3^{org}(t) - c_3^{org^*}(t) \right] \tag{A10}$$

$$\frac{dc_2^{aq}(t)}{dt} = \frac{F_2^{aq}(t)}{V_{mix,3}(t)} \cdot \left[c_0^{aq}(t) - c_2^{aq}(t)\right] - K_3\left[c_3^{org}(t) - c_3^{org^*}(t)\right] \quad \text{(A11)}$$

Organic–electrolyte balance in S1H:

$$\begin{aligned} \frac{dc_4^{org}(t)}{dt} &= \frac{F_4^{org}(t)}{V_{mix,4}(t)} \cdot \left[c_3^{org}(t-t_3) - c_4^{org}(t)\right] - K_4\left[c_1^{el}(t) - c_1^{el^*}(t)\right] \\ \frac{dc_1^{el}(t)}{dt} &= \frac{F_1^{el}(t)}{V_{mix,4}(t)} \cdot \left[c_0^{el}(t) - c_1^{el}(t)\right] + K_4\left[c_1^{el}(t) - c_1^{el^*}(t)\right] \end{aligned} \quad \text{(A12)}$$

From the described balance equations, the following relationships are established:

$$FLO = F_1^{org} = F_2^{org} = F_3^{org} = F_4^{org} \quad \text{(A13)}$$

$$c(RE) = c_1^{el}(t-t_4) \quad \text{(A14)}$$

$$c(RaffP) = c_1^{aq}(t-t_1) \quad \text{(A15)}$$

$$c(RaffS) = c_3^{aq}(t-t_2) \quad \text{(A16)}$$

$$c(LO) = c_3^{org}(t-t_3) \quad \text{(A17)}$$

$$c(BO) = c_4^{org}(t-t_4) \quad \text{(A18)}$$

Theoretical equilibrium values for extraction and re-extraction are determined from the McCabe–Thiele diagram, presented in Figure A1. Isotherms assume that extraction and re-extraction are constant, while the operating lines are constantly changing according to the organic/aqueous volume ratio in each tank. The equilibrium value in a tank is the point of coincidence of the equilibrium isotherm and the inverse operating line, weighted by the efficiency coefficient α.

In extraction, the equilibrium isotherm is not linear:

$$c^{org} = Ac^{aq}/(B + c^{aq}) \quad \text{(A19)}$$

and in re-extraction it is linear:

$$c^{org} = C \cdot c^{aq} + D \quad \text{(A20)}$$

In extraction, the theoretical equilibrium point (100% efficiency) is determined by:

$$y = 1/2a\left(-(Ba - A + b) - \sqrt{(Ba - a + b)^2 - 4aBb}\right) \quad \text{(A21)}$$

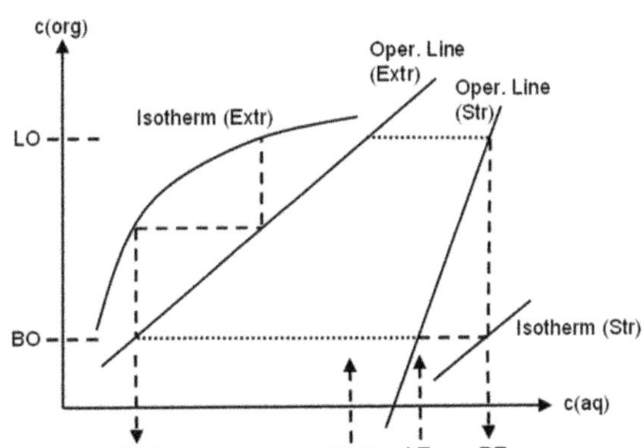

Figure A1. McCabe–Thiele diagram: two extraction stages and one re-extraction stage (adapted from [56]).

The equilibrium value for aqueous and organic concentrations is the weighted efficiency of the theoretical value:

$$c^{aq*} = \alpha_i x^* + (1-\alpha_i)\cdot c_0^{aq} \tag{A22}$$

$$c^{org*} = \alpha y^* + (1-\alpha)\cdot c_0^{org} = \alpha\cdot(ax^* + b) + (1-\alpha)\cdot c_0^{org} \tag{A23}$$

Here, a and b are parameters of the equilibrium isotherm, where the slope a of the operating line is:

$$a = -F^{aq}/F^{org} \tag{A24}$$

while b is the linear term that combines the input concentrations of the organic and aqueous phases:

$$b = c_0^{org} - a\cdot c_0^{aq} \tag{A25}$$

In re-extraction, the isotherm parameters are C and D and a and b; in the same way, the equilibrium point is solved by:

$$y = Cx + D = ax + b \tag{A26}$$

resulting in the theoretical equilibrium concentration of the aqueous phase x^*:

$$x^* = (b-D)/(C-a) \tag{A27}$$

References

1. Liu, G.; Wang, Z.; Mei, C.; Ding, Y. A Review of Decoupling Control Based on Multiple Models. In Proceedings of the 2012 24th Chinese Control and Decision Conference (CCDC), Taiyuan, China, 23–25 May 2012; pp. 1077–1081.
2. Xiong, Q.; Cai, W.J.; He, M.J. A Practical Loop Pairing Criterion for Multivariable Processes. *J. Process Control* **2005**, *15*, 741–747. [CrossRef]
3. Vaes, D.; Swevers, J.; Sas, P. Optimal Static Decoupling to Simplify and Improve Multivariable Control Design. *Mediterr. J. Meas. Control* **2006**, *2*, 48–62.
4. Liu, L.; Tian, S.; Xue, D.; Zhang, T.; Chen, Y.Q.; Zhang, S. A Review of Industrial MIMO Decoupling Control. *Int. J. Control Autom. Syst.* **2019**, *17*, 1246–1254. [CrossRef]
5. Hanuma, R.; Ashok, D.; Anjaneyulu, K. Control Configuration Selection and Controller Design for Multivariable Processes Using Normalized Gain. *Int. J. Electr. Comput. Eng.* **2014**, *8*, 1–5. [CrossRef]
6. Castaño Arranz, M.; Birk, W.; Nikolakopoulos, G. A Survey on Control Configuration Selection and New Challenges in Relation to Wireless Sensor and Actuator Networks. In Proceedings of the IFAC-PapersOnLine, Touluse, France, 10–14 July 2017; Volume 50, pp. 8810–8825.
7. Jain, A.; Babu, B.V. Relative Response Array: A New Tool for Control Configuration Selection. *Int. J. Chem. Eng. Appl.* **2015**, *6*, 356–362. [CrossRef]
8. Niederlinski, A. A Heuristic Approach to the Design of Linear Multivariable Interacting Control Systems. *Automatica* **1971**, *7*, 691–701. [CrossRef]
9. Zhu, Z.X. Variable Pairing Selection Based on Individual and Overall Interaction Measures. *Ind. Eng. Chem. Res.* **1996**, *35*, 4091–4099. [CrossRef]
10. Kinnaert, M. Interaction Measures and Pairing of Controlled and Manipulated Variables for Multiple-Input Multiple-Output Systems: A Survey. *Journal A* **1995**, *16*, 15–23.
11. Mc Avoy, T.; Arkun, Y.; Chen, R.; Robinson, D.; Schnelle, P.D. A New Approach to Defining a Dynamic Relative Gain. *Control Eng. Pract.* **2003**, *11*, 907–914. [CrossRef]
12. Salgado, M.E.; Conley, A. MIMO Interaction Measure and Controller Structure Selection. *Int. J. Control* **2004**, *77*, 367–383. [CrossRef]
13. Wittenmark, B.; Salgado, M.E. Hankel-Norm Based Interaction Measure for Input-Output Pairing. In Proceedings of the IFAC Proceedings Volumes (IFAC-PapersOnline), Barcelona, Spain, 21–26 July 2002; Volume 15, pp. 429–434.
14. Hanzon, B. The Area Enclosed by the (Oriented) Nyquist Diagram and the Hilbert-Schmidt-Hankel Norm of a Linear System. *IEEE Trans. Autom. Control* **1992**, *37*, 835–839. [CrossRef]
15. Birk, W.; Medvedev, A. A Note on Gramian-Based Interaction Measures. In Proceedings of the European Control Conference (ECC), Cambridge, UK, 1–4 September 2003; pp. 2625–2630.
16. Halvarsson, B.; Carlsson, B.; Wik, T. A New Input/Output Pairing Strategy Based on Linear Quadratic Gaussian Control. In Proceedings of the 2009 IEEE International Conference on Control and Automation (ICCA), Christchurch, New Zealand, 9–11 December 2009; pp. 978–982.

17. Jain, A.; Babu, B.V. Sensitivity of Relative Gain Array for Processes with Uncertain Gains and Residence Times. In Proceedings of the IFAC-PapersOnLine, Tiruchirappalli, India, 1–5 February 2016; Volume 49, pp. 486–491.
18. Coetzee, L.C.; Craig, I.K.; Kerrigan, E.C. Robust Nonlinear Model Predictive Control of a Run-of-Mine Ore Milling Circuit. *IEEE Trans. Control Syst. Technol.* **2010**, *18*, 1–86. [CrossRef]
19. Craig, I.K.; MacLeod, I.M. Specification Framework for Robust Control of a Run-of-Mine Ore Milling Circuit. *Control Eng. Pract.* **1995**, *3*, 621–630. [CrossRef]
20. Olivier, L.E.; Craig, I.K. Model-Plant Mismatch Detection and Model Update for a Run-of-Mine Ore Milling Circuit under Model Predictive Control. In Proceedings of the Journal of Process Control, Cape Town, South Africa, 24–29 August 2014; Volume 23, pp. 100–107.
21. Olivier, L.E.; Huang, B.; Craig, I.K. Dual Particle Filters for State and Parameter Estimation with Application to a Run-of-Mine Ore Mill. *J. Process Control* **2012**, *22*, 710–717. [CrossRef]
22. Karelovic, P.; Putz, E.; Cipriano, A. Dynamic Hybrid Modeling and Simulation of Grinding-Flotation Circuits for the Development of Control Strategies. *Miner. Eng.* **2016**, *93*, 65–79. [CrossRef]
23. Bauer, M.; Brooks, K.; Burchell, J.; Coetzee, L.; le Roux, D.; McCoy, J.; Miskin, J.; Winter, D. The Use of a Semi-Rigorous SAG Mill Model for a Hands-on Workshop. In Proceedings of the IFAC-PapersOnLine, Berlin, Germany, 11–17 July 2020; Volume 53.
24. Komulainen, T.; Doyle III, F.J.; Rantala, A.; Jämsä-Jounela, S.-L. Control of an Industrial Copper Solvent Extraction Process. *J. Process Control* **2009**, *19*, 2–15. [CrossRef]
25. Maldonado, M.; Desbiens, A.; Villar, D.R. Decentralized Control of a Pilot Flotation Column: A 3×3 System. *Can. Metall. Q.* **2008**, *47*, 377–385. [CrossRef]
26. Guo, N.; Zheng, H.; Zou, T.; Jia, Y. Integration of Numerical Simulation and Control Scheme for Energy Conservation of Aluminum Melting Furnaces. *IEEE Access* **2019**, *7*, 114659–114669. [CrossRef]
27. Saltelli, A.; Ratto, M.; Andres, T.; Campolongo, F.; Cariboni, J.; Gatelli, D.; Saisana, M.; Tarantola, S. *Global Sensitivity Analysis. The Primer*; John Wiley & Sons, Ltd.: Chichester, UK, 2007; ISBN 9780470725184.
28. Saltelli, A.; Annoni, P.; Azzini, I.; Campolongo, F.; Ratto, M.; Tarantola, S. Variance Based Sensitivity Analysis of Model Output. Design and Estimator for the Total Sensitivity Index. *Comput. Phys. Commun.* **2010**, *181*, 259–270. [CrossRef]
29. Homma, T.; Saltelli, A. Importance Measures in Global Sensitivity Analysis of Nonlinear Models. *Reliab. Eng. Syst. Saf.* **1996**, *52*, 1–17. [CrossRef]
30. Lucay, F.A.; Gálvez, E.D.; Salez-Cruz, M.; Cisternas, L.A. Improving Milling Operation Using Uncertainty and Global Sensitivity Analyses. *Miner. Eng.* **2019**, *131*, 249–261. [CrossRef]
31. Mellado, M.; Cisternas, L.; Lucay, F.; Gálvez, E.; Sepúlveda, F. A Posteriori Analysis of Analytical Models for Heap Leaching Using Uncertainty and Global Sensitivity Analyses. *Minerals* **2018**, *8*, 44. [CrossRef]
32. Helton, J.C.; Burmaster, D.E. Treatment of Aleatory and Epistemic Uncertainty in Performance Assesments for Complex Systems. *Reliab. Eng. Syst. Saf.* **1996**, *54*, 91–94. [CrossRef]
33. Zio, E.; Pedroni, N. Methods for Representing Uncertainty—A Literature Review. *Found. Ind. Saf. Cult. (FonCSI)* **2013**, *5*, 77–88.
34. Saltelli, A. Sensitivity Analysis for Importance Assessment. *Risk Anal.* **2002**, *22*, 579–590. [CrossRef]
35. Reuter, U.; Liebscher, M. Global Sensitivity Analysis in View of Nonlinear Structural Behavior. *LS-Dyna Users Conf.* **2008**, *56*, 903–915.
36. Lilburne, L.; Tarantola, S. Sensitivity Analysis of Spatial Models. *Int. J. Geogr. Inf. Sci.* **2009**, *23*, 151–168. [CrossRef]
37. Sepúlveda, F.D.; Lucay, F.; González, J.F.; Cisternas, L.A.; Gálvez, E.D. A Methodology for the Conceptual Design of Flotation Circuits by Combining Group Contribution, Local/Global Sensitivity Analysis, and Reverse Simulation. *Int. J. Miner. Process.* **2017**, *164*, 56–66. [CrossRef]
38. Lucay, F.A.; Cisternas, L.A.; Gálvez, E.D. An LS-SVM Classifier Based Methodology for Avoiding Unwanted Responses in Processes under Uncertainties. *Comput. Chem. Eng.* **2020**, *138*, 106860. [CrossRef]
39. Calisaya-Azpilcueta, D.; Herrera-Leon, S.; Lucay, F.A.; Cisternas, L.A. Assessment of the Supply Chain under Uncertainty: The Case of Lithium. *Minerals* **2020**, *10*, 604. [CrossRef]
40. Lucay, F.A.; Lopez-Arenas, T.; Sales-Cruz, M.; Gálvez, E.D.; Cisternas, L.A. Performance Profiles for Benchmarking of Global Sensitivity Analysis Algorithms. *Rev. Mex. Ing. Quim.* **2020**, *19*, 423–444. [CrossRef]
41. Saltelli, A.; Aleksankina, K.; Becker, W.; Fennell, P.; Ferretti, F.; Holst, N.; Li, S.; Wu, Q. Why so Many Published Sensitivity Analyses Are False: A Systematic Review of Sensitivity Analysis Practices. *Environ. Model. Softw.* **2019**, *114*, 29–39. [CrossRef]
42. Velez, C. Global Sensitivity and Uncertainty Analysis (GSUA). Available online: https://www.mathworks.com/matlabcentral/fileexchange/47758-global-sensitivity-and-uncertainty-analysis-gsua (accessed on 5 May 2022).
43. Pianosi, F.; Beven, K.; Freer, J.; Hall, J.W.; Rougier, J.; Stephenson, D.B.; Wagener, T. Sensitivity Analysis of Environmental Models: A Systematic Review with Practical Workflow. *Environ. Model. Softw.* **2016**, *79*, 214–232. [CrossRef]
44. Austin, L.G.; Barahona, C.A.; Weymont, N.P.; Suryanarayanan, K. An Improved Simulation Model for Semi-Autogenous Grinding. *Powder Technol.* **1986**, *47*, 265–283. [CrossRef]
45. Austin, L.G.; Menacho, J.M.; Pearcy, F. A General Model for Semi-Autogenous and Autogenous Milling. In Proceedings of the Twentieth International Symposium on the Application of Computers and Mathematics in the Mineral Industries APCOM 87, Johannesburg, South Africa, 19–23 October 1987; pp. 107–126.

46. Barani, K.; Balochi, H. First-Order and Second-Order Breakage Rate of Coarse Particles in Ball Mill Grinding. *Physicochem. Probl. Miner. Process.* **2016**, *52*, 268–278. [CrossRef]
47. Fuerstenau, D.W.; De, A.; Kapur, P.C. Linear and Nonlinear Particle Breakage Processes in Comminution Systems. *Int. J. Miner. Process.* **2004**, *74*, S317–S327. [CrossRef]
48. Magne, L.; Améstica, R.; Barría, J.; Menacho, J. Modelización Dinámica de Molienda Semiautógena Basada En Un Modelo Fenomenológico Simplificado. *Rev. Metal.* **1995**, *31*, 97–105. [CrossRef]
49. Magne, L.; Barría, J.; Améstica, R.; Menacho, J. Evaluación de Variables de Operación En Molienda Semiautógena. In Proceedings of the Segundo Congreso en Metalurgia e Ingeniería de Materiales, IBEROMET II, Ciudad de Mexico, Mexico, 8–14 November 1992; Available online: https://www.revistas.usach.cl/ojs/index.php/remetallica/article/view/1701/1581 (accessed on 5 May 2022).
50. Tripathy, S.K.; Murthy, Y.R.; Singh, V.; Srinivasulu, A.; Ranjan, A.; Satija, P.K. Performance Optimization of an Industrial Ball Mill for Chromite Processing. *J. S. Afr. Inst. Min. Met.* **2017**, *117*, 75–81. [CrossRef]
51. Xu, Y. The Fractal Evolution of Particle Fragmentation under Different Fracture Energy. *Powder Technol.* **2018**, *323*, 337–345. [CrossRef]
52. Helton, J.C.; Oberkampf, W.L. Alternative Representations of Epistemic Uncertainty. Engineering and System Safety. *Reliab. Eng. Syst. Saf.* **2004**, *85*, 1–10. Available online: https://www.sciencedirect.com/science/article/abs/pii/S0951832004000481?via%3Dihub (accessed on 5 May 2022). [CrossRef]
53. Salazar, J.L.; Valdés-González, H.; Vyhmesiter, E.; Cubillos, F. Model Predictive Control of Semiautogenous Mills (Sag). *Miner. Eng.* **2014**, *64*, 92–96. [CrossRef]
54. César, G.Q.; Daniel, S.H. Multivariable Model Predictive Control of a Simulated SAG Plant. *IFAC Proc. Vol.* **2009**, *42*, 37–42. [CrossRef]
55. Kämpjärvi, P.; Jämsä-Jounela, S.L. Level Control Strategies for Flotation Cells. *Miner. Eng.* **2003**, *16*, 1061–1068. [CrossRef]
56. Komulainen, T.; Pekkala, P.; Rantala, A.; Jämsä-Jounela, S.L. Dynamic Modelling of an Industrial Copper Solvent Extraction Process. *Hydrometallurgy* **2006**, *81*, 52–61. [CrossRef]

minerals MDPI

Article

Studying Flotation of Gold Microdispersions with Carrier Minerals and Pulp Aeration with a Steam–Air Mixture

Sergei Ivanovich Evdokimov [1], Nikolay S. Golikov [2,*], Alexey F. Pryalukhin [2,*], Viktor V. Kondratiev [3], Anatolii Mishedchenko [4], Alexandra Vl. Kuzina [5], Natalia Nikolaevna Bryukhanova [3] and Antonina I. Karlina [6]

1 Department of Mineral Processing, North Caucasian Institute of Mining and Metallurgy, State Technological University, 362021 Vladikavkaz, Russia; eva-ser@mail.ru
2 Department of Mechanical Engineering, Saint-Petersburg Mining University, 199106 St. Petersburg, Russia
3 Laboratory of Geochemistry of Ore Formation and Geochemical Methods of Prospecting, A. P. Vinogradov Institute of Geochemistry of the Siberian Branch of the Russian Academy of Sciences, 664033 Irkutsk, Russia; v.kondratiev@igc.irk.ru (V.V.K.); nnb@igc.irk.ru (N.N.B.)
4 Faculty of Urban Studies and Urban Economy, Moscow Polytechnic University, 107023 Moscow, Russia; a.a.mishedchenko@mospolytech.ru
5 Department of Technique and Technology of Mining and Oil and Gas Production, Moscow Polytechnic University, 107023 Moscow, Russia; 1314598@mail.ru
6 Stroytest Research and Testing Center, Moscow State University of Civil Engineering, 129337 Moscow, Russia; karlinat@mail.ru
* Correspondence: golikov_ns@pers.spmi.ru (N.S.G.); pryalukhin_af@pers.spmi.ru (A.F.P.)

Abstract: This work is aimed at obtaining new knowledge in the field of interactions of polydisperse hydrophobic surfaces in order to increase the extraction of mineral microdispersions via flotation. The effect of high velocity and the probability of aggregating fine particles with large ones are used to increase the extraction of finely dispersed gold in this work. Large particles act as carrier minerals, which are intentionally introduced into a pulp. The novelty of this work lies in the fact that a rougher concentrate is used as the carrier mineral. For this purpose, it is isolated from three parallel pulp streams by mixing the rougher concentrate, isolated from the first stream of raw materials, with an initial feed of the second stream; accordingly, the rougher concentrate of the second stream is mixed with the initial feed of the third stream, and the finished rougher concentrate is obtained. In this mode of extracting the rougher concentrate, the content of the extracted metal increases from stream to stream, which contributes to the growth in its content in the end product. Moreover, in order to supplement forces involved in the separation of minerals with surface forces of structural origin in the third flotation stream, the pulp is aerated for a short time (about 15%–25% of the total) with air bubbles filled with a heat carrier, i.e., hot water vapor. Within this accepted flotation method, the influence that the surface currents occurring in the wetting film have on its thinning and breakthrough kinetics is proposed to be in the form of a correction to a length of a liquid slip in the hydrophobic gap. The value of the correction is expressed as a fraction of the limiting thickness of the wetting film, determined by the condition of its thickness invariability when the streams are equal in an interphase gap: outflowing (due to an action of the downforce) and inflowing (Marangoni flows and a thermo-osmotic stream). Gold flotation experiments are performed on samples of gold-bearing ore obtained from two deposits with conditions that simulate a continuous process. Technological advantages of this developed scheme and a flotation mode of gold microdispersions are shown in comparison with the basic technology. The purpose of this work is to conduct comparative tests on the basic and developed technologies using samples of gold-bearing ore obtained from the Natalka and Olimpiada deposits. Through the use of the developed technology, an increase in gold extraction of 7.99% and in concentrate quality (from 5.09 to 100.3 g/t) is achieved when the yield of the concentrate decreases from 1.86 to 1.30%, which reduces the costs associated with its expensive metallurgical processing.

Keywords: gold microdispersions; flotation; carrier minerals; wetting film; stability; slip; correction

Citation: Evdokimov, S.I.; Golikov, N.S.; Pryalukhin, A.F.; Kondratiev, V.V.; Mishedchenko, A.; Kuzina, A.V.; Bryukhanova, N.N.; Karlina, A.I. Studying Flotation of Gold Microdispersions with Carrier Minerals and Pulp Aeration with a Steam–Air Mixture. *Minerals* **2024**, *14*, 108. https://doi.org/10.3390/min14010108

Academic Editors: Fardis Nakhaei, Ahmad Hassanzadeh and Luis A. Cisternas

Received: 5 January 2024
Revised: 15 January 2024
Accepted: 16 January 2024
Published: 19 January 2024

1. Introduction

The main peculiarity of ores currently involved in processing is the increased content of small fractions of extracted valuable components in them [1]. For example, the proportion of disseminated gold contained in sulfide minerals (such as pyrite or arsenopyrite) can reach up to 15%–35%. Taking into account peculiarities of the material composition of currently processed ores, a strategy for the development of the mineral resource base of the Russian Federation until 2035 has been established. It provides for improvements in the theory and methods of extracting and enriching solid mineral raw materials of natural and technogenic origin [2]; in particular, flotation technology [3] represents one of the main processes for the primary processing of the majority of ore [4] and non-metallic minerals.

In the case of the traditional approach to the problem of effectively extracting fine particles (whose size does not exceed 10–30 μm) via flotation, a flotation complex formation has been investigated using a thermodynamics apparatus, which considers the physical and chemical patterns of the action of applied flotation reagents [5]. This approach is the most rational when solving technological problems, for example, water treatment, when the completeness of a collective extraction of all solid particles present in an aqueous phase is a necessary and sufficient condition. During ore flotation, a requirement of the completeness of extracting valuable components is supplemented by a condition of selective separation of minerals [6,7].

During the selective flotation of minerals, the flotation complex is formed as a result of the inertial collision of a particle with a bubble surface. The thinning (when a particle slips along a bubble surface from its upper pole to equator) of a wetting (interphase) film separates a particle and a bubble up to a thickness at which surface adhesion forces acquire a noticeable intensity [8,9] and the formation of the contact angle [10]. Moreover, contact angles should reach values at which the continuous transportation of a particle, adhering to a bubble surface, into a foam layer is ensured under dynamic (turbulent) conditions of the flotation process. A strong adhesion of a particle to a bubble surface is possible with a sufficiently extended wetting perimeter and a large contact angle. Fine particles are fixed in an aft part of a bubble under the action of a hydrodynamic downforce without forming a contact angle in a distant potential pit. When the particle size decreases from 100 to 1 μm, the releasing force decreases 10^6 times [11]. Selective flotation is possible only with forming a flotation complex according to the first mechanism: when particle sizes decrease, the probability of their non-selective fixation on the bubble surface increases without the contact angle formation.

Increasing the completeness of the extraction of fine (inertialess) particles is achieved by enlarging them, due to introducing polymers or high-molecular flocculants into the flotation system. It happens less often by recharging bubbles and using flotation machines of a special design. Hydrophobic glass beads and crushed copper ore have been used as carrier materials in order to extract graphite microdispersions [12]. When the flotation carrier platform, in the form of hydrophobic chalcopyrite and malachite microdispersions, was approached at a distance of up to 8.6–12.5 nm, significant attractive forces appeared between them due to hydrophobic interactions [13–19].

In this work, to increase the completeness and selectivity of the extraction of microdispersions of minerals, large particles, or a solid wall, are used, and the rate of adhesion (as well as induction time [20,21]) is two orders of magnitude higher than the rate at which fine particles aggregate among themselves [14]. When studying an aggregation of hydrophobic polydisperse particles, basic mechanisms of the DLVO theory (van der Waals (dispersion) forces and electrostatic interaction) are complemented by "non-DLVO" forces [22–27], i.e., surface forces of a structural origin associated with a difference between a water structure of boundary layers and a liquid volume [28–32]. A sweep effect contributes to an increase in the rate of liquid removal from an interphase gap (and, as a result, an increase in an adhesion probability) between hydrophobic particles [32–37].

The aim of this work is to study the possibility of increasing the extraction of fine metal fractions from gold-bearing ores using new technological solutions. These are the introduction of carrier minerals and additives to air, supplied for aeration of the flotation system, and a coolant into the pulp, which, from the viewpoint of industrial and environmental safety, fully fits into the context of the Agenda for the field of sustainable development up to 2030, which was adopted by the UN [38–40].

2. Methods and Materials

Field experiments on flotation using the developed method were performed with ore samples from the Natalka deposit (Magadan region) and the Olimpiadinsky deposit (Krasnoyarsk region).

To determine the chemical composition of the samples (Table 1), a spark mass spectrometer of the JMS-BM2 type (JEOL, Tokyo, Japan) equipped with inductively coupled plasma (ICP-MS analysis) was used.

Table 1. Chemical composition of ore samples from the Natalka deposit.

Name of the Element, Component	Content, %	Name of the Element, Component	Content, %
SiO_2	63.5	$Fe_{gen.}$	4.17
Al_2O_3	14.2	Fe_{ok}	2.95
K_2O	3.27	Fe_s	1.22
CaO	2.65	$S_{gen.}$	0.84
MgO	1.95	$S_{sulph.}$	<0.25
MnO	0.12	Zn	0.0070
P_2O_5	0.22	Pb	0.0018
TiO	9.56	Cu	0.0020
Na_2O	2.33	As	0.253
C_{gen}	2.48	Sb	<0.005
$CO_{2\ carb}$	4.55	Au	1.47
C_{org}	1.24	Ag	0.60

According to the scintillation spectral analysis, gold in an ore sample of the Natalka deposit is randomly distributed by size classes. Table 2 shows the results of the scintillation spectral analysis of the gold distribution by size classes in the sample of ores from the Natalka deposit.

Table 2. Gold distribution by size classes in the sample of ores from the Natalka deposit.

Size Class, μm	Gold Distribution, %	Size Class, μm	Gold Distribution, %
0–5	29.7	38–71	1.4
5–9	19.4	71–100	12.0
9–12	4.0	100–150	4.4
12–15	0.7	150–250	12.8
15–25	7.0	250–500	2.8
25–38	5.8		

Seven ore samples, obtained from the Natalka deposit, were taken. For each of these samples, the purity of gold in that ore was determined by an assay test. An average value of the gold purity was 801 (values in the studied samples ranged from 795 to 807).

In the sample of ores from the Olimpiadinsky deposit, sulfides are represented by arsenopyrite, pyrrhotite and antimonite, placed in a form of an insignificant impurity in a mass of non-metallic minerals: calcite (35%–40%), quartz (30%–43%), muscovite (8%–10%) and biotite (10%–15%), chlorite, sericite and a number of other minerals.

Free gold in ores is represented by finely dispersed and pulverized fractions (Table 3).

Table 3. Distribution of gold by size classes in the sample of ores of the Olimpiadinsky deposit.

Grain-Size Class, μm	Gold Distribution, %	Grain-Size Class, μm	Gold Distribution, %
3–5	31.7	12–15	9.9
5–9	29.0	15–25	14.3
9–12	15.1		

Native gold is associated with quartz (35%), arsenopyrite (35%), pyrite and marcasite (15%), pyrrhotite (5%), berthierite and antimonite (5%), carbonates (5%), as well as (single grains) with chalcopyrite, tetrahedrite and muscovite. When grinding ores to a size of 74 μm, the content of free gold is 15%.

A bulk consists of gold grains, whose surface has been subjected to a technogenic impact. Gold is represented by rolled thin plates, whose contours are torn; gold grains of an oblong shape are indicated. A gold color is bright yellow and straw yellow. A surface of individual grains is covered with oxide films. A grain surface texture is mainly bumpy, porous and rough. High-grade gold is 950‰; it contains mercury, silver, copper, iron, antimony, arsenic, sulfur in the form of impurities. A phase analysis of gold in an ore sample was performed (Table 4).

Table 4. Results of a phase analysis of a gold sample of initial ore (when a grinding size of 98% of a class is 74 μm).

Form of Finding Gold	Natalka Deposit	Olimpiadinsky Deposit	Natalka Deposit	Olimpiadinsky Deposit
	Au Content, g/t		Au Distribution, %	
Free gold and in accretions, cyaninated	1.04	1.43	70.70	47.70
Gold associated with sulfides	0.28	0.86	19.15	28.70
Gold in films and "jackets" of iron hydroxides	0.036	0.37	2.48	12.40
Gold encased in quartz	0.11	0.34	7.67	11.20
Initial sample	1.47	3.00	100.00	100.00

In the flotation experiments, a countercurrent column ($D \times L = 0.6 \times 7.4$ m) was used when an initial feed was supplied between a purification zone (1.6 m high) and a mineralization zone (5.8 m high) and the airlift unloading of tailings. During hydraulic transport of tailings, 0.045 $m^3/(min \times m^3)$ of air under a pressure of 0.14 MPa was supplied to an airlift. The reduced pulp velocity (a ratio of a productivity to a column cross-section) was 1.03 cm/s. The steam consumption was 1.07×10^{-2} $kg{\cdot}s^{-1}{\cdot}m^{-2}$ (under an overpressure of 0.15 MPa) and the air consumption was 2.61 $kg{\cdot}s^{-1}{\cdot}m^{-2}$. The flotation machine was equipped with a confuser–diffuser type aerator when water vapor was supplied (an electric steam generator of an EPG-50 type) to a low-pressure zone created by a working airflow (a compressor of a 4BU1-5/9 type). The gas content in the flotation column was 14%–15%. No rinsing water was applied to the foam layer.

2.1. Flotation of Ores Using the Basic Technology

A schematic block diagram of ore flotation using the basic technology is shown in Figure 1.

Gold was extracted from the ores using a gravity–flotation technology, similar to that adopted at an operating gold extracting factory. An experiment was performed according to a scheme simulating a closed enrichment cycle. A total of 65.43% of Au was extracted into a "golden head" (817.88 g/t of Au) by gravity. Gold was extracted from gravity tailings using a flotation method. The flotation flowsheet included a rougher flotation operation, whose rougher concentrate was recleaned three times; tailings of the rougher flotation operation were subjected to two control flotation operations. In the experiment that simulated a closed flotation cycle, intermediate flotation products (tailings of recleaning and foam products of control flotation operations) were sent from subsequent flotation operations to previous ones. At a rate of 500 g/t, starch was used to suppress organic carbon, which allows a reduction in the carbon content in a concentrate from 6.1 to 1.4% and the extraction from 47.0 to 1.4%. Gold-bearing sulfides were activated with copper sulfate when loading sodium carbonate up to pH 8.5–8.7. Another 15.28% of Au was extracted into a flotation concentrate, using butyl xanthogenate and a foam agent T-92.

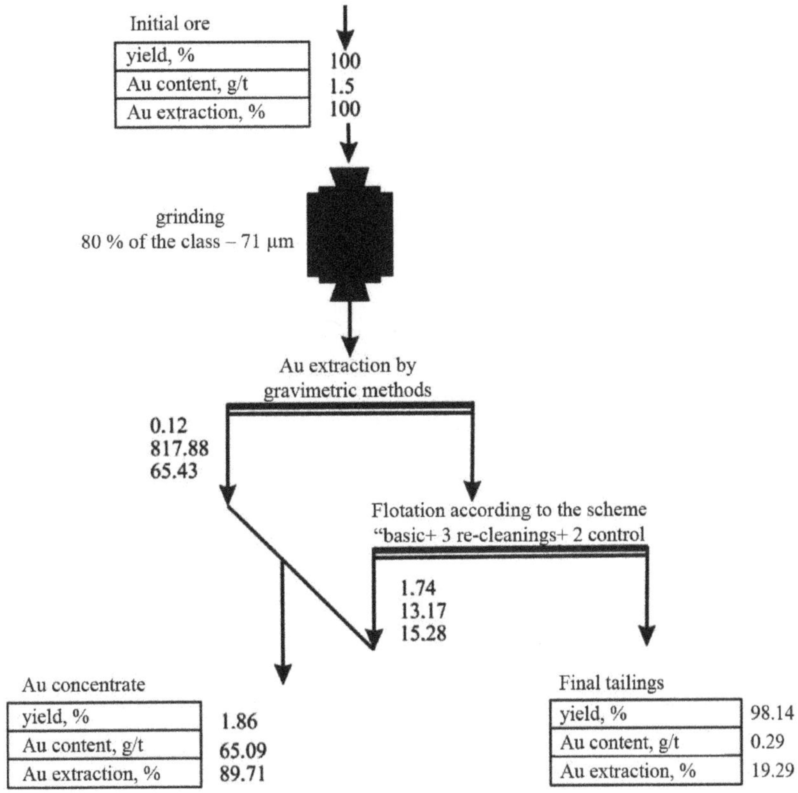

Figure 1. Basic block diagram of flotation of ores from the Natalka deposit using the basic technology.

2.2. *Flotation of Ores Accompanied by Isolating the Rougher Concentrate in "Three Steps"*

Figure 2 shows the results of flotation of ores from the Natalka deposit (a cycle of extracting gold by gravitational methods and a cycle of basic flotation), using the rougher concentrate as carrier minerals to extract gold microdispersions. The results were obtained in the experiment modulating a closed flotation cycle.

A structural and technological scheme of ore flotation (and a reagent mode of mineral separation), using the rougher concentrate as carrier minerals, intended for extracting gold microdispersions, was similar to that used in the basic technology (Section 2.1, Figure 1). The finishing of the rougher concentrate consisted in its three re-cleanings; the finishing of tailings of three main flotation operations consisted of two control flotation operations.

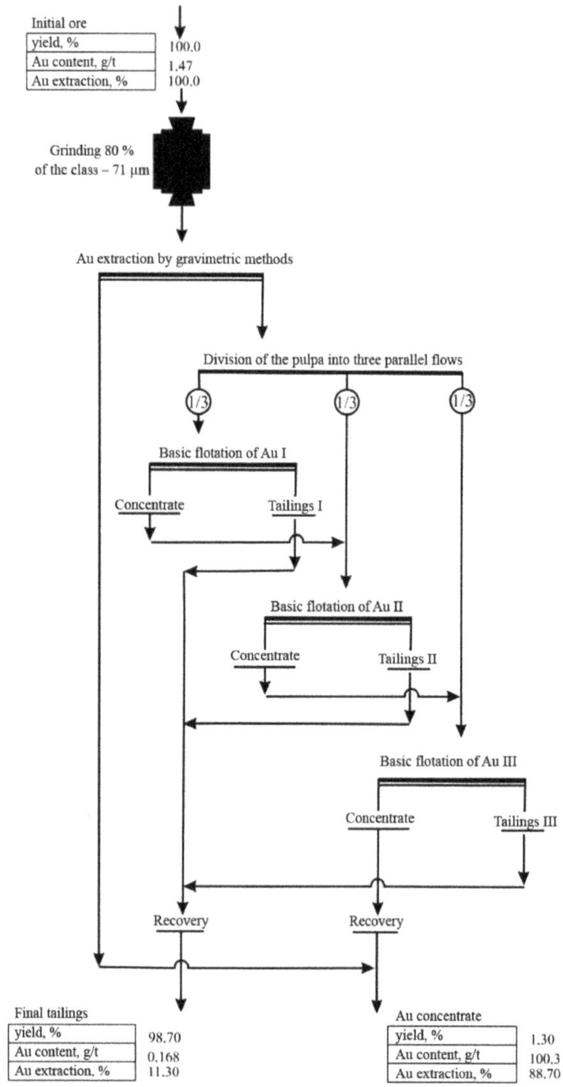

Figure 2. Basic block diagrams of flotation of ores from the Natalka deposit accompanied by isolating the rougher concentrate in "three steps".

3. Theoretical Provisions

Upon approaching a bubble, large particles collide with a bubble surface, deviating under the action of inertia forces from fluid flow lines, flowing around a bubble. Inertia forces decrease along with the decreasing particle size, which makes it difficult for them to collide with the bubble surface. Therefore, with large particles, a process of flotation complex formation is limited by an adhesion stage, while with fine particles it is limited by their transfer to the bubble surface [41]. Based on peculiarities of the interaction of fine particles with an air bubble, the completeness of extracting fine particles by flotation can be increased by their preliminary aggregation with large particles.

Based on the geometric probability of a collision, we can prove the predominant aggregation of large R_i and fine R_j particles in a polydisperse mineral system.

Polydisperse particles move in a laminar stream along an Ox axis between a surface $y = 0$ and $y = h$; a field of external forces is oriented across streams (Figure 3).

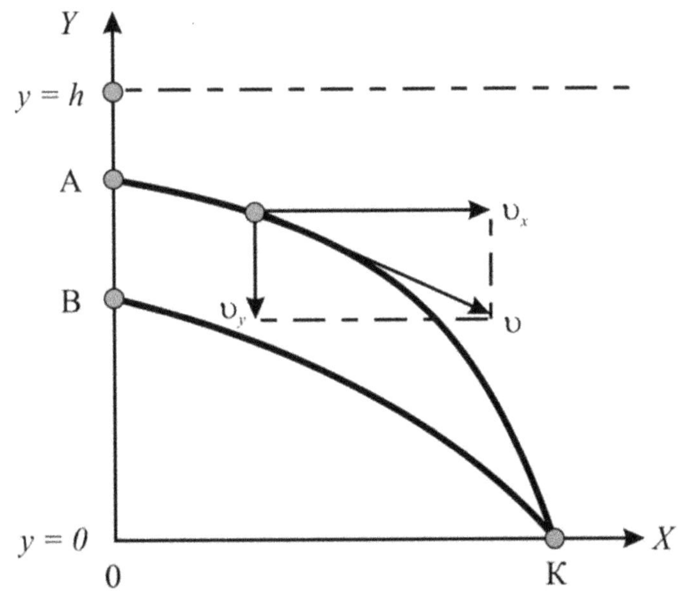

Figure 3. Scheme used for calculating the geometric probability of the collision of polydisperse particles.

According to the diagram (Figure 1), a particle with a size of $R_j < R_i$ begins to move along with a stream at point B having an ordinate:

$$y_B = \phi\ (y_h\ ,\ R_i\ ,\ R_j) = \psi\ (y_h) \tag{1}$$

Specifying the size of the large particle R_i and an ordinate of its entry into the y_A, stream allows for the determination of an AK path and an abscissa of a point of its collision with a wall—$x_K = 0\ K$. This means that particles of an R_i size, when moving along an AK trajectory, can only collide with particles that have crossed an AB segment and have a size of R_j $(R_j < R_i)$.

Provided that particles of j-size are separated from each other by a distance of:

$$\delta_j = \frac{R_j}{\sqrt[3]{c\ f\ (R_j)\ \Delta\ R_j}} \tag{2}$$

The geometric probability of a collision of polydisperse particles can be determined by the following formula:

$$p_{i\,j} = \begin{cases} (R_i + R_j)^2 / \delta_j^2 & when\ (R_i + R_j) \leq \delta_j \\ 1 & when\ (R_i + R_j) > \delta_j \end{cases} \tag{3}$$

where $f\ (R_j)$—mass differential function of particle size distribution; c—fraction of a large particle surface occupied by fine particles (a density factor of packing fine particles on the large particle surface).

We will divide a segment of $0\ y_1$ into $n = h\ /\ \Delta\ y$ parts, and a BK segment will be divided into $l\ (k)$ parts (where $\Delta\ y = \delta_j, l\ (k) = (y_h - y_1)\ /\ \delta_j)$.

Then, the geometric probability of a collision of polydisperse particles will be:

$$p_{i\,j\,h} = 1 - \left[1 - (R_i + R_j)^2 / \delta_j^2\right]^{l\,(h)} \tag{4}$$

If a volumetric unit of a suspension contains $C'\, f\,(R_i)\,\Delta\, R_i\, /\, R_i^2$ particles of i-th size, then the number of particles of the same size crossing a ΔS strip having an area of $\Delta\, S = 1 \cdot \delta_j$ per unit time at the entrance is equal to:

$$\Delta\, n_i|_{\delta_j} = c\, Q\, \delta_j\, f\,(R_i)\,\Delta\, R_i\, /\, \left(R_i^3\, h\right) \tag{5}$$

Following [42], the expression for determining the mathematical expectation of the number of collisions of large R_i and fine R_j particles in a suspension of volume V can be written as:

$$\begin{aligned} M\,(R_i\,/\,R_j) &= \sum_{k=1}^{n} p_{i\,j\,k}\,\Delta\, n_i = \\ &= \frac{C'\, V\, f\,(R_i)\,\Delta\, R_i}{R_i^3\, h} \sum_{k=1}^{n} \left\{1 - \left[1 - \left(\frac{R_i + R_j}{\delta_j}\right)^2\right]^{l\,(k)}\right\} \delta_j \end{aligned} \tag{6}$$

Applying Expressions (4) and (5), as well as taking into account the fact that $\Delta\, y = \delta_j << 1\,,\, n >> 1$, Expression (6) can be rewritten as:

$$\begin{aligned} M\,(R_i\,/\,R_j) &= \sum_{k=1}^{n} p_{i\,j\,k}\,\Delta\, n_i = \\ &= \frac{c\, V\, f\,(R_i)\,\Delta\, R_i}{R_i^3\, h} \sum_{k=1}^{n} \left\{1 - \left[1 - \left(\frac{R_i + R_j}{\delta_j}\right)^2\right]^{\frac{h\,Y - \psi\,(h\,Y)}{\delta_j}} d\,Y\right\} \end{aligned} \tag{7}$$

where $Y = y\,/\,h$.

If $R_j < R_i$, the mathematical expectation of the number of collisions of particles, whose size is $R_i \in \left[R_i', R_i''\right]$ and $R_i' \in \left[R'_i\,,\, R_i''\right]$ when $\Delta\,\delta_i \to 0$, equals:

$$\begin{aligned} M\,(R_i' \le R_i \le R_i''\,|\,R_j) &= c\, V \sum_{i=i'}^{i=i''} \frac{f\,(R_i)}{R_i^3} \times \\ &\times \left\{1 - \int_0^1 \left[1 - \left(\frac{R_i + R_j}{\delta_j}\right)^2\right]^{\alpha} d\,Y\right\} \Delta\, R_i \end{aligned} \tag{8}$$

where $\alpha = \dfrac{h\,Y - \phi\,(h\,Y\,,\, r\,,\, R_j)}{\delta_j}$

As a result, we can find the ratio of the number of collisions of large particles R_i with fine particles R_j to the number of large particles n:

$$M\,(R_i\,|\,R_j)\,/\,n_i = 1 - \int_0^1 \left\{1 - \left(\frac{R_i + R_j}{\delta_j}\right)^2\right\}^{\frac{h\,Y - \psi\,(h\,Y)}{\delta_j}} d\,Y \tag{9}$$

Equation (16) allows for the conclusion that particles of the same size moving along the same trajectory practically do not collide with each other. On the contrary, $\psi\,(hY) \approx 0\,,\, (R_i + R_j)^2\,/\,\delta_j^2 < 1$ and a relative collision frequency of large and fine particles is close to 1.

In this work, the effect of a high velocity and the probability of aggregating fine particles with large ones is used to increase the extraction of finely dispersed gold: it is extracted after the preliminary aggregation with carrier minerals. Large particles that act as carrier minerals are intentionally introduced into the pulp. Moreover, as carrier minerals (solid walls), a material as homogeneous as possible with extracted fine particles is used, i.e., a rougher concentrate, which is isolated for this purpose in "three steps" (Figure 4).

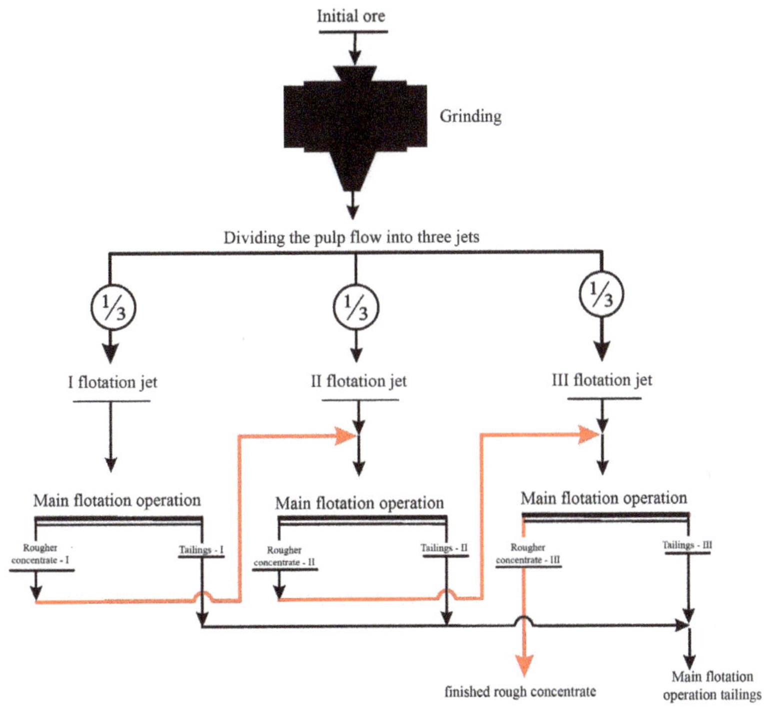

Figure 4. The structure of a basic flotation operation under study. Red line—the means of obtaining the finished rough concentrate.

For this purpose, the rougher concentrate is isolated from three parallel streams (jets) of the pulp. A rougher concentrate isolated in the first flotation jet is mixed with an initial feed of the second flotation jet; accordingly, the rougher concentrate of the second jet is mixed with an initial feed of the third jet, releasing the finished rougher concentrate.

In this mode of extracting the rougher concentrate, the content of extracted metal increases from jet to jet. Consequently, when using the developed flotation scheme, the achievement of a "synergetic" technological effect is possible due to the aggregation of fine particles with carrier minerals, as well as because of increasing a valuable component content in the basic flotation operation, which contributes to an increase in its extraction.

In order to supplement forces, involved in the separation of minerals by surface forces of structural origin in the third flotation stream, a pulp was aerated during a short time (about 15%–25% of the total) with an air mixture containing hot (≥104 °C) water vapor (steam–air mixture).

It is possible to analyze the relationship between the content of a valuable component α in an initial feed and an amount of its extraction ε.

We can assume that along an enrichment curve $\varepsilon_0\ (\alpha)$, with a planned content of an extracted metal in an initial ore (equal to α_0), a level of achieving its extraction into concentrate (Figure 5) is $\varepsilon_0\ (\alpha_0)$. Let us choose other arbitrary values of a metal content in the initial ore α and an operating mode of a flotation unit, at which a metal extraction level along the enrichment curve is $\varepsilon_0\ (\alpha)$. For the selected mode, an extraction of a metal into a concentrate ε, reduced to α_0, will be:

$$\varepsilon\ (\alpha_0) = \varepsilon\ (\alpha) - \varepsilon_0\ (\alpha) + \varepsilon_0\ (\alpha_0) \tag{10}$$

since

$$\varepsilon\ (\alpha)\ -\ \varepsilon_0\ (\alpha)\ =\ \varepsilon\ (\alpha_0)\ -\ \varepsilon_0\ (\alpha_0) \tag{11}$$

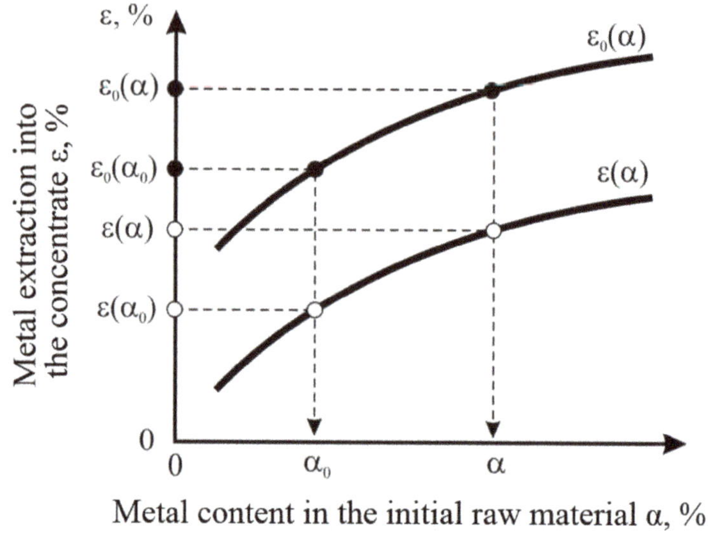

Figure 5. Enrichment curves for the derivation of Equation (10).

The value $\varepsilon\ (\alpha_0)$ does not depend on the position of the curve $\varepsilon_0\ (\alpha)$ in the graph in Figure 5, based on the fact that:

$$\varepsilon\ (\alpha_0)\ = \varepsilon\ (\alpha)\ -\ \varepsilon_1\ (\alpha)\ +\ \varepsilon_1\ (\alpha_0) \tag{12}$$

Figure 6 shows the dependence of extracting gold into a concentrate on its content in the initial feed for ores of the Olimpiadinsky deposit, obtained during the study of ores for enrichment.

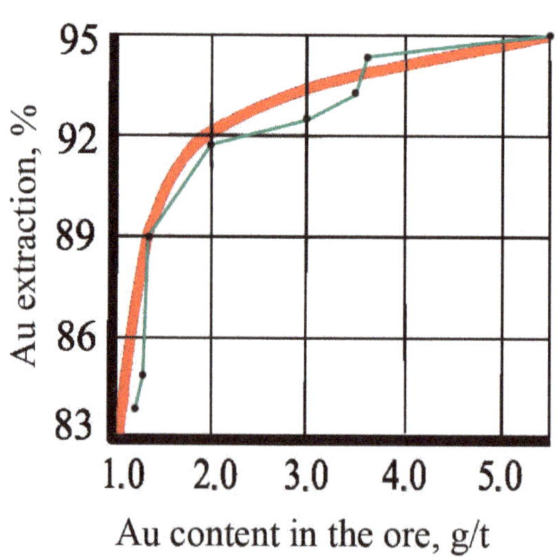

Figure 6. Extraction of gold into a concentrate as a function of its content in the initial ore of the Olimpiadinsky deposit. The red line—middle line; green line—line of experimental values.

The results provided in Figures 4 and 5 do not contradict data of other researchers [10] and the practices of the operation of enriching factories [11].

In practice, a necessary increase in the content of an extracted metal in a basic flotation operation is achieved by returning middling products to it, i.e., tailings I of the re-cleaning and the concentrate of the control flotation operation. When the initial feed is mixed with rich middling products, the content of an extracted metal in the basic flotation operation increases. However, a uniformity condition for the separation of the resulting mixture is not fulfilled, because an equality of metal contents in mixed products is not equivalent to their identity in terms of a separation ability. With such a discrepancy, rapidly and slowly floated fractions of extracted and suppressed minerals appear in a flotation cascade. When middling products are mixed with an initial feed, an increase in the metal content in it is achieved mainly via an increase in the concentration of the most difficult-to-float forms of the extracted mineral. A consequence is a decrease in the contrast of a material in terms of flotation properties due to the appearance of extracted minerals of a wide range of floatability. A ratio of flotation rate constants of separated minerals (Beloglazov's selectivity index) turns out to be close to 1: the flotation rates of a slowly floated fraction of an extracted mineral and a rapidly floated fraction of a suppressed mineral are aligned.

A jet movement of an initial feed and a rougher concentrate ensure a high content of an extracted metal in a basic flotation operation without the appearance of an undesirable distribution of minerals by floatability in the scheme.

Therefore, two possible reasons for increasing the completeness of an extraction of microdispersions of minerals were identified. One reason may be a high rate and a probability of an aggregation of fine minerals with minerals of a rougher concentrate (containing carrier minerals). Another reason may be an increase in the content of an extracted metal in the basic flotation operation when using a rougher concentrate in circulation.

In the general case, the number of parallel streams depends on the content of an extracted metal in ore. Enrichment processes are characterized by a monotonous increase in the extraction of the metal into the concentrate when decreasing the metal content in it. At the same time, when controlling the process, the most important limitation is imposed on the content of the base metal in the concentrate: as a rule, the metal content in the concentrate must correspond to (or exceed) a planned indicator. Based on the limitations imposed on the quality of the concentrate, when determining the number of streams, the goal is to achieve maximum extraction at a given concentration degree in a basic flotation operation.

Isolation of a finished rougher concentrate in the third flotation jet.

Based on the analysis of the density functional of a liquid in a thin gap, the authors demonstrated [42] that a non-uniform decrease in the density of a film thickness led to the appearance of attractive forces, whose magnitude in the case of thin films could significantly exceed van der Waals forces. A distinctive feature of such forces is not a decrease, as was assumed in [43], but the increase in the absolute magnitude of forces when the temperature increases. It also involves a decrease in minimum time required for fixing a particle on the surface of a bubble (induction time) [44] and an increase in the contact angle [45]. Therefore, to complement forces, participating in a separation of minerals, with surface forces characterized by high sensitivity to temperature, it is sufficient to heat a layer of water 4 to 8 nm thick near a bubble, inside which an action of surface forces is localized. In this work, water in the boundary layer of bubbles is heated due to heat of condensation of hot water vapor when a bubble contacts cold water. For this purpose, a pulp was aerated with an air mixture containing hot ($\geq$104 °C) water vapor (steam–air mixture). Steam–air flotation is carried out in a laboratory countercurrent column (Figure 7).

The initial feed is supplied from a mechanical agitator (3) into a column (1) from above under a foam layer (at a distance of ~1/3 of a column height) towards air bubbles generated by a pneumatic atomizer (2) installed in the bottom part of the column. A chamber product is unloaded using an airlift (5), and during emergency unloading it is unloaded through a gate valve (6). There is a system (4) that supplies water to a foam layer (with a secondary concentration of minerals by washing out non-float particles from the foam layer).

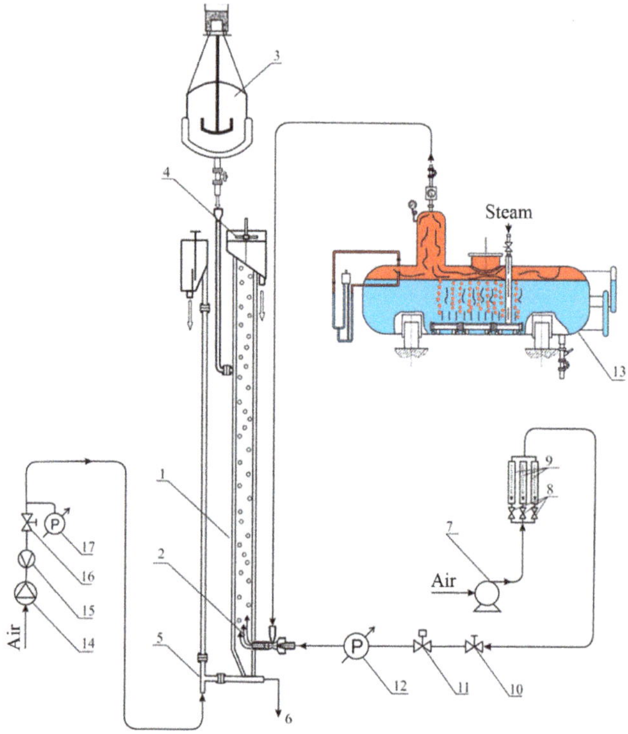

(**a**)

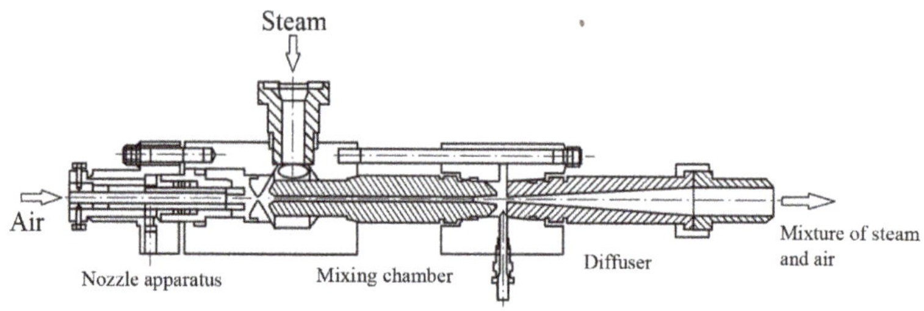

(**b**)

Figure 7. A laboratory-scale plant of column flotation equipped with an aerator: (**a**)—general column flotation scheme; (**b**)—steam and air mixer scheme.

The concurrent streams of air and water vapor are created using a steam generator (13) equipped with a steam line and a pneumatic system assembled on the basis of an air blower (7). It is supplied with fine-adjustment valves (8) and a flow meter (9) at an outlet; a pressure gauge (12) with shut-off and control valves (10, 11) is installed in front of the atomizer to control the air pressure. To supply air to an airlift, there is an independent pneumatic system based on a compressor (14) equipped with a flow meter (15), a gas valve (16) and a pressure gauge (17).

A study of the effect of water temperature in boundary layers of bubbles on the results of flotation consisted in analysis of the temperature dependence of the hydrophobic slip and the associated limiting thickness h_{lim} of the interphase film. Physically, h_{lim} corresponds to a limiting distance that a particle can cover to approach a bubble surface under given conditions of their hydrodynamic interaction.

Slipping (inaccurate observance of a boundary condition of the sticking–slipping of a liquid stream over a solid surface) is associated with the presence on a hydrophobic surface of a layer of reduced viscosity (formed due to a stratification of a saturated gas–liquid solution during thinning an interphase gap (Figure 8a) [46–49]) or microcavities (submicrocavities) containing gas bubbles 10^{-4}–10^{-2} cm in size retained by surface tension forces (with a rough surface (Figure 8b) [50,51]). Macroscopic slipping on such a surface is a consequence of reduced friction between a liquid and a gas.

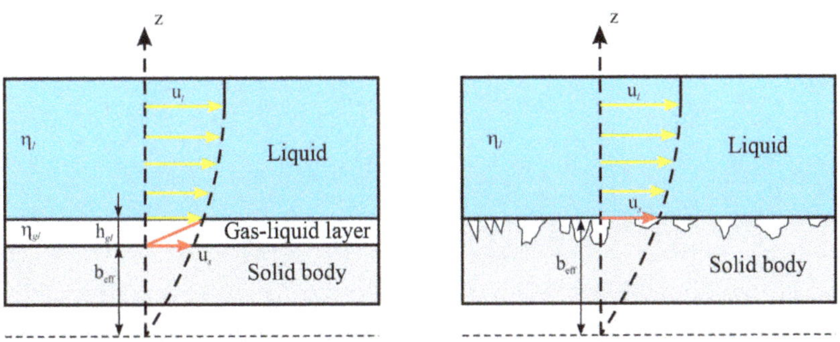

(a) (b)

Figure 8. A scheme of the liquid slipping along a wall-mounted gas–liquid layer of reduced viscosity (**a**) and on a rough surface (**b**).

The Navier slipping condition (with a zero normal component of a velocity vector $u_n = 0$, a tangent component of a velocity u_s is proportional to a tangent stress τ) is as follows [52]:

$$u_s = b_{eff} \left(\frac{\partial u}{\partial n} \right) \tag{13}$$

where b_{eff}—apparent slipping length, whose value with a thickness h_{gl} of a wall-mounted gas–liquid layer can be found from the following ratio [53]:

$$b_{eff} = h_{gl} \left(\frac{\eta_l}{\eta_{gl}} \right) \tag{14}$$

During flotation with a steam–air mixture [52–55], a slip length should be taken into account by a correction k_E in the form of:

$$k_E = \frac{\Delta \sigma_\Gamma}{\Delta \sigma_T}, \tag{15}$$

considering the appearance of tangential stresses associated with the adsorption layer of a surface active agent (SAA):

$$\Delta \sigma_\Gamma = (\partial \sigma / \partial \Gamma) \, \Gamma \tag{16}$$

and a temperature gradient:

$$\Delta \sigma_T = (\partial \sigma / \partial T) \, \Delta T. \tag{17}$$

Moreover, a rate of a heat transfer is higher than a rate of a mass transfer (Schmidt number is three orders of magnitude higher than Prandtl number) and when $k_E < 1$, an adsorption convection is completely suppressed by an oncoming Marangoni thermal flow towards a periphery of a wetting film.

When $k_E \sim 1$, there is a dynamic balance between counter-tangential stresses, in which a generated concentration and heat fluxes have approximately the same intensity. Under these conditions, at a tangential velocity of a liquid outflow from an interphase gap of $u\,(r, z)$, a wetting film is formed, whose limiting thickness h_{lim} is given by an expression in [56,57]. An expression for determining a velocity u in a shear flow of a liquid at a point x, located in a gap of a thickness L (along an Ox axis) between two parallel plates, can be obtained in the form of:

$$u = u_S + u_{h_{gl}} \frac{x}{h_{gl}} \text{ when } 0 < x < h_{gl} \tag{18}$$

$$u = u_S + u_{h_{gl}} + \left(\Delta u - 2\, u_{h_{gl}}\right) \frac{x - h_{gl}}{L - 2\, h_{gl}} \; when\; h_{gl} < x < \left(L - h_{gl}\right) \tag{19}$$

$$u = u_S + \Delta u + u_{h_{gl}} \frac{x - L}{h_{gl}} \; when \left(L - h_{gl}\right) < x < L \tag{20}$$

Here, u_S, $u_{h_{gl}}$—velocity of a liquid on a surface of a solid and at a distance of a thickness of a liquid layer h_{gl} with a structure and properties changed compared to volume; Δu—relative velocity of a liquid flow relative to an Oy axis. Consequently:

$$\eta_{gl} \frac{u_{h_{gl}}}{h_{eff}} = \eta_l \frac{\Delta u - 2\, u_{h_{eff}}}{L - 2\, h_{eff}} \tag{21}$$

and

$$\frac{u_{h_{eff}}}{h_{eff}} = \frac{\eta_l}{\eta_{gl}\, L - 2\, h_{gl}\left(\eta_l - \eta_{h_{eff}}\right)} = \frac{\eta_l}{\eta_{gl}} \frac{\Delta u}{L + 2\, h_{eff}\left(\frac{\eta_l}{\eta_{gl}} - 1\right)} \tag{22}$$

Based on a sticking condition (13), an expression for a slipping length is obtained in the form of (Figure 9):

$$b_{eff} = h_{eff}\left(\frac{\eta_l}{\eta_{gl}} - 1\right). \tag{23}$$

Here, a value h_{eff} is given by the following expression:

$$h_{\text{eff}} = h_{gl}\,(1 + k_h) \tag{24}$$

where k_h—dimensionless thickness of an interphase film. A limiting thickness of the film is used as a characteristic value h_{cr}:

$$k_h = \frac{h_{lim}}{h_{cr}} \tag{25}$$

The relationship between a limiting thickness of the interphase film h_{limb} and the time of contact of a particle with a bubble surface is as follows:

$$t_{bp} = R_b \,/\, u_b \tag{26}$$

and relaxation time to a state of a new equilibrium of the film (defined as the time after which a volume of an outflowing liquid differs by no more than 1% from the total volume change during the transition of the film from one equilibrium state to another) [58]:

$$t_{rel} = \left(2\,\eta\, R_0^2\, h_0^{m-2}\right) / \, m\, A \approx a^2 \,/\, D_s \tag{27}$$

where R_0—radius of a curvature of an undisturbed film (meniscus); A—Hamaker constant; m = 1 for a film on a cylinder and m = 2—on a sphere) was obtained in the form of [59]:

$$h_{lim} = \frac{\gamma}{\eta}\frac{9\,\pi\,R}{48}\frac{a}{b}\,tg^2\frac{\theta}{2}\sum_{n=1}^{\infty}\frac{1}{\mu_n^2\left(1+\frac{1}{2}\mu_n^2\,\chi\right)} \tag{28}$$

where

$$\chi = \frac{t_{bp}}{t_{rel}}\frac{t_K - t_\Sigma}{\left|\varepsilon \times \cos\theta_{eff} \times t_{rel}\right|} = \frac{t_{ind}}{\left|\varepsilon \times \cos\theta_{eff} \times t_{rel}\right|} \tag{29}$$

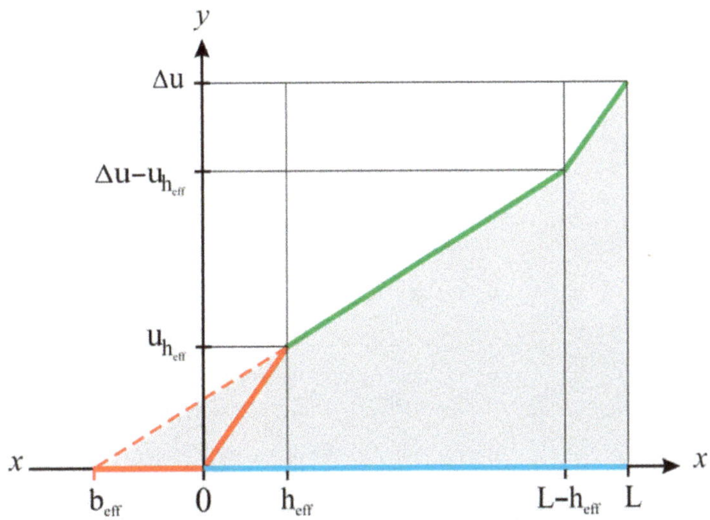

Figure 9. A scheme for the geometric determination of a slipping length b in a shear fluid flow.

Here, t_c—time of contact of a particle with a surface of a bubble. $t_\Sigma = \frac{b}{v_\Gamma} + \frac{b}{v_T} + \frac{b}{v_\chi}$—correction for fluid slippage; $v_\Gamma = R_b\,\eta_l^{-1}\,(\partial\,\sigma\,/\,\partial\,C)\,\Delta\,C$—capillary condensation flow rate [56–58]; $v_T = R_b\,\eta_l^{-1}\,(\partial\,\sigma\,/\,\partial\,T)\,\Delta\,T$—thermocapillary flow rate [60–62]; $v_\chi = 2\,h\,\chi\,R_b\,T\,/\,(\partial\,T\,/\,\partial\,R_b)$—thermosmotic flow rate [58]; θ_{eff}—effective receding contact angle; ΔT, ΔC—difference in a length of an interphase film of temperature and SAA concentration; t_{ind}—induction time; ε—proportion of stuck particles at a given contact time t_c.

A value of a contact angle θ ($\theta \in 90 \div 180^0$) is determined by the physicochemical properties of a solid surface, and in a mode of a heterogeneous wetting of a rough Cassie—Baxter surface, a relationship of an effective receding contact angle θ_{eff} with a contact angle determined by a Young equation θ_0 is provided by the following ratio [63,64]:

$$\cos\theta_{eff} = f\,(r \times \cos\theta_0 + 1) - 1 \tag{30}$$

where θ_{eff}—effective receding contact angle, determined in a wetting mode, when a surface is heterogeneous, composed of areas having a Young's contact angle θ_0 and depressions partially or completely filled with air. $f \in 0 \div 1$—proportion of a surface area wetted in a homogeneous Wenzel–Deryagin regime; r—coefficient considering a deviation in Young's contact angle from a contact angle measured on a rough surface. When $f \rightarrow 1$, the Wenzel–Deryagin regime of the homogeneous wetting corresponding to a boundary minimum of free energy, which replaces a local minimum corresponding to a heterogeneous Cassie–Baxter wetting. Conversely, when $f \rightarrow 0$, minimum free energy becomes global (less than a boundary one, which leads to decreasing the wettability of a solid surface).

4. Results and Discussion

The results of calculating (in the Maple 2021) the values of limiting thickness h_{lim} of an interphase film with Formula (33) with various flotation factors (Figure 10) are presented.

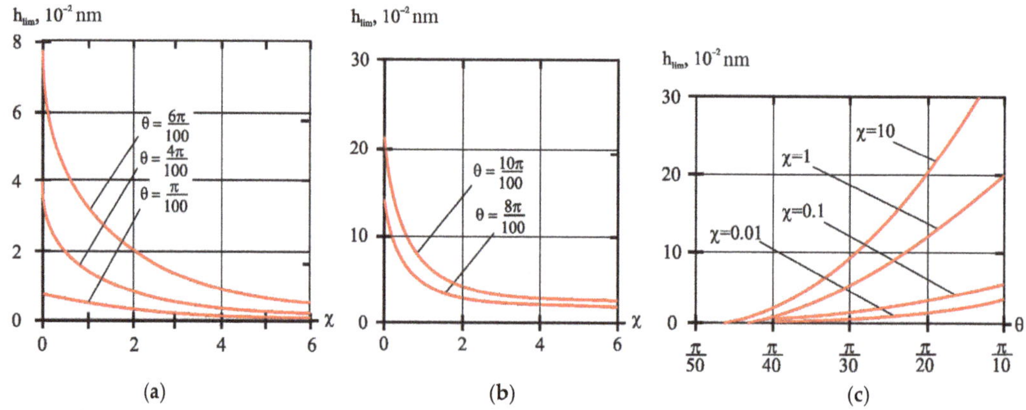

Figure 10. Results of analyzing the dependence of a limiting thickness h_{lim} of an interphase film on flotation factors: the dependence of h_{liim} on a magnitude of correction for stabilization of an interphase film χ (when a crossing angle θ of a particle and a bubble ranges from $\pi/100$ to $6\pi/100$ (**a**) and the crossing angle θ varies from $8\pi/100$ to $10\pi/100$ (**b**)) and on the crossing angle θ of a particle and a bubble (when changing the value of correction for stabilizing an interphase film χ from 0.01 to 10) (**c**).

The fixation of a particle on a bubble surface is possible when a wetting film thickness is $h = h_{cr} \approx 10$ nm, at which the hydrophobic interaction intensity (structural attraction forces) begins to increase, and the hydrophilic interaction (structural repulsion forces) weakens, which, along with the subsequent thinning of the wetting film, changes the sign of a structural component of wedging pressure. Based on this condition, a high probability of a particle deposition on a bubble surface exists only near its upper pole, when a particle trajectory deviates from a vertical axis of a bubble by an insignificant angle θ_{cr}—$\theta < \pi/100 = \theta_{cr}$ (Figure 10a). When counter flows formed by a pressure drop, created by an inertia force of a particle approaching a bubble, and a Marangoni convection are aligned in an interphase gap, a capture efficiency becomes negligible (Figure 10b).

At low values of a $\chi << 1$ ratio, an intensity of the surface current, caused by shear stress associated with a concentration SAA gradient, is limited by the diffusion process. And during the time of contact of a particle with a bubble surface, it does not have time to significantly influence, as follows from Formula (23), a limiting thickness of a wetting film.

On the contrary, under a condition of $\chi >> 1$, the driving force for reducing the limiting thickness h_{lim} of the wetting film is a flow, caused by surface SAA diffusion and directed from the center to the periphery of the film (Figure 10c).

In laboratory conditions, studies were conducted to assess the efficiency of a technology of flotation enrichment of gold-bearing ores using a rougher concentrate isolated from a part of ore as carrier minerals to increase the extraction of gold microdispersions.

In the first experiment, tests were performed on a sample of ores from the Natalka deposit. Gold was extracted from ores using two technologies. The first basic technology is described in Section 2.1. The second modified technology is described in Section 2.2.

The difference between the basic technology (Section 2.1, Figure 1) and the modified technology (Section 2.2, Figure 2) is that, in a cycle of a basic flotation operation, a rougher concentrate is isolated in "three steps" from three parallel pulp streams: the finished rougher concentrate is obtained through mixing a recycled rougher concentrate twice with an initial feed. Such countercurrent movement of the recycled rougher concentrate and the initial feed is performed for two purposes. Firstly, a material of the recycled rougher concentrate acts as carrier minerals, whose presence allows for the extraction of gold microdispersions by means of the most effective flotation mechanism of inertialess particles, i.e., sticking to a hydrophobic "solid wall" extremely related to them in genesis and physico-chemical properties. In addition, when recycling the rougher concentrate, the extracted metal content in the feed of the basic flotation operation increases, which enables separating the mixture via flotation of a greater technological efficiency. The effects of introducing carrier minerals and increasing the content of an extracted valuable component are complemented by a flotation mode with a steam–air mixture, whose duration is approximately 15%–25% of the flotation time in an operation [62–64].

Concentrates obtained in the first two experiments were analyzed for the distribution of gold in them by size classes using sedimentometric analysis (Figure 11).

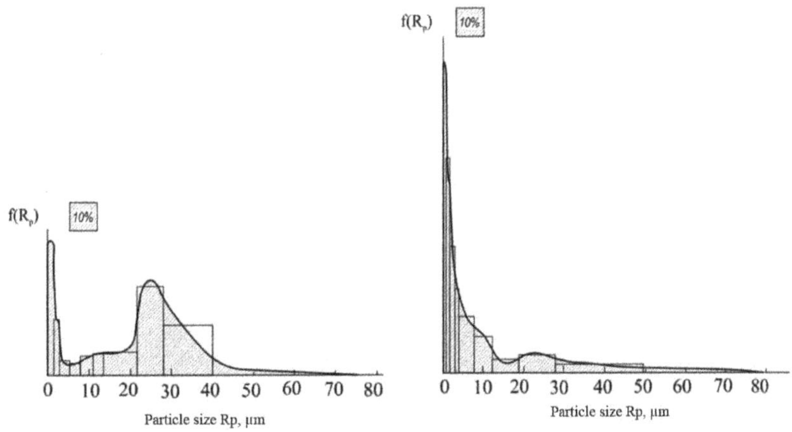

(a) (b)

Figure 11. Histograms and differential functions of the particle size distribution f(R_p) for concentrates obtained from gold-bearing ores of the Natalka deposit using basic (**a**) and newly developed (**b**) technologies.

A comparison of results obtained during flotation of ores from the Natalka deposit, using two competing technologies, allows for the conclusion that one of the factors contributing to an increase in gold extraction (Figures 8 and 9) is an increase in the extraction of small fractions of a gold-bearing material into a concentrate (Figure 10).

Selective enrichment of a feed of a basic flotation operation contributes to a gold extraction increase, which changes the ratio of useful and polluting minerals in a favorable direction: the content of an extracted metal in a concentrate obtained using the newly developed technology was 100.3 g/t versus 65.09 g/t obtained applying the basic technology. When the feed is enriched with a flotation-active rougher concentrate, a load on a surface of air bubbles is increased with an extracted mineral, and their mineralization with rock particles and other minerals decreases. A consequence of a "competition for space" on an air bubble surface is an improvement in the concentrate quality. As a result of improving the concentrate quality, its yield decreased from 1.86 to 1.30%, which enables reducing costs associated with its expensive metallurgical preparation before leaching gold.

Studying the new technology in laboratory conditions was continued on the basis of a sample of sulfide ores from the Olimpiadinsky deposit. As in the previous experiment, indicators obtained during ore flotation performed according to basic (adopted at an operating gold extracting factory) and newly developed technologies were compared (Figure 12).

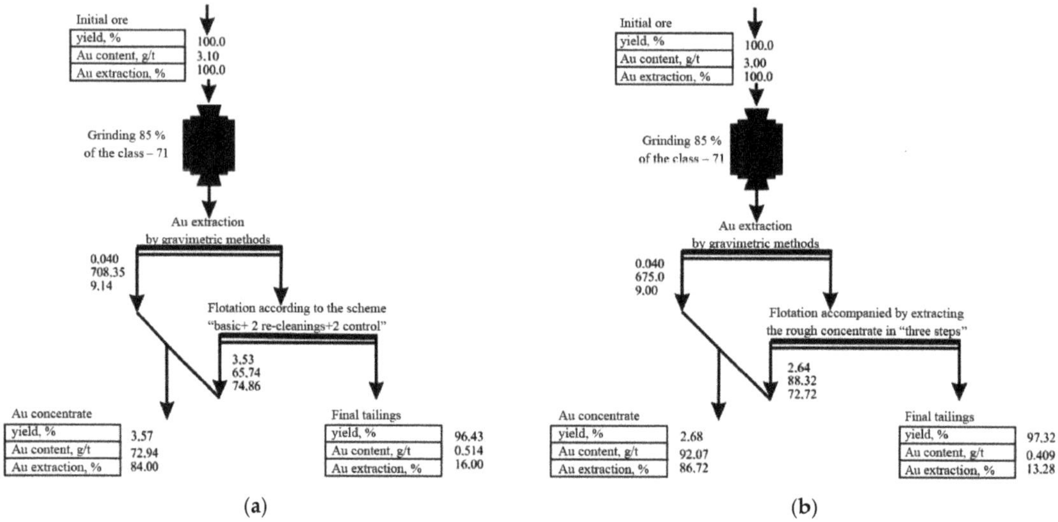

Figure 12. Basic block diagrams of flotation of ores from the Olimpiadinsky deposit according to the basic technology (**a**) and that accompanied by a rougher concentrate isolation in "three steps" (**b**).

The basic scheme of ore flotation (Figure 12a) included a flotation cycle of primary ore and a flotation cycle of industrial products. The first cycle consisted of the main flotation operation involving two re-cleanings of an isolated rougher concentrate, as well as two control flotation operations on tailings of a basic flotation operation. Gold reflotation was realized from an industrial product consisting of tailings I of re-cleaning and concentrate I of control flotation.

After optimizing the reagent flotation mode according to the basic scheme, an experiment on six parallel ore charges was conducted on the principle of a continuous process (when a movement direction of intermediate products was adopted in a canonical flotation scheme). The gold content in tailings of five–six ore charges was stabilized (after some growth observed in previous charges), which indicated that a closed cycle mode had been reached. In the control experiment, gold extraction into a concentrate from six ore charges was 84.0% when the concentrate quality was 65.74 g/t of Au.

When switching to the technology of isolating a rougher concentrate in "three steps" (Figure 12b), the increase in gold extraction was 2.72% (the gold content in the tailings decreased from 0.514 to 0.409 g/t) when the concentrate quality increased from 72.94 to 97.07 g/t. An important technological effect of using the new technology is a 24.9% decrease, thus relatively a concentrated yield.

The gold distribution by size classes of concentrates obtained during the experiment was established by sedimentometric and optical–geometric analysis methods (Figure 13).

The results of sedimentometric and optical–geometric analyses of flotation concentrates (Figure 13) allow for the conclusion that the set goal, i.e., to increase the extraction of gold microdispersions, has been achieved.

Improving the efficiency of flotation provides a solution to the problem of rational environmental management, which allows not only a technical and economic effect to be obtained, but also respect for the environment [65].

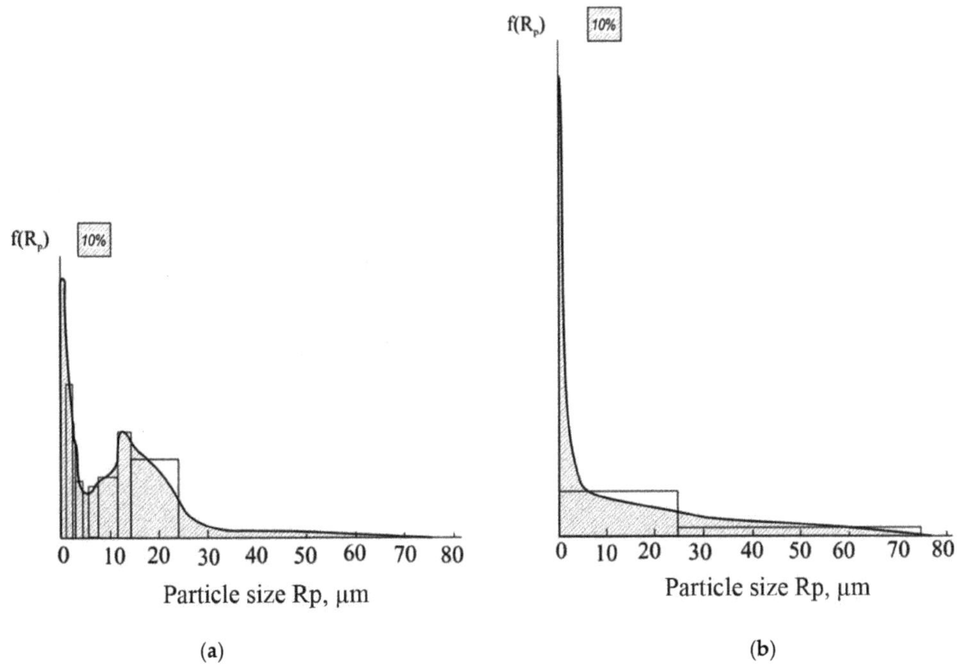

Figure 13. Results of determining dispersity of concentrates isolated by means of basic (**a**) and newly developed (**b**) technologies.

5. Conclusions

This work is focused on increasing the extraction of microdispersions of minerals by flotation.

Employing the example of flotation of two gold-bearing ores, the possibility of using high velocity and the probability of the sticking of fine particles to large ones to increase the extraction of finely dispersed gold, when introducing carrier minerals into a pulp in the form of a rougher concentrate, has been shown.

Carrier minerals are isolated from three parallel pulp streams, and the initial feed of the subsequent pulp stream is mixed with the rougher (recycled) concentrate isolated from the previous pulp stream. The ready-made rougher concentrate is isolated from the third pulp stream in a mode of flotation with the air mixture, containing hot water vapor, and is sent for recleaning.

During flotation with the vapor–air mixture, the stability of wetting films is determined by the surface flow of the liquid caused by a temperature-dependent surface-tension gradient (and the shear stress associated with it). The influence of such flows on the thinning rate of wetting films has been proposed to be considered in the form of correction to a sliding length of a liquid in a hydrophobic gap. A correction value is expressed in fractions of a critical thickness of a wetting film: at these distances between a particle and a bubble, the influence of surface forces of structural origin begins to manifest itself: forces of hydrophobic attraction and those of hydrophilic repulsion.

Author Contributions: Conceptualization, S.I.E. and A.I.K.; methodology, V.V.K. and A.M.; formal analysis, V.V.K. and A.M.; investigation, V.V.K. and A.M.; data curation, A.V.K. and N.N.B.; writing—original draft preparation, S.I.E., N.S.G. and A.F.P.; writing—review and editing, S.I.E. and A.I.K.; supervision, S.I.E.; project administration, N.S.G. and A.F.P.; visualization—A.V.K. and N.N.B. All authors have read and agreed to the published version of the manuscript.

Funding: This research received no external funding.

Institutional Review Board Statement: Not applicable.

Informed Consent Statement: Not applicable.

Data Availability Statement: The data presented in this study are available from the corresponding authors upon reasonable request.

Conflicts of Interest: The authors declare no conflicts of interest.

References

1. Ivanik, S.A.; Ilyukhin, D.A. Flotation extraction of elemental sulfur from gold-bearing cakes. *J. Min. Inst.* **2020**, *242*, 202–208. [CrossRef]
2. Alexandrova, T.N. Complex and deep processing of mineral raw materials of natural and technogenic origin: State and prospects. *J. Min. Inst.* **2022**, *256*, 503–504.
3. Alexandrova, T.N.; Romashev, A.O.; Kuznetsov, V.V. Development of a methodological approach to determining the flotation capacity of thinly impregnated sulfides. *Obogashchenie Rud* **2020**, *2*, 9–14. [CrossRef]
4. Mitrofanova, G.V.; Marchevskaya, V.V.; Taran, A.E. Flotation extraction of the titanium concentrate from apatite-nepheline-titanite ores of anomalous zones of Khibiny deposits. *J. Min. Inst.* **2022**, *256*, 560–566. [CrossRef]
5. Li, D.; Wang, H.; Li, C.; Liang, Y.; Yan, X.; Zhang, H. Determination and modulation of the typical interactions among dispersed phases relevant to flotation applications: A review. *Adv. Colloid Interface Sci.* **2021**, *288*, 102359. [CrossRef] [PubMed]
6. Pan, L.; Jung, S.; Yoon, R.-H. A fundfmental study on the role of collector in the kinetics of bubble-particle interaction. *Int. J. Miner. Process.* **2012**, *106–109*, 37–41. [CrossRef]
7. Wang, J.; Yoon, R.-H.; Morris, J. AFM surface force measurements conducted between gold surface treated in xanthate solutions. *Int. J. Miner. Process.* **2013**, *122*, 13–21. [CrossRef]
8. Huang, K.; Yoon, R.-H. Surface forces in the thin liquid films (tlfs) of water confined between n-alkane drops and hydrophobic gold surfaces. *Langmuir* **2019**, *35*, 15681–15691. [CrossRef]
9. Dziadkowiec, J.; Javadi, S.; Royne, A.; Bratvold, J.E.; Nilsen, O. Surface forces apparatus measurements of interactions between rough and reactive calcite surfaces. *Langmuir* **2018**, *34*, 7248–7263. [CrossRef]
10. Abreu, S.B.; Skinner, W.; Addai-Mensah, J.; Zanin, M. The influence of pyriite content on the flotation of chalcopyrite/pyrite mixtures. *Miner. Eng.* **2014**, *55*, 87–95.
11. Xie, L.; Wang, J.; Lu, Q.; Hu, W.; Zeng, H. Surface interaction mechanisms in mineral flotation: Fundamentals, measurements, and perpectives. *Adv. Colloid Interface Sci.* **2021**, *295*, 102491. [CrossRef] [PubMed]
12. Arriagada, S.; Acuna, C.; Vera, M. New technology to improve the recovery of fine particles in froth flotation based on using hydrophobized glass bubbles. *Miner. Eng.* **2020**, *156*, 106364. [CrossRef]
13. Liu, S.; Xie, L.; Liu, G.; Zhang, H.; Zeng, H. Understanding the hetero-aggregation mechanism among sulfide and oxide mineral particles driven by bifuctional surfactants: Intensification flotation of oxide minerals. *Miner. Eng.* **2021**, *169*, 106928. [CrossRef]
14. Teisala, H.; Butt, H.J. Hierarchical structures for superhydrophobic and superoleophobic surfaces. *Langmuir ACS J. Surf. Colloids* **2019**, *35*, 10689–10703. [CrossRef] [PubMed]
15. Mishchuk, N.A. The model of hydrophobic attraction in the framework of classical DLVO forces. *Adv. Colloid Interface Sci.* **2011**, *168*, 149–166. [CrossRef] [PubMed]
16. Christenson, H.K.; Claesson, P.M. Direct measurements of the force between hydrophobic surfaces in water direct measurements of the force between hydrophobic surfaces in water. *Adv. Colloid Interface Sci.* **2001**, *91*, 391–436. [CrossRef]
17. Krasowska, M.; Malysa, K. Wetting films in attachment of the colliding bubble. *Adv. Colloid Interface Sci.* **2007**, *134–135*, 138–150. [CrossRef]
18. Englert, A.H.; Krasowska, M.; Formasiero, D.; Raiston, J.; Rubio, J. Interaction force between an air bubble and a hydrophilic spherical particle in water, measured by the colloid probe technique. *Int. J. Miner. Process.* **2009**, *92*, 121–127. [CrossRef]
19. Nguyen, A.; Nalaskowski, J.; Miuller, J.D.; Butt, H.-J. Attraction between hydrophobic surfaces studied by atomic force microscopy. *Int. J. Miner. Process.* **2003**, *72*, 215–225. [CrossRef]
20. Verrelli, D.I.; Koh, P.T.L.; Bruckard, W.J.; Schwarz, M.P. Variations in the induction period for particle–bubble attachment. *Miner. Eng.* **2012**, *36–38*, 219–230. [CrossRef]
21. Verrelli, D.I.; Albijanic, B. A comparison of methods for measuring the induction time. *Miner. Eng.* **2015**, *80*, 8–13. [CrossRef]
22. Smith, A.M.; Borkovec, M.; Trefalt, G. Forces between solid surfaces in aqueous electrolyte solutions. *Adv. Interface Sci.* **2020**, *275*, 102078. [CrossRef] [PubMed]
23. Sedev, R.; Exerowa, D. DLVO and non-DLVO surface forces in foam films from amphiphilic block copolymers. *Adv. Colloid Interface Sci.* **1999**, *83*, 111–136. [CrossRef]
24. Pshenin, V.; Zaripova, N.; Zaynetdinov, K. Modeling of the crude oil (or petroleum products) vapor displacement during rail tanks loading. *Pet. Sci. Technol.* **2019**, *37*, 2435–2440. [CrossRef]
25. Guo, H.; Kovscek, A.R. Investigation of the effects of ions on short-range non-DLVO forces at the calcite/brine interface and implications for low salinity oil-recovery processes. *J. Colloid Interface Sci.* **2019**, *552*, 295–311. [CrossRef]

26. Grasso, D.; Subramaniam, K.; Butkus, M.; Strevett, K.; Bergendahl, J. A review of non-DLVO interactions in environmental colloidal systems. *Rev. Environ. Sci. Biotechnol.* **2002**, *1*, 17–38. [CrossRef]
27. Bal, V. Stability characteristics of nanoparticles in a laminar linear shear flow in the presence of DLVO and non-DLVO forces. *Langmuir ACS J. Surf. Colloids* **2019**, *35*, 11175–11187. [CrossRef]
28. Sendner, C.; Horinek, D.; Bocquet, L.; Netz, R.R. Interfacial Water at Hydrophobic and Hydrophilic Surfaces: Slip, viscosity, and diffusion. *Langmuir* **2009**, *25*, 10768–10781. [CrossRef]
29. Miller, J.D.; Wang, X.; Jin, J.; Shrimali, K. Interfacial water stucture and the wetting of mineral surfaces. *Int. J. Miner. Process.* **2016**, *156*, 62–68. [CrossRef]
30. Boinovich, L.B.; Emelyanenko, A.M. Forces due to dynamic structure in thin liquid films. *Adv. Colloid Interface Sci.* **2002**, *96*, 37–58. [CrossRef]
31. Gunko, V.M.; Turov, V.V.; Bogatyrev, V.M.; Zarko, V.I.; Chuiko, A.A. Unusual properties of water at hydrophilic/hydrophobic interfaces. *Adv. Colloid Interface Sci.* **2005**, *118*, 125–132. [CrossRef]
32. McKee, C.; Walz, J.Y. Interaction forces between colloid particles in a solution of like-charged, adsorbing nanoparticles. *J. Colloid Interface Sci.* **2012**, *365*, 72–80. [CrossRef] [PubMed]
33. Feuillebois, F.; Bazant, M.Z.; Vinogradova, O.I. Effective slip over superhydrophobic surfaces in thin channels. *Phys. Rev. Lett.* **2009**, *102*, 026001. [CrossRef] [PubMed]
34. Yakubov, G.E.; Butt, H.-J.; Vinogradova, O.I. Interaction forces between hydrophobic surfaces. Attractive jump as an indication of formation of "stable" submicrocavities. *J. Phys. Chem. B* **2000**, *104*, 3407–3410. [CrossRef]
35. Ya Malkin, A.; Patlazhan, S.A. Wall slip for complex liquids—Phenomenon and its causes. *Adv. Colloid Interface Sci.* **2018**, *257*, 42–57. [CrossRef] [PubMed]
36. Ke, S.; Xiao, W.; Quan, N.; Zhang, L.; Hu, J.; Dong, Y. Formation and stability of bulk nanobubbles in different solutions. *Langmuir ACS J. Surf. Colloids* **2019**, *35*, 5250–5256. [CrossRef] [PubMed]
37. Esteso, V.; Carretero-Palacios, S.; Míguez, H.; Thiyam, P.; Parsons, D.F.; Brevik, I.; Boström, M. Trapping of gas bubbles in water at a finite distance below a water-solid interface. *Langmuir ACS J. Surf. Colloids* **2019**, *35*, 4218–4223. [CrossRef]
38. Pashkevich, M.A.; Danilov, A.S. Environmental safety and sustainable development. *J. Min. Inst.* **2023**, *260*, 153–154.
39. Malyukova, L.S.; Martyushev, N.V.; Tynchenko, V.V.; Kondratiev, V.V.; Bukhtoyarov, V.V.; Konyukhov, V.Y.; Bashmur, K.A.; Panfilova, T.A.; Brigida, V. Circular Mining Wastes Management for Sustainable Production of *Camellia sinensis* (L.) O. Kuntze. *Sustainability* **2023**, *15*, 11671. [CrossRef]
40. Fedotov, P.K.; Senchenko, A.E.; Fedotov, K.V.; Burdonov, A.E. Studies of enrichment of sulfide and oxidized ores of gold deposits of the Aldan shield. *J. Min. Inst.* **2022**, *242*, 218. [CrossRef]
41. Yamilev, M.Z.; Azat Masagutov, M.; Nikolaev, A.K.; Pshenin, V.V.; Zaripova, N.A.; Plotnikova, K.I. Modified equations for hydraulic calculation of thermally insulated oil pipelines for the case of a power-law fluid. *Sci. Technol. Oil Oil Prod. Pipeline Transp.* **2021**, *11*, 388–395. [CrossRef]
42. Boinovich, L.B. Long-range surface forces and their role in the development of nanotechnology. *Adv. Chem.* **2007**, *76*, 510–528.
43. Wang, J.; Yoon, R.-H.; Eriksson, J.C. Excess thermodynamic properties of thin water films confined between hydrophobized gold surfaces. *J. Colloid Interface Sci.* **2011**, *364*, 257–263. [CrossRef] [PubMed]
44. Albijanic, B.; Ozdemir, O.; Nguyen, A.V.; Bradshaw, D. A review of induction and attachment times of wetting thin films betweem air bubbles and particles and its relevance in the separation of particles by flotation. *Adv. Colloid Interface Sci.* **2010**, *159*, 1–21. [CrossRef] [PubMed]
45. Esipova, N.E.; Rusanov, A.I.; Sobolev, V.D. Temperature dependence of the contact angle of a sessile bubble at the water-silicon interface. *Colloid. J.* **2020**, *82*, 569–575. [CrossRef]
46. Semenov, E.V. Calculation of the probability of collision of suspended particles in a suspension stream. *Colloid. J.* **1981**, *VXLIII*, 912–917.
47. Denisov, E.F.; Yamilev, M.Z.; Tigulev, E.A.; Pshenin, V.V. Analysis of the current level of technologies for determining the location of non-metallic underground services. *Neft. Khozyaystvo—Oil Ind.* **2022**, *2022*, 121–125. [CrossRef]
48. Li, C.; Li, D.; Zhang, H. Surface nanobubbles on the hydrophobic surface and their implication to flotation. *Int. J. Miner. Metall. Mater.* **2022**, *29*, 1491–1492. [CrossRef]
49. Jadhav, A.; Barigou, M. Bulk nanobubbles or not nanobubbles: That is the question. *Langmuir* **2020**, *36*, 1699–1708. [CrossRef]
50. Huang, W.; Huang, J.; Guo, Z.; Liu, W. Icehobic/anti-icing properties of superhydrophobic surfaces. *Adv. Colloid Interface Sci.* **2022**, *304*, 102658. [CrossRef]
51. Li, M.; Xing, Y.; Zhu, C.; Liu, Q.; Gui, X. Effect of roughness on wettability and flotability: Based on wetting film drainage between bubbles and solid surface. *Int. J. Miner. Sci. Technol.* **2022**, *32*, 1389–1396. [CrossRef]
52. Kondratiev, S.A.; Khamzina, T.A. Assessment of collecting activity of physically sorbed reagents on the example of easily floatable coking coal sludge. *J. Min. Inst.* **2022**, *256*, 549–559. [CrossRef]
53. Evdokimov, S.I.; Golikov, N.S.; Zadkov, D.A.; Voitovich, E.V.; Kondratiev, V.V.; Petrovskiy, A.A.; Konyukhov, V.Y.; Gladkikh, V.A. Studying the Flotation of Gold-Bearing Ores Using Carrier Minerals. *Minerals* **2024**, *14*, 88. [CrossRef]
54. Evdokimov, S.I.; Panshin, A.M.; Solodenko, A.A. Minerallurgy. In *2 Volumes, V. 2. Advances of Flotation*; OOO NPKP "MAVR": Vladikavkaz, Russia, 2010; 992p.

55. Evdokimov, S.I.; Gerasimenko, T.E. Determination of rational steam consumption during flotation of apatite-nepheline ores with a steam-air mixture. *J. Min. Inst.* **2022**, *256*, 567–578. [CrossRef]
56. Vinogradova, O.I. Drainage of a thin liquid film confined between hydrophobic surfaces. *Langmuir* **1995**, *11*, 2213. [CrossRef]
57. Andrienko, D.; Dunweg, B.; Vinogradova, O.I. Boundary slip as a result of a prewetting transition. *J. Chem. Phys* **2003**, *119*, 13106. [CrossRef]
58. Strateichuk, D.M.; Martyushev, N.V.; Klyuev, R.V.; Gladkikh, V.A.; Kukartsev, V.V.; Tynchenko, Y.A.; Karlina, A.I. Morphological Features of Polycrystalline $CdS_{1-x}Se_x$ Films Obtained by Screen-Printing Method. *Crystals* **2023**, *13*, 825. [CrossRef]
59. Bosikov, I.I.; Martyushev, N.V.; Klyuev, R.V.; Tynchenko, V.S.; Kukartsev, V.A.; Eremeeva, S.V.; Karlina, A.I. Complex Assessment of X-ray Diffraction in Crystals with Face-Centered Silicon Carbide Lattice. *Crystals* **2023**, *13*, 528. [CrossRef]
60. Alexandrova, T.N.; O'Connor, C. Processing of platinum group metal ores in Russia and South Africa: Current state and prospects. *J. Min. Inst.* **2020**, *244*, 462–473. [CrossRef]
61. Boldyrev, D.V.; Anufrieva, S.D. Approximation of the temperature dependence of the liquid viscosity. *Innov. Sci.* **2015**, *11*, 18–20.
62. Alexandrova, T.N.; Elbendari, A.M. Increasing the efficiency of phosphate ore processing using flotation method. *J. Min. Inst.* **2021**, *248*, 260–271. [CrossRef]
63. Yelemessov, K.; Baskanbayeva, D.; Martyushev, N.V.; Skeeba, V.Y.; Gozbenko, V.E.; Karlina, A.I. Change in the Properties of Rail Steels during Operation and Reutilization of Rails. *Metals* **2023**, *13*, 1043. [CrossRef]
64. Boinovich, L.B.; Emelianenko, A.M. Hydrophobic materials and coatings: Principles of creation, properties and application. *Adv. Chem.* **2008**, *77*, 619–638.
65. Antoninova, N.Y.; Sobenin, A.V.; Usmanov, A.I.; Shepel, K.V. Assessment of the possibility of using iron-magnesium production waste for wastewater treatment from heavy metals (Cd^{2+}, Zn^{2+}, Co^{2+}, Cu^{2+}). *J. Min. Inst.* **2023**, *260*, 257–265. [CrossRef]

minerals

MDPI

Article

Studying the Flotation of Gold-Bearing Ores Using Carrier Minerals

Sergei Ivanovich Evdokimov [1], Nikolay S. Golikov [2,*], Denis A. Zadkov [2,*], Elena V. Voitovich [3], Viktor V. Kondratiev [4], Aleksey A. Petrovskiy [4], Vladimir Yu. Konyukhov [5] and Vitaliy A. Gladkikh [6]

1 Department of Mineral Processing, North Caucasian Institute of Mining and Metallurgy, State Technological University, 362011 Vladikavkaz, Russia; eva-ser@mail.ru
2 Department of Mechanical Engineering, Saint-Petersburg Mining University, 199106 St. Petersburg, Russia
3 Department of Industrial and Civil Engineering, Moscow Polytechnic University, 107023 Moscow, Russia; e.voitovich@mail.ru
4 Laboratory of Geochemistry of Ore Formation and Geochemical Methods of Prospecting, A. P. Vinogradov Institute of Geochemistry of the Siberian Branch of the Russian Academy of Sciences, 664033 Irkutsk, Russia; imz@mail.ru (V.V.K.); begimotus@mail.ru (A.A.P.)
5 Department of Automation and Control, Irkutsk National Research Technical University, 664074 Irkutsk, Russia; konyukhov_vyu@mail.ru
6 Stroytest Research and Testing Center, Moscow State University of Civil Engineering, 26, Yaroslavskoye Shosse, 129337 Moscow, Russia; vgladkich_87@mail.ru
* Correspondence: golikov_ns@pers.spmi.ru (N.S.G.); zadkov_da@pers.spmi.ru (D.A.Z.)

Abstract: This work is aimed at the analysis of the development of flotation technology by applying carrier minerals. Based on the concepts of continuum mechanics, a theoretical analysis of the influence of the carrier minerals (wall) on the motion of a single solid particle is provided, taking into account their hydrodynamic interaction (in the case of low Reynolds numbers). A correction was obtained in the form of a ratio of the particle size to its distance from the wall to take into account the influence of the wall on the hydrodynamic force acting on the particle. The influence of the wall is manifested through a rapid approximation of the liquid vortex flow in the gap between the solid wall and the particle to the steady-state mode, accompanied by the suppression of the transverse movement of particles. When the liquid slides along a wall-mounted gas–liquid layer with a reduced viscosity, the liquid flow increases in the interfacial gap, which can be analyzed by a dimensionless correction that includes values describing the properties of a continuous medium (dynamic viscosity) and a disperse phase (geometric particle size). The reason for the decrease in the induction time when gold grains adhere to each other is assumed to be due to the forces of hydrophobic attraction (when the grains have a mirror-smooth surface) and the sliding of the flow along the hydrophobic surface of the particles along the gas layer (when the grains have a rough surface). When polydisperse particles are aggregated, the threshold energy of the fast coagulation was established to be lower than that arising during the interaction of monodisperse particles, whose aggregation requires a large depth of the potential pit. Performing natural experiments on the ore using a rougher concentrate as a carrier material showed that the concentrate yield decreases by 20.52% rel. In the second case, the gold extraction was higher by 4.69% abs. While maintaining the achieved level of gold extraction, the double mixing of the rougher concentrate and the initial feed increased the gold content in the rougher concentrate from 4.97 to 6.29 g/t.

Keywords: gold-bearing ore; flotation; fine particles; flotation scheme; carrier minerals; wall correction; slip correction; field experiments

Citation: Evdokimov, S.I.; Golikov, N.S.; Zadkov, D.A.; Voitovich, E.V.; Kondratiev, V.V.; Petrovskiy, A.A.; Konyukhov, V.Y.; Gladkikh, V.A. Studying the Flotation of Gold-Bearing Ores Using Carrier Minerals. *Minerals* **2024**, *14*, 88. https://doi.org/10.3390/min14010088

Academic Editors: Ahmad Hassanzadeh, Luis A. Cisternas and Fardis Nakhaei

Received: 3 November 2023
Revised: 3 January 2024
Accepted: 8 January 2024
Published: 11 January 2024

1. Introduction

Modern gold mining companies in Russia and worldwide are facing problems with optimizing production costs. These are caused by intense price competition on the one hand and deteriorating the mining and geological conditions on the other hand, subsequently decreasing the gold content of the ore, increasing the depth of development, and reducing unallocated reserves [1]. One of the current trends of sustainable economic development existing in the country, taking into account the need to increase the reserves and the consumption of mineral resources, is the introduction of innovative technologies that use a wide range of minerals [2].

A significant part of gold ore resources are represented by sulfide ores, which are difficult to obtain via extraction and require the application of highly efficient new technologies for processing low-grade and refractory materials [3]. Extracting the ores and concentrates, in which fine gold is associated with pyrite or arsenopyrite, needs much more complex technological schemes due to the fact that, during grinding, such gold is only slightly revealed, and the bulk of it remains in the sulfides. It is not dissolved during cyanidation, and in the processes of gravity and flotation enrichment, it is extracted together with the carrier minerals [4].

In recent decades, the gold content of ores has significantly decreased, and the share of poor and rather difficult-to-uncover, refractory, gold-bearing raw materials involved in processing has increased to 40% [5].

Improving the efficiency of processing hard-to-process gold-bearing ores and concentrates, which are characterized by complex mineralogical and geochemical compositions, submicroscopic grain sizes, heterogeneous textures, and a variety of genetic processes of ore formation, leads to a fine dispersion of gold particles in waste rock minerals. This also requires the assessment of structural and textural parameters [6] and the modeling of technological processes [7], the possible use of selective disintegration and separation technologies [8], and reducing the amount of gold in waste rock minerals.

The main enrichment process for such ores is flotation, which is a complex, multiphase process, and the works of many researchers have focused on optimizing this operation as a key link in enrichment technology [9].

The ores in which finely disseminated gold is associated with sulfide minerals, such as pyrite, arsenopyrite, or antimonite, are processed using complex technological schemes. In the case of fine dissemination of gold in sulfide minerals, the very fine grinding of ores is required. However, in the case of a significant decrease in the particle size, the hydrodynamic field of the bubbles that flow up significantly reduces the efficiency and selectivity of the flotation process [10–12]. Due to inertia forces, coarse particles are known to approach the surface of bubbles along a rectilinear trajectory, either by the impact or by the effect followed by sliding along the surface of the bubbles. When the particle size decreases, its hydrodynamic interaction with the bubbles depends on the fact that, during the contact time (the time of the particle's movement from the upper pole of a bubble to its equator and below), the trajectory of the particle's movement and the liquid flow lines flowing around the bubble curve become twisted and coincide. The inertia-free interaction with a bubble, i.e., displacement along with a liquid flow without contact with the surface of a bubble up to its lower hemisphere, is the main reason for the loss of fine particles during flotation, including gold [13,14]. Regardless of the nature of the liquid flow near the surface of the bubble (Reynolds number is $Re = (2\,R_b\,v_b/\nu) >> 1$) or viscous Stokes ($Re < 1$), there is a quadratic dependence of the collision efficiency E on the value of the R_p/R_b ratio. In this case, v_b is the velocity of the bubble with radius R_b, ν is the kinematic viscosity of the liquid, and R_p is the particle size.

If the high losses of the fine particles are caused by a decrease in the collision efficiency and a decrease in the particle size due to hydrodynamic interactions, then the deterioration in the selectivity of their separation is associated with the developing balance of the surface forces [15–19] acting on the separated particles.

An obvious (but technically far from being simple) solution to the problem of fine particle flotation is the use of nanobubbles [20–23]. Another promising direction in the flotation of fine particles is their preliminary aggregation [24,25].

The adhesion of the fine particles to the surface of the bubble is possible under the condition of their convergence to distances coinciding with the long-range potential minimum.

But the collision followed by the subsequent adhesion of the particle to the surface of the bubble without the formation of a contact angle is indiscriminate, and, in the case of the selective flotation of ores, it reduces the efficiency of the mineral separation process.

To increase the flotation efficiency of the microdispersions of minerals, conditions are created for their preliminary aggregation with coarse minerals [26,27]. These technologies are based on the fact that the rate of the adhesion of the fine particles to coarse ones is 400–500 times higher than the rate of the aggregation of the fine particles among themselves [28,29]. And the frequency of the collision of the particles with significantly different sizes tends to unity, while in the case of fine particles, it is negligible [30].

We can assume that during the flotation processes, the particles move in an unlimited liquid, since the value of the ratio is $R/l >> 1$. Here, R and l are the particle size and the distance from the center of the particle to the wall of the apparatus or to a single coarse particle in the case of a polydisperse system, respectively. The liquid flow lines induced by the motion of the single particles in the unlimited liquid are close to infinity [31,32]. However, when several particles move together, their hydrodynamic interaction manifests itself in the fact that the movement of each particle in the group is influenced by the movement of neighboring particles [33–35]. As a result, a single particle is influenced by a greater (than that determined by the Stokes formula) viscous drag force than that which influences each particle in the group. When the value of the R/l ratio decreases, the effect of the volume substitution and the alternating motion of the liquid caused by it, in which the particle participates, begins to manifest itself. The change in the hydrodynamic resistance during the transition from the movement of a single particle in the unlimited liquid to the movement of a group of the particles (or near the wall) was quantitatively described by Brenner [36,37]. The convergence resistance is lower in the case of fine particles, which may be the reason for their effective adhesion to coarse particles [26–28].

During the hydrodynamic interaction of the hydrophobic particles, the sliding of the liquid decreases the hydrodynamic resistance to the liquid flow in the gap between the particles. This effect of increasing (against the expected one by calculation when meeting the adhesion condition) the liquid flow is a consequence of its sliding along the gas layer (or nanosized gas bubbles) owing to a large difference in the dynamic viscosity of water and gas (η_l/η_g~50) [38–40].

Gold is effectively recovered by flotation after loading xanthogenate [41–46]. Valderrama L. demonstrated [23,47] that, due to the adhesion to coarse particles, the extraction of fine gold by xanthogenate is increased by 24%, and the retention is increased by 50% when the flotation rate is increased 3–4 times. Two flotation peaks were revealed when the shear energy was 0.5–2.0 and 3.0–4.0 kW/m^3.

The overall positive effect of using analogue technologies based on the aggregation of polydisperse particles is an increase in the extraction of fine-dispersed fractions of the target component. For example, hydrophobic glass beads [48], paraffin, organic polymers, magnetized iron (or magnetite) isolated from ores, and specially prepared monominerals are used [27,49]. However, the development of the attractive forces between polydisperse particles is possible only at a high concentration of coarse particles [50]; in the presence of a large number of microdispersions, the conditions established for the extraction of coarse particles become critical. Therefore, up to 200% (of the mass of the fine particles) of carrier minerals must be introduced into the flotation system [51]. Such flotation carrier platforms are expendable. The organization of the regeneration of their surface for the purpose of reuse in most cases is inefficient and increases capital expenditures and operating costs for conducting the flotation process.

This work is focused on studying the techniques of improving the technology of fine gold flotation using carrier minerals. To improve the technical and economic performance of this technology, the efficiency of using a maximally homogeneous material, including fine gold, i.e., the rougher concentrate isolated from a part of the ore, as carrier minerals is proven [51,52]. The rougher concentrate is the most flotation-active part of the raw material enriched with gold. When it is mixed with the initial feed, an increase in the fine gold extraction is possible due to two effects: the adhesion of the fine particles to the coarse ones and an increase in the gold content in the flotation feed, which is unattainable when using inert materials [30].

The fact that the flotation of the fine particles is a qualitatively new process is proven by using special terms, such as "microflotation" [26] and the flotation of "inertialess particles" [30], when discussing the results of researching it. In terms of modern colloidal chemistry, microflotation is an orthokinetic heterocoagulation. This allows for interpreting the experimental data on the aggregation of hydrophilic particles using two approaches that form the basis of the DLVO theory: dispersion (van der Waals) forces and ion-electrostatic interactions. However, in the case of the aggregation and the flotation of hydrophobic particles, in order to harmonize the theory and the experiment, it is necessary to apply new mechanisms of long-range surface forces. They are collectively called "non-DLVO" forces in the literature [53–55] and are taken into account by the extended DLVO theory (XDLVO [56–58]). The relationship between the particle adhesion to the bubble during flotation and the interaction forces caused by the altered structure of the liquid present in the wetting film, i.e., hydrophobic attractive forces, is shown in [41,59], including during gold flotation with xanthogenate [44,60,61].

The achieved level of understanding the physical regularities of the flotation of fine particles has allowed for developing the technologies for extracting minerals from ores, reaching high technical and economic indicators. However, the problem of reducing the losses of mineral microdispersions involving flotation waste is still relevant.

The purpose of the work is to develop a flotation technology for gold-bearing ores based on the revealed patterns of the hydrodynamic interaction of polydisperse hydrophobic surfaces used as carrier minerals of fine gold.

2. Materials and Equipment

2.1. Research Object

The research object was composite ore samples obtained from two sites of the Bereznyakovsky gold ore field. The samples were composed of a witness core of the wells and ditches (77 samples 50 mm in size, whose total weight was 143 kg) located at various hypsometric levels. When they were processed at the gold recovery factory, the gold size in the flotation tailings was characterized by the histograms shown in Figure 1.

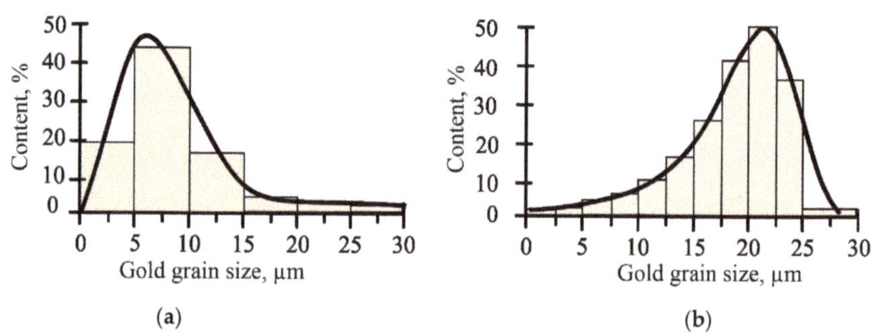

Figure 1. Experimental curves of the grain size distribution of gold in the flotation tailings during the flotation of the ores of the first (**a**) and second (**b**) samples obtained at the gold recovery factory.

The histograms of the size distribution of gold grains are based on the results of measuring 100 grains in each sample. The histograms of the size distribution of the gold grains in the flotation tailings of both ore samples have a pronounced asymmetric nature.

The mineralogical studies were conducted using a Nikon Eclipse LV 100 Pol microscope. The X-ray diffraction phase analysis (DRON-3M, Cu-K$_{\alpha}$ radiation) and the mineralogical studies (Nikon Eclipse Lv 100n Pol microscope, Nikon Instruments Inc., Tokyo, Japan) allowed for revealing the fact that 35%–40% of the material of the two samples belonged to quartz and about 50% of the material was represented by pyrophilite, illite, sericite, and paragonite. The bulk of the sulfides that were present in the samples was composed of pyrite (up to 10% by weight). Tahr ores, secondary copper sulfides, and sulfosalts together amounted to no more than 0.6%, while sphalerite, chalcopyrite, and galena accounted for 0.10%–0.45% (sphalerite predominated, amounting to 0.4%). Pyrite grains (of idiomorphic, hypidiomorphic, and framboidal shapes) had a size ranging from 1–5 µm to 0.05–0.10 mm.

The gold in the ores was mainly accompanied by pyrite, recovered at the operating gold processing plant by flotation into a concentrate, which is subjected to cyanidation after the autoclave gold extraction.

2.2. Flotation Equipment

Full-scale tests of the flotation technology using the rougher concentrate as a carrier mineral were performed on a laboratory bench (Figure 1) [30].

The flotation method of extracting gold was chosen followed by constructing a flow diagram according to the jet principle [28–30]. In the experiments on flotation, a laboratory-scale plant with a square cross-section of 47 × 47 mm in size was used for the column flotation (Figure 2).

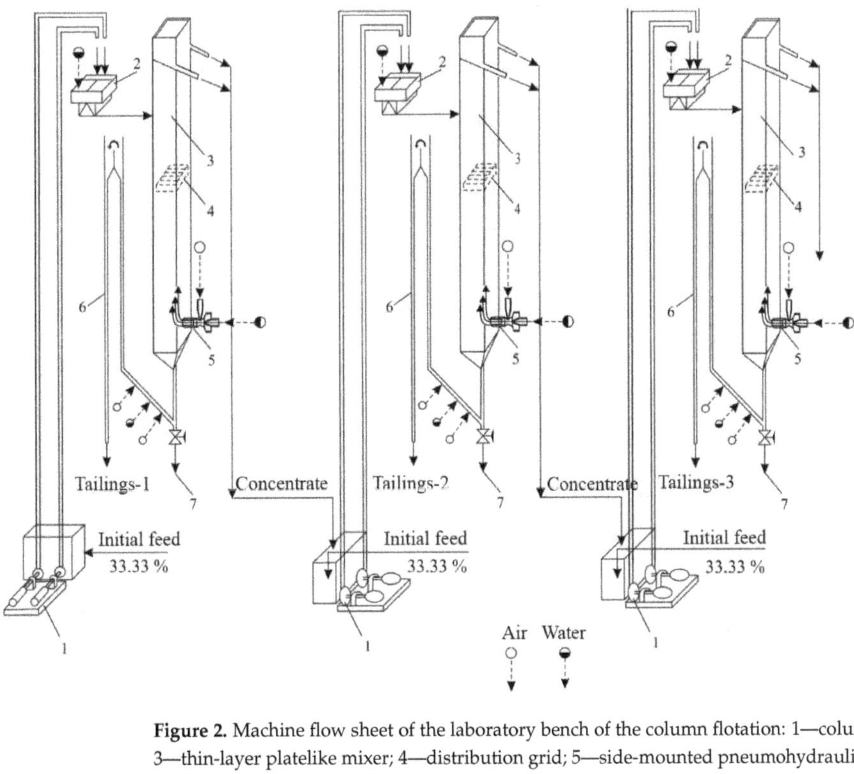

Figure 2. Machine flow sheet of the laboratory bench of the column flotation: 1—columns; 2—pump; 3—thin-layer platelike mixer; 4—distribution grid; 5—side-mounted pneumohydraulic aerator of the confusor–diffuser; 6—airlift; 7—Teflon gate.

The bench was based on three column-type flotation machines, each of which received 1/3 of the initial feed. The columns were interconnected by a rougher concentrate flow. The concentrate isolated on the first flotation column was mixed with the initial feed of the second flotation column. Then, the concentrate of the second flotation column was mixed with the initial feed of the third flotation column, and a ready-made rougher concentrate was obtained. Therefore, the ready-made rougher concentrate was isolated in three steps, using the concentrate material of the first two flotation columns as the carrier minerals.

The main part of the bench (Figure 1) consisted of three columns (1) made of corrosion-resistant steel (of the 8X18H10 type) with a square cross-section (47 × 47 mm) that was 2070 mm in height. The feeding depth was 767 mm. The initial feeding system included a sand pump (2) equipped with a thin-layer platelike mixer (3) of the Jones sampler type. The column was countercurrent and was supplied with the initial feed from above the distribution grid (4) and the gas phase from below. The air was supplied by a side-mounted pneumohydraulic aerator of the confusor–diffuser type (5). An airlift (6) served to unload the chamber product; during the emergency unloading of the column, the pulp was discharged through a Teflon gate (7).

Monominerals were floated in a countercurrent column-type flotator that was 64 mm in diameter and 1.7 m in height when the initial feed was supplied under the cleaning zone to a depth of 0.46 m. The flotator was equipped with a pneumohydraulic aerator of the confusor–diffuser type. The cylindrical mixing chamber was placed between the confusor (Ø was 1 mm) and the conical expanding diffuser (Ø was 5.1 mm). The diffuser outlet was closed with a polyurethane mesh, which allowed for calculating the size of the formed air bubbles. The total length of the aerator was 152 mm. Air was supplied to the mixing chamber by a compressor through a vertically mounted connecting pipe with a nozzle of Ø 1.47 mm at a rate of 1.35 m^3/min per 1 m^2 of the chamber cross-section. The aerator was installed in the bottom part of the column on its outer side and on the side.

The column capacity in terms of the initial feed was 1.5 m^3/min per 1 m^2 of the chamber section.

The column water balance allowed for the conclusion that at a washing flow rate of 0.34 m^3/min per 1 m^2 of the column section, the water flux flowing into the tailings exceeded the water flux flowing into the feed by 7%–8%, which made it possible to suppress the mechanical removal of non-floating minerals into the concentrate.

Potassium butyl xanthate was supplied at a rate of 25 mg/dm^3 for recleaning.

The air pressure at the entrance to the airlift was 0.14 MPa at a flow rate of 0.045 m^3 per 1 m^3 of the pulp removed into the tailings.

2.3. Flotation Mode

The base ore, which was 3 mm in size, was ground in a rod mill at a ratio of S:W = 1:1 to a grain size of 80% of a 71 μm class in the presence of sodium sulfide (112 g/t). After being activated with copper sulfate (15 g/t), the sulfides were floated with potassium butyl xanthogenate (85 g/t) and a foaming agent T-92 (35 g/t).

When performing the flotation, the reduced air velocity was 1.75×10^{-2} m/s at a flow rate of 3.85×10^{-5} m^3/s. The airflow rate was measured by a diaphragm rheometer (sequentially connected to the pneumatic system) or a gasometer of the UGIMETERS type (in some cases, by the volumetric method); the air pressure at the inlet to the ejector was measured by a mercury pressure gauge. The excess air pressure in the pneumatic system could be adjusted in the range from 1.1×10^{-2} to 1.4×10^{-2} MPa. The air supply to the ejector was controlled by an adjustable clamp. The indicators of the gas flowmeters were verified by calculating the static pressure drop of the flotation chamber atmosphere, as measured by a U-shaped meter.

The liquid workflow into the ejector was fed from an overflow tank (to maintain a constant level of filling with water) by a water pump through a flow meter equipped with fine-adjustment valves and pressure control implemented by pressure gauges installed at the inlet and outlet. A laboratory shut-off-and-control valve was mounted on the feed pipe; in the working jet, the water pressure was changed in the range from 20.01×10^{-2} to 25.20×10^{-2} MPa.

The washing water consumption was 0.4 m^3/min per 1 m^2 of the chamber section, which provided a 7%–8% excess of the water flow into the tailings compared to the amount of water flow into the feed and the removal of rock minerals that were mechanically trapped from the foam layer.

Air was supplied to the airlift by a compressor through a flow meter, a shut-off armature, and a pressure gauge.

During the flotation, the volumetric capacity of the flotation column was maintained at 1.59×10^{-5} m^3/s when the reduced pulp velocity was ~1×10^{-2} m/s.

3. Results and Discussion

3.1. Studying the Hydrodynamic Interaction of Polydisperse Solid Particles with Carrier Minerals under Flotation Conditions

During flotation using carrier minerals, the formation of a flotation complex should be preceded by the aggregation of polydisperse particles; the accumulation of fine gold occurs on its coarse particles. In order to preliminarily aggregate the polydisperse gold particles, the rougher concentrate isolated from a part of the initial ore is mixed with another part of the ore, and only then flotation is implemented.

At a sufficiently close distance from each other, the particles enter into a hydrodynamic interaction, which manifests itself in the perturbation of the fields of local liquid flows occurring near them. The factors influencing this interaction are, first of all, the sizes of the interacting particles, their velocity, the forces causing their movement, the orientation relative to each other, etc.

The subject of this study is the hydrodynamic interaction of polydisperse particles under flotation conditions, i.e., the adhesion of fine gold onto the coarse particles of a rougher concentrate (carrier minerals). The interaction of a coarse particle and a single solid particle moving along it (a micron-sized mineral) in the case of its Stokes flow is investigated.

Let us assume that the flow of the liquid is symmetrical with respect to the 0z axis and that, therefore, the resultant of forces F of the viscous resistance ($r_{r\theta}$) and pressure (t_{rr}) applied to the particle are also directed along the 0z axis and coincide with the direction of the liquid flow (Figure 3).

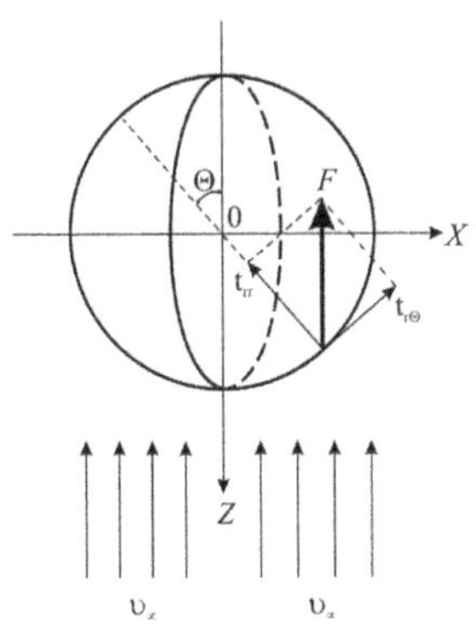

Figure 3. Components of the resistance force acting on the solid particle in the viscous liquid flow.

Integrating the sum of the projections of the pressure forces (t_{rr}) and viscous friction ($t_{r\theta}$) on the 0z axis over the entire surface of the particle provides:

$$\begin{aligned} F &= \int_S (-t_{rr}\cos\theta + t_{r\theta}\sin\theta)\, dS = \\ &= 2\pi \int_0^{\pi} (-t_{rr}\cos\theta + t_{r\theta}\sin\theta)\, R^2 \sin\theta\, d\theta = 3\pi\mu v_\infty R \int_0^{\pi} \sin\theta\, d\theta + \\ &+ 2\pi R^3 \rho g \int_0^{\pi} \cos^2\theta \sin\theta\, d\theta \end{aligned} \quad (1)$$

or

$$F = 6\pi\mu v_\infty R + \frac{4}{3}\pi R^3 \rho g. \quad (2)$$

Therefore, when:

$$Re = v_\infty R_p / \nu << 1 \quad (3)$$

from the side of the flow of the viscous unlimited liquid, the resistance force acts on a solid particle moving in it [36], which consists of two forces. First, this is the Stokes drag force:

$$F_{Stk} = 6\pi\eta v_\infty R_p \quad (4)$$

which tends to compensate for the velocity difference between the surrounding liquid and the particle, and, second, the inertial buoyancy force arises:

$$F_{Arh} = 4/3\,\pi R^3 \rho g. \quad (5)$$

In the case under consideration, the system of basic differential equations of fluid hydrodynamics, taking into account external volumetric forces (external mass forces are neglected) acting on the liquid, consists of the continuity equation:

$$\frac{d\rho}{dt} + \rho\, div\, \vec{v} = 0 \quad (6)$$

and the Stokes equation:

$$-\frac{1}{\rho}\nabla p + \nu\Delta\vec{v} = 0 \quad (7)$$

provided that the hydrodynamic force acts on a solid particle whose dimensions R_p are small compared to its distance from the wall l:

$$R/l << 1. \quad (8)$$

In this case, *Re*—Reynolds number; v_∞—typical velocity scale (an absolute value of the vector of the velocity of the incident flotation of the liquid); $\nu = \eta / \rho$, η—kinematic and dynamic viscosity; ρ—density; R_p—size of the body, streamlined by the liquid or the inner radius of the pipe through which the liquid is flowing; g—free-fall acceleration; t—time; $\vec{v}$—velocity vector; and p—hydrostatic pressure. When writing Equations (6) and (7), the following notations were used:

$$\nabla = \vec{i}\frac{\partial}{\partial x} + \vec{j}\frac{\partial}{\partial y} + \vec{k}\frac{\partial}{\partial z};\ \Delta\vec{v} = i \rightarrow \Delta v_x + \vec{j}\,\Delta v_y + \vec{k}\,\Delta v_z \quad (9)$$

where $\vec{i}$, $\vec{j}$, $\vec{k}$ are unit vectors along the directions of the *Ox*, *Oy*, *Oz* axes of the Cartesian coordinate system and Δ is the Laplace operator.

The inequality (8) corresponds to the condition of the near-hydrodynamic interaction between the coarse (wall) particle and the fine one moving at a speed of:

$$\vec{u} = \vec{i}\,u \quad (10)$$

where $i \rightarrow$ is the unit vector along the *Ox* axis.

Provided that the surface *S* of the fine particle is on the surface of the coarse particle (wall) Σ, we can write the boundary conditions as follows:

$$\vec{v}^{(1)}\Big|_S = \overline{u};\ \vec{v}^{(1)} \rightarrow 0\ when\ r \rightarrow \infty; \quad (11)$$

$$\vec{v}^{(2)}\Big|_\Sigma = -\vec{v}^{(1)};\ \vec{v}^{(2)} \rightarrow 0\ when\ r \rightarrow \infty; \quad (12)$$

$$\vec{v}^{(3)}\Big|_S = -\vec{v}^{(2)};\ \vec{v}^{(3)} \to 0\ when\ r \to \infty; \tag{13}$$

Taking into account the linearity of Equations (6) and (7), we may represent the fields of local pressures as the sum of the fields:

$$\vec{v}\,\beta^* = p = \vec{v}^{(1)}\beta^* + \vec{v}^{(2)}\beta^* + \vec{v}^{(3)}\beta^* + \cdots = p^{(1)} + p^{(2)} + p^{(3)} + \cdots \tag{14}$$

where β is the dimension sliding factor and velocities:

$$\vec{v} = \vec{v}^{(1)} + \vec{v}^{(2)} + \vec{v}^{(3)} + \cdots, \tag{15}$$

each member of which satisfies the boundary conditions (11)–(13).

From the side of the unlimited extent of the liquid, the particle moving in it is influenced by the initial field of the velocities $\vec{v}^{(1)}$ and the force corresponding to it:

$$\vec{F}^{(1)} = \overleftarrow{F}_\infty. \tag{16}$$

The boundary condition (12) is equivalent to the field of the velocities $\vec{v}^{(2)}$ neutralizing the initial field $\vec{v}^{(1)}$ on the surface of the carrier mineral Σ. The determination of $\vec{v}^{(2)}$ allows for finding the field $\vec{v}^{(3)}$ that, in accordance with the boundary condition (16), neutralizes the field $\vec{v}^{(2)}$ on the surface S of the fine particle.

The calculation of the individual contributions of the fields to the local field of the velocity $\vec{v}$ makes it possible to find the force $\vec{F}$ acting from the side of the liquid, bounded by the carrier mineral, on the fine particle moving along it. At the same time, we should note that, in accordance with the boundary conditions (11)–(13), only the velocity fields with odd indices contribute to the magnitude of the force $\vec{F}$:

$$\vec{F} = \vec{F}^{(1)} + \vec{F}^{(3)} + \ldots \tag{17}$$

In accordance with the selected conditions (16) and (17), the total force $\vec{F}$ acting on the particle from the side bounded by the liquid wall and the force $\vec{F}_\infty$ determining the first contribution to it are opposite to the movement direction of the particle:

$$\begin{array}{l} \vec{F}_\infty == -\vec{i}\,F_\infty \\ \vec{F} = -\vec{i}\,F \end{array}. \tag{18}$$

Provided that the fine particle moves along the surface Σ or in the plane that is perpendicular to its symmetry, it is possible to write:

$$\begin{aligned} \vec{v}^{(2)} &= -\vec{i}\,v^{(2)} \\ \vec{F}^{(3)} &= F_\infty \frac{\vec{v}^{(2)}}{u} \\ \vec{F}^{(3)} &= -\vec{i}\,F^{(3)} \\ F^{(3)} &= F_\infty \frac{\vec{v}^{(2)}}{u} \end{aligned} \tag{19}$$

Since the conditions are considered according to G. Lamb, its influence on the hydrodynamic interaction occurring in the system of polydisperse particles consists in the coincidence of the fields created by the moving particle and the force $\vec{F}_\infty$, located in its center, which can be considered by the following expressions:

$$\vec{v}^{(1)}\beta = p^{(1)} = \frac{1}{4\pi}\left(\vec{F}_{\infty}\cdot\nabla\right)\frac{1}{r} + o\left(\frac{1}{r^2}\right) \tag{20}$$

$$\vec{v}^{(1)} = -\frac{\vec{F}_{\infty}}{6\,\pi\,\mu\,r} - \frac{r^2}{24\,\pi\,\mu}\left(\vec{F}_{\infty}\cdot\nabla\right)\nabla\frac{1}{r} + o\left(\frac{1}{r}\right). \tag{21}$$

Using (17), we can obtain an expression for the total resistance force acting on the fine particle from the side of the resistance force limited by the liquid wall:

$$F = F_{\infty} \times \left(1 - \frac{v^2}{u}\right)^{-1} \cdot \beta \tag{22}$$

where β is a nondimensional sliding coefficient.

Proceeding from (22), between the magnitude of the velocity $\vec{v}^{(1)}$ and the magnitude of the force F_{∞} acting on the particle in the unlimited extent of the liquid, there is a directly proportional dependence. On the other hand, in accordance with the conditions (14) and (15), the velocity $\vec{v}_S^{(1)}$ on the fine particle surface and the velocity $\vec{v}_{\Sigma}^{(2)}$ on the carrier–mineral surface (wall) are interrelated linearly. The force $\vec{F}$ introduced into the expression (22), is proportional to the liquid viscosity μ and the value $\vec{v}^{(2)} \to 0$, provided that $l \to \infty$, i.e., when the distance between the particle and the wall is increasing unlimitedly.

Then, the analysis of the dimensionalities leads to the conclusion that the above dimensional quantities are related by the following ratio:

$$\vec{v}^{(2)} = \frac{F_{\infty}}{6\,\pi\,\mu\,l}. \tag{23}$$

By substituting the expression (23) into (21), we obtain an expression for the correction to the hydrodynamic resistance force acting on the fine particle from the liquid side bounded by the wall (the surface of the carrier mineral):

$$F = \frac{F_{\infty}}{1 - \frac{F_{\infty}}{6\,\pi\,\mu\,l\,u}} \cdot \beta. \tag{24}$$

Taking into account [36], the expression for the force acting at the time moment t on the solid particle moving in the liquid at the velocity $\vec{u}$ can be written as:

$$\frac{F}{6\,\pi\,\eta\,R\,u} = 1 + \frac{R}{\sqrt{\pi\,v\,t}} + \frac{9}{16}\frac{R}{l}K\left(\frac{l}{\sqrt{v\,t}}\right). \tag{25}$$

Let $\phi = l\,/\,\sqrt{v\,t}$, and when $\varphi < 1$, the decomposition for the function $K(\varphi)$:

$$K(\phi) = 1 - \frac{16}{9\sqrt{\pi}}\phi + \frac{8}{9\sqrt{\pi}}\phi^3 - \frac{1}{6}\phi^4 + O\left(\phi^5\right) \tag{26}$$

will be substituted into the expression for the resistance force (25), and for steady flow when $t \to \infty$, we can obtain:

$$\frac{F}{6\,\pi\,\eta\,R\,u} = 1 + \frac{9}{16}\frac{R}{l} + \frac{1}{2\sqrt{\pi}}\frac{R}{l}\left(\frac{l}{\sqrt{v\,t}}\right)^3 \tag{27}$$

At the initial moment of time t, the particle is located at a sufficiently large distance from the wall, so that $\varphi > 1$ and, substituting the decomposition of the function $K(\varphi)$, we have:

$$K(\phi) = \frac{1}{3}\phi^{-2} + \frac{4}{3\sqrt{\pi}}\phi^{-3} + O\left(\phi^{-4}\right) \tag{28}$$

In the ratio (27), we have an expression for the force in the absence of the wall:

$$\frac{F}{6\pi\eta R u} = 1 + \frac{R}{\sqrt{\pi \nu t}} + \frac{9}{16}\frac{R}{l}\left(\frac{\nu t}{3 l^2}\right)^3 \tag{29}$$

According to the expression (27), near the coarse particle, the velocity of the approaching of the vortex (nonstationary) liquid flow to the steady state will change in the same way as the value $t^{-3/2}$ does. Based on the expression (29), at large distances between the particles (in the case of the particle motion in the unlimited liquid), the correction conditioned by the unsteadiness of the liquid flow will be changed in the same way as the value $t^{-1/2}$ will be. In the case of the Stokes flow, the correction to the hydrodynamic force acting on the particle is of the R/l order. Consequently, during the flotation involving the carrier minerals in the gap between the fine particle and the coarse one, the velocity of the approaching of the liquid flow to the steady-state mode is increased.

During the flotation, the particles hydrodynamically interact with plane (two-dimensional) and axisymmetric (two-dimensional and three-dimensional) fluid flows, for which the components of the velocity vector and the pressure field are determined based on the expressions:

$$\begin{aligned} v_r(r,\theta) &= -v_\infty \cos\theta\left(1 - \tfrac{3}{2}\tfrac{R}{r} + \tfrac{1}{2}\tfrac{R^3}{r^3}\right); \\ v_\theta(r,\theta) &= v_\infty \sin\theta\left(1 - \tfrac{3}{4}\tfrac{R}{r} - \tfrac{1}{4}\tfrac{R^3}{r^3}\right); \\ p(r,\theta) &= \left(\tfrac{3}{2}\mu\tfrac{v_\infty R}{r^2} + \rho g r\right)\cos\theta \end{aligned} \tag{30}$$

In the case of these two flows, the distribution of the velocity and the pressure depends on two coordinates (r, θ are polar coordinates), and the continuity equation contains the sum of two derivatives, which allows for introducing a scalar Stokes current function depending on these two coordinates, i.e., the current function $\psi(r,\theta)$. In this case, the constituent velocities v_r and v_θ of the liquid are determined based on the current function in accordance with the equalities:

$$v_r = v_\infty \cos\theta = -\frac{1}{r^2 \sin\theta}\frac{\partial\psi}{\partial\theta};\; v_\theta = -v_\infty \sin\theta = \frac{1}{r\sin\theta}\frac{\partial\psi}{\partial r}. \tag{31}$$

The hydrodynamic interaction of the polydisperse particles can be estimated by the value of the current function for the current line (surface), the movement along which for the deposition of the fine particle on the surface of the coarse one is the boundary ψ_{cr}:

$$E = \psi_{cr} / v_0 R_p(+) \tag{32}$$

where v_0 is the velocity of the incoming undisturbed flow, provided that:

$$\begin{aligned} r &= R_p(+) + R_p(-);\; \theta = \pi/2; \\ D &= \left[R_p(+) - R_p(-)/R_p(+)\right] << 1 \\ D_0 &= R_p(-)/R_p(+) \end{aligned} \tag{33}$$

Equation (32) can be obtained in the form of:

$$E = 2\frac{\varpi}{v_0}\left[\frac{R_p(-)}{R_p(+)}\right]^2. \tag{34}$$

Here,

$$\begin{aligned} \varpi &= \frac{v_0}{2(2.00 - \ln Re)}; \\ Re &= 2R_p(+)\, v_0/\nu \end{aligned} \tag{35}$$

where ν is the liquid kinematic viscosity.

The expression for the particle collision efficiency (34) is obtained under the assumption that the condition of the liquid adhesion to the hydrophilic surface of the mineral is observed. In the flotation conditions, during the hydrodynamic interaction of hydrophobic polydisperse particles, when determining the collision efficiency, it is necessary to take into account an inaccurate compliance with the adhesion condition, i.e., the sliding of the liquid along the hydrophobic surface. The change in the hydrodynamic mode of the interaction during the transition from hydrophilic solid particles to hydrophobic solid polydisperse particles can be considered by obtaining an expression for the current function when taking into account the liquid sliding.

Based on Oseen's equation:

$$\left(\vec{v}_\infty \cdot \nabla\right)\vec{v} = -\frac{1}{\rho}\nabla p + \nu \Delta \vec{v}, \tag{36}$$

the velocity components are:

$$\begin{aligned} v_r &= -\frac{A_0}{r^2} + \frac{2A_1 \cos\theta}{r^3} - \frac{C_0 e^{-kr(1-\cos\theta)}}{2kr^2}\left[1 + kr(1-\cos\theta)\right] + v_\infty \cos\theta \\ v_\theta &= \frac{A_1 \sin\theta}{r^3} + \frac{C_0 \sin\theta}{2r} e^{-kr(1-\cos\theta)} - v_\infty \sin\theta \end{aligned} \tag{37}$$

when $\delta r << 1$ (where $\delta = v_\infty / 2\nu$) has the following form:

$$\begin{aligned} v_r &= \frac{A_0}{r} - A_1 \frac{\cos\theta}{r^2} + v_0 \cos\theta - \\ &- \frac{C_0}{2}\left[\frac{1}{\delta r} + \cos\theta - \left(\gamma + \ln\frac{1}{2}\delta r\right)\cos\theta\right], \end{aligned} \tag{38}$$

$$v_\theta = -A_1 \frac{\sin\theta}{r^2} - v_0 \sin\theta - \frac{C_0}{2}\left(\gamma + \ln\frac{1}{2}\delta r\right)\sin\theta. \tag{39}$$

Here, γ is the Euler constant, and A_0, A_1 and C_0 are the constants whose values must be obtained from the boundary conditions.

Based on the boundary condition that $v_r = 0$, when $r = R_p(+)$ by equating the coefficient to zero, when $\cos\theta$, and when a member does not contain $\cos\theta$, the following is true:

$$A_0 = \frac{C_0}{2\delta};\; A_1 = v_0 R_b^2 \frac{C_0 R_p^2(+)}{2}\left(1 - \gamma - \ln\frac{1}{2}\delta r\right). \tag{40}$$

If the condition $v_\theta = 0$ corresponds to the adhesion of the liquid, then the condition of the equality of the tangential stress $p_{r\theta}$ to the tangential force takes into account the sliding of the liquid along the surface of the coarse particle $r = R_p(+)$:

$$v_0 \beta^* = p_{r\theta} = \eta\left(\frac{\partial v_\theta}{\partial r} + \frac{\partial v_r}{r\partial\theta} - \frac{v_\theta}{r}\right)_{r = R_p(+)}. \tag{41}$$

Using (38) and (39), based on (41), when $r = R_p(+)$, we have:

$$A_1\left(\frac{\beta^*}{\eta} + \frac{4}{R_p(+)}\right) = -\frac{\beta^*}{\eta} R_p^2(+)\left[v_0 + \frac{C_0}{2}\left(\lambda + \ln\frac{1}{2}\delta r\right)\right], \tag{42}$$

from which:

$$\begin{aligned} C_0 &= \frac{2v_0}{\left(1 - \gamma - \ln\frac{1}{2}\delta R_p(+)\right) - \frac{1}{2}\left(1 + \frac{2\eta}{\beta^* R_p(+)}\right)^{-1}}; \\ A_1 &= -\frac{C_0 R_p^2(+)}{4\left(1 + \frac{2\eta}{\beta^* R_p^2}\right)} \end{aligned} \tag{43}$$

Proceeding from the condition that:

$$\begin{cases} d\psi = r\,v_r\,d\theta - v_\theta\,d\,r \\ \psi = 0\,ghb\,r = R_p\,(+)\ , \\ \frac{\eta}{\beta^*\,R_p\,(+)} << 1 \end{cases} \tag{44}$$

we will obtain:

$$E = 2\frac{\omega}{v_0}\left(\frac{R_p(-)}{R_p\,(+)}\right)^2\left(1 + \frac{2\,\eta}{\beta^*\,R_p\,(-)}\right). \tag{45}$$

The comparison of the expressions (34) and (45) allows for stating that the correction for slip is given by the expression:

$$f_E = \left(1 + \frac{2\,\eta}{\beta^*\,R_p\,(-)}\right). \tag{46}$$

In view of this, taking into account the sliding, the efficiency of the collision the fine particle with the coarse one is higher by correction, whose value is determined by the ratio (46).

A comparison of the expressions (20), (22) and (26), (37) allows for stating that during the flotation of the microdispersions of minerals using carrier minerals, the adhesion of fine particles to coarse ones is facilitated by the rapid approach of the vortex liquid flow in the interphase gap to the steady-state mode accompanied by suppressing transverse particle movements. In the case of the Stokes flow, this fact can be considered by the correction having the R/l order to the hydrodynamic force acting on the particle. The coefficient value of the capturing by a bubble of an aggregate consisting of hydrophobic particles is higher by the slip correction value in the form of:

$$\left(1 + \frac{2\,\eta}{\beta^*\,R_p\,(-)}\right).$$

3.2. *Flotation of Gold Microdispersions Using Carrier Minerals*

Based on the obtained estimates, the influence of the polydispersity of the particles and the conditions of adhesion/sliding of the liquid along the hydrophobic surface on the result of their hydrodynamic interaction was analyzed. This was analyzed according to the method related to the preliminary adhesion of hard-to-extract forms of the minerals (their microdispersions) to the carrier minerals that were specially introduced into the flotation system. The purpose of this section of the work was to obtain experimental evidence for the technological effectiveness of the flotation technology using the rougher concentrate as carrier minerals.

The aggregation process of the minerals in the flotation cells occurs in the turbulent mode of their movement. Therefore, the methods used for studying the aggregation in a calm suspension, even after mixing, cannot fully reflect the ongoing changes in the particle size. The method of studying the aggregation of the fine particles during their mixing should more fully reflect the change in the dispersed composition of the mineral as compared to static methods.

In order to study the kinetics of the thinning and breakthrough of the symmetric interphase films formed during the interaction of the polydisperse grains of gold, the induction time was measured when the grains adhered to each other. The influence of temperature and the gold samples on the induction time was studied. To solve this problem, a change was introduced to the device design: the system used for generating a gas bubble was replaced by a cantilever beam (probe) containing a grain of gold, glued according to the method that was developed and tested in [62]. When assembling the probe, the most flattened grains were selected: thin plates, scales, and leaves of native gold. The selected gold grains, when mounted on the probe, were oriented towards the material placed in the cuvette by a larger surface, i.e., a face (wall). Providing a predetermined grain position on the probe, the induction time was measured when individual grains adhered to the "gold wall" and not to its edge or corner. Nevertheless, the lack of geometric similarity of the interphase gap during the interaction of the flat wall with the polyhedral gold grains in the material layer was the main reason for the measurement error.

Gold grains of a given coarseness were isolated by sedimentometric analysis.

For the first experiment (Figure 4a, curve 1), native gold grains with a mirror-smooth surface without any relief, corrosive shells on the periphery of the grains, or signs of hypergenic transformation of the gold faces were selected. On the contrary, for the second experiment (Figure 4a, curve 2), the research object was non-rounded gold grains with a shagreen (rough), pitted, and bumpy surface. When using the high-resolution electron microscope, randomly (mosaic) located micron-sized depressions (ranging from fractions of a micron to 1–3 microns in size) of a geometrically regular shape in the form of "honeycombs" were marked on the faces of the gold grains.

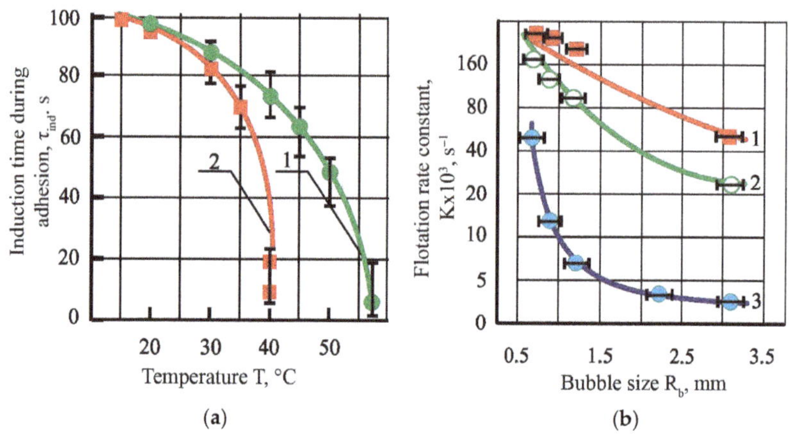

Figure 4. Dependence of the induction time τ_{ind} on the temperature T (**a**) and the constant of the flotation rate and the size of the gas bubble R_b (**b**).

The native gold grains were treated with a 10^{-3} M solution of ethyl xanthogenate.

In the first case (Figure 4a, curve 1), the induction time decrease when the temperature increased can be explained by an increase in the hydrophobic interaction forces, i.e., an endothermic process associated with the difference in the structure and properties of the water located in the boundary layer and in the volume [19,59,63]. This also includes the cases of adsorption of ethyl xanthogenate on the gold surface [42,44,64]. A sharp increase in the dependence of the induction time on temperature when the rough surface particles interact (Figure 4a, curve 2) can be associated with the manifestation of the effect of the water flow sliding along the hydrophobic surface of the particles as a result of stratifying the wall-mounted gas–liquid layer [65,66] or separating stable nanobubbles from the surface relief irregularities [39,67].

The experimental results were processed along with rejecting the measurement runs; if the empirical dispersion of the measurement runs was noticeably greater, then the significance of its difference from the rest was checked by comparing it according to Cochran's test (*G*-criterion). The critical (tabular) values of the *G*-criterion were determined at a significance level of 0.95. The points shown in Figure 4 belong to the midpoints of the confidence intervals constructed using Student's *t*-distribution.

The induction time measurement was supplemented by an experiment conducted on the flotation of native gold of different sizes and samples. The experiments were carried out in a counterflow flotation column that was 64 mm in diameter when the xanthogenate concentration was 15.6 mg/L and the foaming agent concentration was 0.025 mg/L. Washwater was not supplied to the foam layer.

To determine the diameter of the air bubbles (d_b, mm), its dependence (approximation accuracy was R^2 = 0.98) on the pressure drop occurring between the phases (p, MPa) and the outlet hole diameter (d, mm) was in the form of [57]:

$$d_b = -33.81 \times 10^{-2} + 60.08 \times p + 18.15 \times d + 430.02 \times p\,d. \tag{47}$$

In the experiments, high-assay copper (870‰–930‰) native gold (of reddish color) admixed with platinum and with inclusions of ilmenite and magnetite was used; no signs of hypergenic transformations were noted on the surface of the angular gold grains. Only individual grains had a low assay (680‰–770‰); their surface differed by a heterogeneous structure, and they had endogeneous deformation signs and corrosion shells. Small classes of coarseness were enriched with lamellar and flake morphotypes of semiangular grains. Native gold of three size classes was used: (−100 + 71) μm (Figure 4b, curve 1), (−71 + 40) μm (Figure 4b, curve 2), and −20 μm (Figure 4b, curve 3).

The flotation rate constant was determined to decrease (Figure 4b) when the bubble size increased and the R_p/R_b ratio value decreased. The obtained result is explained by the influence of the hydrodynamic interaction of the particles with the bubbles on the flotation complex formation. The result can be explained by the low efficiency of the collision of the fine particles with the large bubbles (the action of viscous forces) and the high efficiency of the collision of the coarse particles with the bubble surface due to the action of inertia forces.

The change in the inertia forces during flotation by the developed method is possible due to the adhesion of gold microdispersions to the coarse particles of the carrier minerals. Moreover, there is an interaction of the particles whose surface is hydrophobized by a collecting reagent. The surface hydrophobization increases the area of the isotherm of the wedging pressure in the region of its negative values (from S_1 to S_2), under which the attraction forces prevail over the repulsion forces (Figure 5).

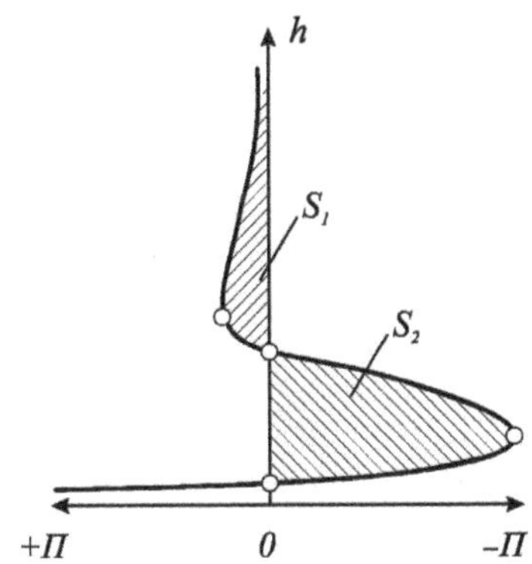

Figure 5. The view of the isotherm of the wedging pressure Π(h) of the wetting film.

The decrease in the surface hydrophilicity was associated with the adsorption of the collector, whose hydrocarbon radicals were the cause of the appearance of the interaction forces caused by a change in the structure of the liquid, i.e., the forces of hydrophobic attraction. When the thickness h of the wetting film decreased, when the particles approached, the energy of their interaction U was determined by the additive contribution of the energies of a different nature [68–70]:

- Molecular attraction V_A

$$V_A = \frac{A_{123}}{6h}\frac{R_{p1}R_{p2}}{R_{p1} + R_{p2}} \tag{48}$$

- Electrostatic repulsion V_R

$$V_R = \pi \varepsilon_0 \frac{R_{p1} R_{p2}}{R_{p1} + R_{p2}} \left(\psi_1^2 + \psi_2^2\right) \left\{ \frac{2 \psi_1 \psi_2}{\psi_1^2 + \psi_2^2} ln \left[\frac{1 + exp(-\kappa h)}{1 - exp(-\kappa h)}\right] + ln\left[1 - exp(-2\kappa h)\right] \right\} \quad (49)$$

- Hydrophobic attraction V_S

$$V_S = 2\pi \frac{R_{p1} R_{p2}}{R_{p1} + R_{p2}} K \lambda exp\left(-\frac{h}{\lambda}\right)$$
$$\lambda = 12.2 k_1;\ K = -\frac{2.51 \times 10^{-3}}{\pi} k_1;\ k_1 = \frac{exp(\theta / 100^0) - 1}{e - 1} \quad (50)$$

The values $|U_{min}| / kT$ were calculated during the interaction of the gold grains, whose sizes were R_{p1} = 1–10 μm and R_{p2} = 100–150 μm (Figure 6b). The calculation conditions were as follows. $A_{123} = 4.1 \times 10^{-12}$ erg is the Hammaker constant for the interaction of gold grains (indices "1" and "2") through a symmetrical film of water (index "3"). $\varepsilon_0 = 8.85 \times 10^{-12}$ F/m is the electrical constant of the dispersion medium. φ_1, $\varphi_2 = -14.7$ mV is the Stern potential of the particles (approximated by their ζ-potential, whose experimental values were determined by the electrophoretic mobility of the particles in water using the Henry equation, a Dispersion DT-310 electroacoustic spectrometer (Dispersion Technology Inc., New York, the United States) and a Zetasizer Nano ZS device, Malvern Instruments Ltd, Malvern, United Kingdom). $\sigma_0 = 3 \times 10^{14}$ charge/cm^2 is the surface-charge density of the gold grains. $\kappa = 0.92 \times 10^6$ cm^{-1} is the parameter of the double electric layer corresponding to the inverse Debye shielding radius (the parameter κ was determined based on the electrical conductivity data). h is the distance between the particles R_{p1} and R_{p2} ($\kappa h < 3$). K is the parameter characterizing the magnitude of surface structural forces, J/m^2 [58,71]. θ = 64^0 is the contact angle. λ is the parameter characterizing the long-range action of the forces, nm [58,71]. e = 2.718 is the transcendental constant. $k = 1.381 \times 10^{-23}$ J/K is the Boltzmann constant. T = 293 K is the absolute temperature. The calculations were performed in the Maple 2021 environment.

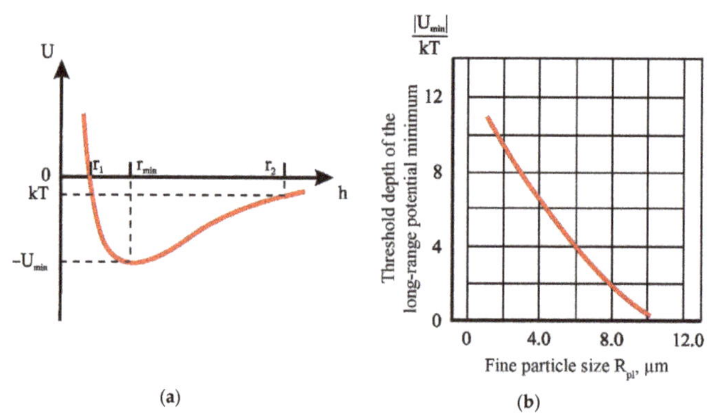

Figure 6. The scheme (**a**) and the calculation results (**b**) of the threshold depth of the long-range potential minimum for the fine particles of various sizes.

Figure 6a shows the change in the potential energy of the interaction of the particles when the distance between them was increased from r_1 to r_2. The selected distance between the particles corresponds to the length of the long-range potential minimum; starting from the distance $r \geq r_1$, the potential energy of the interaction of the particles is $U < 0$; r_2 is the distance between the particles, at which the energy of their interaction is $|U| = kT$.

When the size of the fine particles increased (Figure 6b), the absolute value of the threshold depth of the long-range potential minimum decreased from 0.56 (when the minimum coordinate was r_{min} = 11.6 nm) to 0.36 (when r_{min} = 15.4 nm). Therefore, when the particle size increased, the area of their interaction increased; the polydisperse particles were aggregated at a smaller depth of the long-range potential minimum.

The results of studying the aggregation and flotation of gold allowed for reaching a conclusion about the probability of efficiently recovering the microdispersions of minerals, including native gold, by flotation using the carrier minerals intended for sticking the fine particles onto the coarse ones. This process is followed by the flotation of the formed aggregates containing the bubbles of a reduced size.

Using the two samples of ores, which differed in gold content, experiments were set up to compare the technological indicators obtained using competing technologies that have been adopted at an operating gold recovery factory and developed using the rougher concentrate as carrier minerals. The set of the experiments differed in the fact that the ready-made concentrate was obtained on the ore of sample-1 according to the full technological scheme, and the ore of sample-2 allowed for releasing only the rougher concentrate. This made it possible to evaluate the technological regularities of the scheme as a whole and the main flotation operation efficiency in particular. The experiments were conducted according to the continuous process principle.

During the flotation of the ores of sample-1 according to the technological scheme adopted at the operating gold extraction factory, 82.93% of the gold was extracted into a sellable concentrate when the concentrate yield was 3.07% and the gold content was 20.80 g/t.

During the flotation and proceeding according to the scheme shown in Figure 6, when the rougher concentrate was first mixed with the base ore, the gold content in the main flotation feed increased from 0.77 to 1.22 g/t and up to 1.62 g/t during the second mixing, i.e., by 58.4 and 110.4% rel. with respect to the base ore. The consequence of increasing the gold content in the initial feed of the rough flotation operation was the operational extraction increase in the metal from 83.74 to 91.28 and 94.40%, respectively. The gold extraction into a sellable concentrate was 87.62% when the Au content was 27.65 g/t and the $\Delta\gamma$ concentrate yield decreased by:

$$\Delta\gamma = \frac{3.07 - 2.44}{3.07} \times 100 = 20.52\%.$$

Table 1 shows the experimental results of extracting the rougher concentrate from the ores of sample-2 in the mode adopted at the operating gold processing factory and the results of testing the flotation mode using the rougher concentrate material as carrier minerals (Figure 2).

Figure 7 shows a qualitative and quantitative scheme of the sample-1 flotation when the rougher concentrate was mixed twice and provided with the initial feed.

Table 1. Results of the experiments on the ore flotation using competing flotation schemes.

No. Item	Product Name	Yield, %	Au Content, g/t	Au Extraction, %
		Sample-2 flotation in the factory mode		
1	Rougher concentrate	14.417	4.97	74.95
2	Flotation tailings	85.583	0.280	25.05
3	Base ore	100.0	0.956	100.0
		Sample-2 flotation using carrier minerals		
4	Rougher concentrate	11.40	6.29	74.95
5	Tailings-1	29.76	0.297	9.24
6	Tailings-2	29.42	0.274	8.43
7	Tailings-3	29.42	0.240	7.38
8	General tailings	88.60	0.270	25.05
9	Base ore	100.0	0.956	100.0

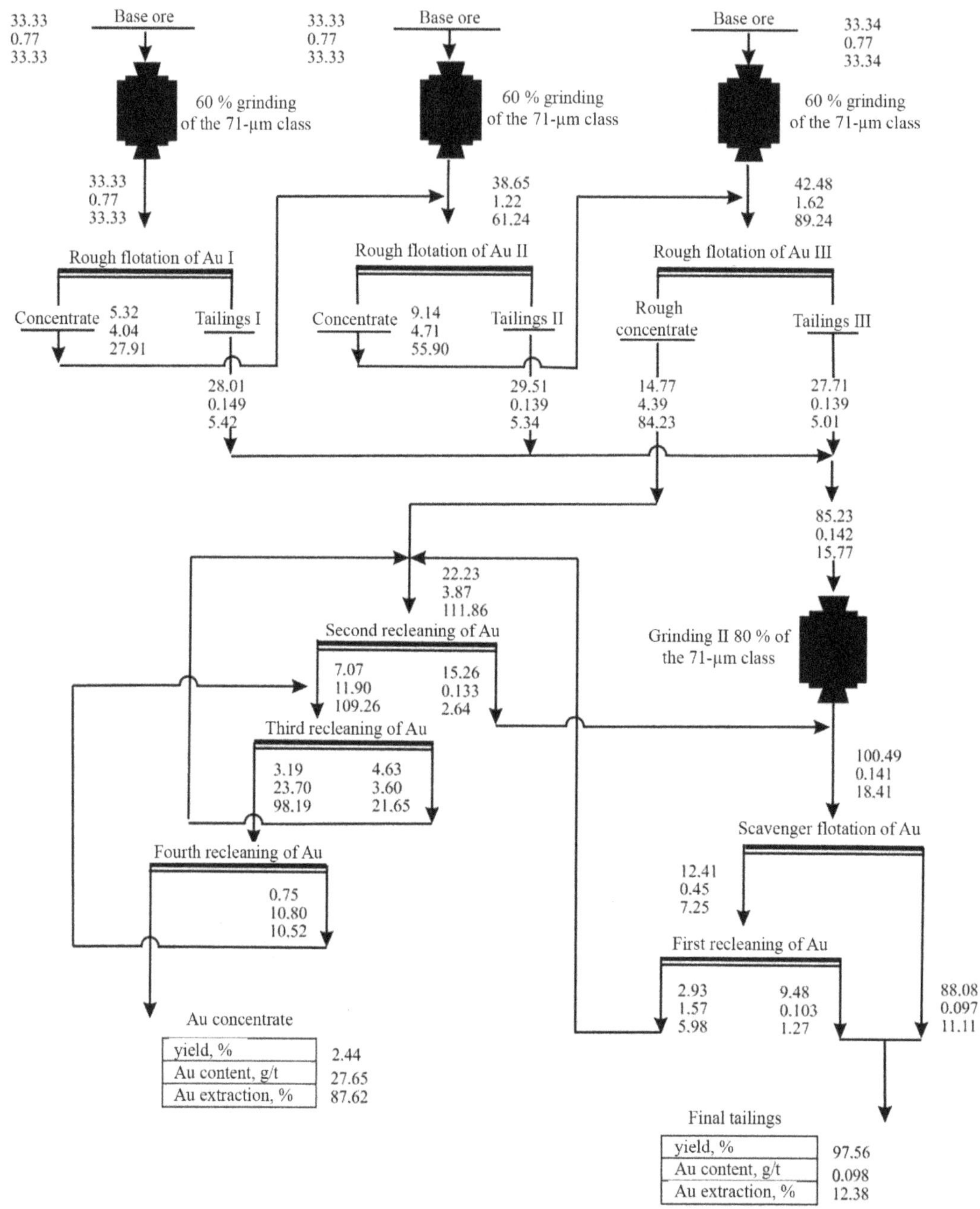

Figure 7. Results of enriching sample-1 in the experiment, modeling a closed cycle.

Table 1 concludes that, while maintaining the achieved level of gold extraction (74.95%), the double-mixing of the rougher concentrate with the initial feed made it possible to increase the gold content in the rougher concentrate from 4.97 to 6.29 g/t (the gold concentration degree increased from 5.199 to 6.579) when the concentrate yield decreased.

Figure 8 demonstrates the dependences of the operational gold extraction on the metal content in the initial feed of the main flotation operation obtained using the material of sample-1 (Figure 8a) and sample-2 (Figure 8b). A similar dependence (Figure 8c) proceeded from the data processing results obtained at the stage of the preliminary research works related to developing the ores of the deposit.

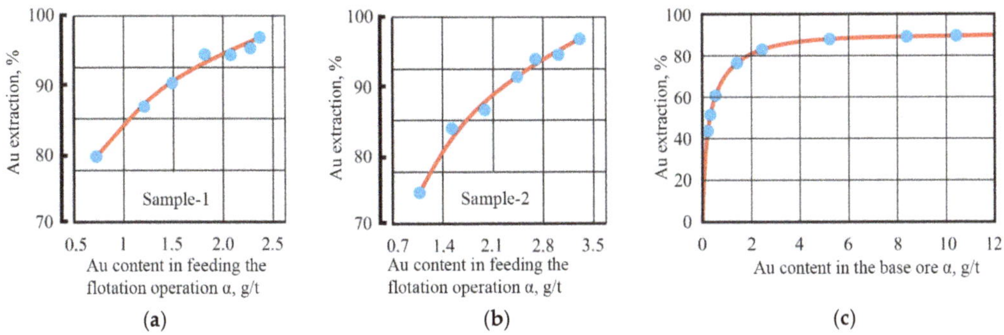

Figure 8. Gold extraction as a function of its content in the initial feed of the flotation operation for sample-1 (**a**), sample-2 (**b**), and according to research works (**c**).

Figure 8 shows that when using the rougher concentrate material as carrier minerals, a mixture was obtained, which, in terms of the extracted mineral content and the flotability, was identical to the base ore with an increased content of the valuable component.

Various methods of processing refractory sulfide gold concentrates are possible; their implementation requires significant operating expenses. Therefore, reducing the gold concentrate amount delivered for metallurgical processing provides a significant economic effect.

The cost of processing the persistent sulfide flotation concentrates by various technologies was calculated using the data given in [72]. When using the developed method of the ore flotation, the possibility of reducing operating costs attributed to 1 ton of the initial ore has been established, which increases the economic efficiency of its processing.

4. Conclusions

A flotation technology of processing gold-bearing ores, which was developed in order to increase the completeness of extracting the fine particles of the valuable component, was studied in this paper.

A correction was obtained for the hydrodynamic drag force acting from the liquid side on the fine particles under conditions of their flotation applying the carrier minerals.

A decrease in the induction time when the temperature increases was revealed when studying the kinetics of the thinning and breakthrough of the symmetrical interphase films formed by the gold grains with a mirror-smooth surface without any relief. The obtained results can be explained by an increase in the forces of hydrophobic interactions. The decrease in the induction time, when the gold grains adhere to the rough surface, can be associated with the manifestation of the effect of the liquid sliding along the gas layer.

The influence of the liquid sliding effect is considered by a dimensionless correction to the magnitude of the collision efficiency of the aggregate of hydrophobic particles with an air bubble. The correction expression includes the values that describe the properties of a continuous medium (dynamic viscosity) and a disperse phase (geometric particle size); the correction value is always greater than one.

We have revealed that the flotation rate constant decreases when the size of the bubbles increases and the ratio of the particle size to the bubble size decreases.

We have found that during the aggregation of polydisperse particles, the threshold energy of the rapid coagulation is lower than that occurring during the interaction of monodisperse particles, whose aggregation requires a large depth of the potential pit.

When performing full-scale experiments using the ore of sample-1, sellable concentrates were obtained according to two complete technological schemes. The first has been adopted at the existing gold recovery factory and the second uses the rougher concentrate as a carrier material. We have established the fact that when the concentrate yield decreased by 20.52% rel. in the second case, the gold extraction was 4.69% abs. higher.

According to the example of the ore of sample-2, when comparing two technologies used for extracting the rougher concentrate (the factory one and the developed one using the rougher concentrate as a carrier material), we have shown that while maintaining the achieved level of gold extraction (74.95%), the double-mixing of the rougher concentrate with the initial feed allowed for obtaining an increase in the gold content in the rougher concentrate from 4.97 to 6.29 g/t. At the same time, the degree of the gold concentration increased from 5.199 to 6.579.

A possible level of increasing the commercial efficiency indicators of production has been shown when using the developed approach to extracting microdispersions of minerals.

Author Contributions: Conceptualization, S.I.E.; methodology, A.A.P.; formal analysis, V.Y.K.; investigation, V.A.G.; data curation, A.A.P., V.Y.K. and V.A.G.; writing—original draft preparation, S.I.E., N.S.G. and D.A.Z.; writing—review and editing, E.V.V. and V.V.K.; supervision, S.I.E.; project administration, E.V.V. and V.V.K.; visualization, N.S.G. and D.A.Z. All authors have read and agreed to the published version of the manuscript.

Funding: This research received no external funding.

Institutional Review Board Statement: Not applicable.

Informed Consent Statement: Not applicable.

Data Availability Statement: The data presented in this study are available from the corresponding authors upon reasonable request.

Conflicts of Interest: The authors declare no conflicts of interest.

References

1. Marinin, M.; Marinina, O.; Wolniak, R. Assessing of Losses and Dilution Impact on the Cost Chain: Case Study of Gold Ore Deposits. *Sustainability* **2021**, *13*, 3830. [CrossRef]
2. Litvinenko, V.S.; Petrov, E.I.; Vasilevskaya, D.V.; Yakovenko, A.V.; Naumov, I.A.; Ratnikov, M.A. Assessment of the role of the state in the management of mineral resources. *J. Min. Inst.* **2023**, *259*, 95–111. [CrossRef]
3. Boduen, A.; Zalesov, M.; Melamud, V.; Grigorieva, V.; Bulaev, A. Combined Bacterial and Pressure Oxidation for Processing High-Sulfur Refractory Gold. *Processes* **2023**, *11*, 3062. [CrossRef]
4. Ivanik, S.A.; Ilyukhin, D.A. Flotation extraction of elemental sulfur from gold-bearing cakes. *J. Min. Inst.* **2020**, *242*, 202–208. [CrossRef]
5. Fedotov, P.K.; Senchenko, A.E.; Fedotov, K.V.; Burdonov, A.E. Studies of enrichment of sulfide and oxidized ores of gold deposits of the Aldan shield. *J. Min. Inst.* **2020**, *242*, 218–227. [CrossRef]
6. Duryagina, A.M.; Talovina, I.V.; Lieberwirth, H.; Ilalova, R.K. Morphometric parameters of sulphide ores as a basis for selective ore dressing. *J. Min. Inst.* **2022**, *256*, 527–538. [CrossRef]
7. Aleksandrova, T.N.; Nikolaeva, N.V.; Lvov, V.V.; Romashev, A.O. Ore processing efficiency improvements for precious metals based on process simulations. *Obogashchenie Rud* **2019**, *2*, 8–13. [CrossRef]
8. Alexandrova, T.N.; Nikolaeva, N.V.; Kuznetsov, V.V. Study of the strength properties of gold ore from the Bam deposit. *Gorn. Zhurnal* **2021**, *11*, 851. [CrossRef]
9. Yakovleva, T.A.; Romashev, A.; Mashevsky, G.N. Digital technologies for optimizing the dosing of flotation reagents during flotation of non-ferrous metal ores. *Min. Informational Anal. Bull.* **2022**, 175–188. [CrossRef]
10. Wang, D.; Lin, Q. Hydrodynamics of froth flotation and its effects on fine and ultrafine mineral particle flotation: A literature review. *Miner. Eng.* **2021**, *173*, 107220. [CrossRef]
11. Xie, L.; Wang, J.; Lu, Q.; Hu, W.; Zeng, H. Surface interaction mechanisms in mineral flotation: Tals, measurements, and perspectives. *Adv. Colloid Interface Sci.* **2021**, *295*, 102491. [CrossRef] [PubMed]
12. Yin, W.Z.; Tang, Y. Interactive effect of minerals on complex ore flotation: A brief review. *Int. J. Miner. Metall. Mater.* **2020**, *27*, 571–601. [CrossRef]

13. Kiventera, J.; Golek, L.; Yliniemi, J.; Ferreira, V.; Dejce, J.; Llikainen, M. Utilization of sulphidictailinds from gold mine as a raw material in geopolymerization. *Int. J. Miner. Process.* **2016**, *149*, 104–110. [CrossRef]
14. Zhang, J.; Zhang, Y.; Richmond, W.; Wang, H.-P. Processing technologies for gold-telluride ores. *Int. J. Miner. Met. Mater.* **2010**, *17*, 1–10. [CrossRef]
15. Long, Q.; Wang, H.; Jiang, F.; Tan, W.; Xu, Z. Enhancing flotation separation of fine copper oxide from silica by microbubble assisted hydrophobic aggregation. *Miner. Eng.* **2020**, *189*, 107863. [CrossRef]
16. Li, Z.; Yoon, R.-H. AFM force measurements between gold and silver surface treated in ethyl xanthate solutions: Effect of applied potentials. *Miner. Eng.* **2012**, *36–38*, 126–131. [CrossRef]
17. Nguyen, A.V.; Nalaskowski, J.; Miller, J.D.; Butt, H.-J. Attraction between hydrophobic surface studied by atomic microscopy. *Int. J. Miner. Process.* **2003**, *72*, 215–225. [CrossRef]
18. Gillies, G.; Kappl, M.; Butt, H.-J. Direct measurements of particle-bubble interactions. *Adv. Colloid Interface Sci.* **2005**, *114–115*, 165–172. [CrossRef]
19. Skvarla, J. Hydrophobic interaction between macroscopic and microscopic surfaces. Unification using surface thermodynamics. *Adv. Colloid Interface Sci.* **2001**, *91*, 335–390. [CrossRef]
20. Capponi, F.; Azevedo, A.; Oliveira, H.; Rubio, J. Column rougher flotation of fine niobium-bearing particles assisted with micro and nanobubbles. *Miner. Eng.* **2023**, *199*, 108119. [CrossRef]
21. Zhang, Z.; Ren, L.; Zhang, Y. Role of nanobubbles in the flotation of fine rutile particles. *Miner. Eng.* **2021**, *172*, 107140. [CrossRef]
22. Li, C.; Li, D.; Zhang, H. Surface nanobubbles on the hydrophobic surface and their implication to flotation. *Int. J. Miner. Metall. Mater.* **2022**, *29*, 1491–1492. [CrossRef]
23. Jadhav, A.J.; Barigou, M. Bulk nanobubbles: That is the question. *Langmuir* **2020**, *36*, 1699–1708. [CrossRef] [PubMed]
24. Sviridov, V.V.; Sviridov, A.V.; Nikiforov, A.F. *Physico-Chemical Bases of Microflotation Processes*; Lan: St. Petersburg, Russia, 2018.
25. Bryk, P.; Bryk, M. Effective interactions in polydisperse colloidal suspensions investigated using Ornstkin-Zernike integral equations. *J. Colloid Interface Sci.* **2009**, *338*, 92–98. [CrossRef]
26. Xue, J.; Ren, D.; Chen, T.; Bu, X.; Wan, X.; Song, Z.; Zhao, C. Hydrophobic agglomeration flotation of oxidized digenite fine particles induced by Na_2S and butyl xanthate. *Miner. Eng.* **2021**, *168*, 106932. [CrossRef]
27. Hu, Y.; Qiu, G.; Miller, J.D. Hydrodynamic interactions between particles in aggregation and flotation. *Int. J. Miner. Process.* **2003**, *70*, 157–170. [CrossRef]
28. Verrelli, D.I.; Albijanic, B. A comparison of methods for measuring the induction time. *Miner. Eng.* **2015**, *80*, 8–13. [CrossRef]
29. Xia, W. Role of surface roughness in the attachment time between air bubble and flat ultra-low-ash coal surface. *Int. J. Miner. Process.* **2017**, *168*, 19–24. [CrossRef]
30. Evdokimov, S.I.; Panshin, A.M.; Solodenko, A.A. Mineralurgy. In *Successes of Flotation*; OOO NPKP "MAVR": Vladikavkaz, Russia, 2010; Volume 2, p. 2109.
31. Chorru, F.; Larrieu, E. Motion of a particle near a rough wall in a viscous shear flow. *J. Fluid Mech.* **2007**, *570*, 431453.
32. Yahiaoui, S.; Feuillebois, F. Lift on a sphere moving near a wall in a parabolie flow. *J. Fluid Mech.* **2010**, *662*, 447–474. [CrossRef]
33. Asmolov, E.S.; Feuillebois, F. Far-field disturbance flow induced by a small non-neutrally buoyant sphere in a linear shear flow. *J. Fluid Mech.* **2010**, *643*, 449–470. [CrossRef]
34. Lebedeva, N.A.; Asmolov, E.S. Migration of settling particles in a horizontal viscous flow through a vertical slot with porous walls. *Int. J. Multiph. Flow.* **2011**, *27*, 453–461. [CrossRef]
35. Maximov, R.N. Interaction of particles with the wall of the concentrator channel with a screw pulp flow. *Min. Inf. Anal. Bull.* **2008**, 344–347.
36. Protodiakonov, I.O.; Lyublinskaya, I.E.; Ryzhkov, A.E. *Hydrodynamics and Mass Transfer in Disperse Liquid-Solid Systems*; Chemistry: Leningrad, Russia, 1987; 336p.
37. Nasyrov, V.V.; Nasyrova, M.G. On the applicability of Stokes' law. *Math. Struct. Model.* **2020**, *2*, 40–48. [CrossRef]
38. Zuo, H.; Javadpour, F.; Deng, S.; Jiang, X.; Li, Z.; Li, H. Reassessing water slippage in hydrophobic nanostructures. *J. Chem. Phys.* **2020**, *153*, 191101. [CrossRef]
39. Malkin, A.Y.; Patlazhan, S.A. Wall slip for complex liquids—Phenomenon and its canses. *Adv. Colloid Interface Sci.* **2018**, *257*, 42–57. [CrossRef]
40. Wu, Y.; Cai, M.; Li, Z.; Song, X.; Wang, H.; Pei, X.; Zhou, F. Slip flow of diverse liquids on robust super omniphobic surface. *J. Colloid Interface Sci.* **2014**, *414*, 9–13. [CrossRef]
41. Pan, L.; Jung, S.; Yoon, R.-H. A fundamental study on the role of collector in the kinetics of bubble-particle interaction. International. *J. Miner. Process.* **2012**, *106–109*, 37–41. [CrossRef]
42. Wang, J.; Yoon, R.-H.; Morris, J. AFM surface force measurements conducted between gold surface treated in xanthate solutions. *Int. J. Miner. Process.* **2013**, *122*, 13–21. [CrossRef]
43. Kemppinen, J.; Aaltonen, A.; Sihvonen, T.; Leppinen, J.; Siren, H. Xanthate degradation occurring in flotation process waters of a gold concentrator plant. *Miner. Eng.* **2015**, *80*, 1–7. [CrossRef]
44. Golik, V.I.; Klyuev, R.V.; Martyushev, N.V.; Kondratiev, V.V.; Tynchenko, V.S.; Gladkikh, V.A.; Iushkova, L.V.; Brigida, V. Reuse and Mechanochemical Processing of Ore Dressing Tailings Used for Extracting Pb and Zn. *Materials* **2023**, *16*, 7004. [CrossRef] [PubMed]

45. Gul, A.; Kangal, O.; Sirkeci, A.A.; Onal, G. Beneficiation of the goldbearing ore le gravity and flotation. *Int. J. Miner. Metall. Mater.* **2012**, *19*, 106–110. [CrossRef]
46. Afanasova, A.V.; Aburova, V.A.; Prokhorova, E.O.; Lushina, E.A. Investigation of the influence of depressors on flotation-active rock-forming minerals in sulphide gold-bearing ore flotation. *Min. Informational Anal. Bull.* **2022**, 161–174. [CrossRef]
47. Valderrama, L.; Rubio, J. High intensity conditioning and the carrier flotation of gold fine particles. *Int. J. Miner. Process.* **1998**, *52*, 273–285. [CrossRef]
48. Arriagada, S.; Acuna, C.; Vera, M. New technology to improve the recovery of fine particles in froth flotation based on using hydrophobized glass bubbles. *Miner. Eng.* **2020**, *156*, 106364. [CrossRef]
49. Liu, S.; Xie, L.; Liu, G.; Zhang, H.; Zeng, H. Understanding the hetero-aggregation mechanism among sulfide and oxide mineral particles driven by bifuctional surfactants: Intensification flotation of oxide minerals. *Miner. Eng.* **2021**, *169*, 106928. [CrossRef]
50. McKee, C.; Walz, J.Y. Interactionforcesbetweencolloidalparticlesinasolutionoflike-charged, adsorbing nanoparticles. *J. Colloid Interface Sci.* **2012**, *365*, 72–80. [CrossRef]
51. Evdokimov, S.I.; Golik, V.I.; Gerasimenko, T.E.; Rubtsov, A.S. Flotationregime of gold-bearing ores with acirculationof therougher concentrate. *News Tula State Univ. Geosci.* **2023**, *1*, 382–404. (In Russian)
52. Evdokimov, S.I.; Gerasimenko, T.E. Development of the flotation mode of gold-bearing ores by an air mixture and water vapor. *Phys.-Tech. Probl. Dev. Miner. Resour.* **2021**, *2*, 162–177.
53. Bal, V. Stability characteristics of nanoparticles in a laminar linear shear flow in the presence of DLVO and non-DLVO forces. *Langmuir* **2019**, *35*, 11175–11187. [CrossRef]
54. Grasso, D.; Subramaniam, K.; Butkus, M.; Strevett, K.; Bergendahl, J. A review of non-DLVO interactions in environmental colloidal systems. *Environ. Sci. Biotechnol.* **2002**, *1*, 17–38. [CrossRef]
55. Li, K.; Wang, W.; Xiao, F.; Ge, Y.; Jin, H.; Yu, Z.; Gong, J.; Gao, W.; Peng, Z. Atomic force microscopy study of non-DLVO interactions between drops and bubbles. *Langmuir* **2021**, *37*, 6830–6837. [CrossRef] [PubMed]
56. Ojaniemi, U.; Riihimaki, M.; Manninen, M.; Pattikangas, T. Wall function model for particulate fouling applying XDLVO theory. *Chem. Eng. Sci.* **2012**, *84*, 57–69. [CrossRef]
57. Brant, J.A.; Childress, A.E. Membrane–colloid interactions: Comparison of extended DLVO predictions with AFM force measurements. *Environ. Eng. Sci.* **2002**, *19*, 413–427. [CrossRef]
58. Gomez-Flores, A.; Solongo, S.K.; Heyes, G.W.; Ilyas, S.; Kim, H. Bubble−particle interactions with hydrodynamics, XDLVO theory, and surface roughness for flotation in an agitated tank using CFD simulations. *Miner. Eng.* **2020**, *152*, 106368. [CrossRef]
59. Wu, Z.; Wang, X.; Liu, H.; Zhang, H.; Miller, J.D. Some physicochemical aspects of water-soluble mineral flotation. *Adv. Colloid Interface Sci.* **2016**, *235*, 190–200. [CrossRef]
60. Pan, L.; Yoon, R.-H. Measurement of hydrophobic forces in thin liquid films of water between bubbles and xanthate-treated gold surfaces. *Miner. Eng.* **2016**, *98*, 240–250. [CrossRef]
61. Liu, J.; Cui, X.; Xie, L.; Huang, J.; Zeng, H. Probing effects of molecular-level heterogeneity of surface hedrophobicity on hydrophobic interactions in air/water/solid systems. *J. Colloid Interface Sci.* **2019**, *557*, 438–449. [CrossRef]
62. Ong, Q.K.; Sokolov, I. Attachment of nanoparticles to the AFM tips for direct measurements of interaction between a single nanoparticle and surface. *J. Colloid Interface Sci.* **2000**, *310*, 385–390. [CrossRef]
63. Miller, J.D.; Wang, X.; Jin, J.; Shrimali, K. Interfacial water structure and the wetting of mineral surfaces. *Int. J. Miner. Process.* **2016**, *156*, 62–68. [CrossRef]
64. Nizkaya, T.V.; Dubov, A.L.; Mourran, A.; Vinogradova, O.I. Probing effective slippage on superhydrophobic stripes by atomic force microscopy. *Soft Matter. R. Soc. Chem.* **2016**, *12*, 6910–6917. [CrossRef] [PubMed]
65. Zhou, J.; Smiatek, J.; Asmolov, E.S.; Vinogradova, O.I.; Schmid, F. Application of Tunable-Slip Boundary Conditions in Particle-Based Simulations. *High Perform. Comput. Sci. Eng.* **2015**, *13*, 19–30.
66. Theodorakis, P.E.; Che, Z. Surface nanobubbles: Theory, simulation, and experiment. A review. *Adv. Colloid Interface Sci.* **2019**, *272*, 47–61. [CrossRef]
67. Sukharev, I.S. Experimental determination of the size of gas bubbles during effluence in the air/water system. *Bull. Volga State Acad. Sports* **2016**, *46*, 198–204.
68. Lu, S.T. On the role of hydrophobic interaction in flotation and flocculation. *Colloid J.* **1990**, *52*, 858–864.
69. Vigdergauz, V.E.; Schrader, E.A.; Sarkisoa, L.M.; Kuznetsova, I.N. Evaluation of long-range interactions between hydrophobic surfaces as applied to the flotation of sulfide minerals. *Phys.-Tech. Probl. Miner. Dev.* **2006**, *5*, 107–114.
70. Zhang, N.; Pang, T.; Han, R.; Chen, S.; Li, Z.; Yu, Y.; Shi, Z.; Liu, L.; Qu, J.; Zhou, A. Interactions between bubble and particles of key minerals of diasporic bauxite through the extended DLVO theory. *Int. J. Min. Sci. Technol.* **2022**, *32*, 201–214. [CrossRef]
71. Churaev, N.V. Surface forces and physicochemistry of surface phenomena. *Success Chem.* **2004**, *73*, 26–38.
72. Ficeriova, J.; Balaz, P. Leaching of gold from a mechanically and mechanochemically activated waste. *Acta Montan. Slovaca* **2010**, *15*, 183–187.

minerals
MDPI

Article

Experimental and Modelling Study of Pt, Pd, and 2E+Au Flotation Kinetics for Platreef Ore by Exploring the Influence of Reagent Dosage Variations

Parisa Doubra *, Candice Carelse, Deshenthree Chetty and Marian Manuel

Mintek, 200 Malibongwe Drive, Randburg, Johannesburg 2194, South Africa; candicec@mintek.co.za (C.C.); deshc@mintek.co.za (D.C.); marianm@mintek.co.za (M.M.)
* Correspondence: parisad@mintek.co.za; Tel.: +27-(11)-709-4032

Abstract: This study investigates the flotation kinetics of individual platinum-group elements (PGEs) and gold, namely Pt, Pd, and 2E+Au (i.e., Pt+Pd+Au), in the context of Platreef ore flotation. Experimental tests were conducted on a Platreef ore feed using various dosages of depressants, frothers, and collectors under controlled agitation and pH conditions. The recoveries of the individual PGEs were analysed using six kinetic models, with the modified Kelsall model identified as the most suitable for accurately describing the flotation kinetics and predicting elemental recovery. Notably, the model incorporates two rate constants (k_{fast} and k_{slow}) to account for the distinct flotation behaviours of the PGEs. The results indicate that Pt has the fastest floatability, followed by Pd and 2E+Au. The modified Kelsall model demonstrates high effectiveness in predicting the recovery of these PGEs. Three empirical correlations for Pt, Pd, and 2E+Au recoveries based on the modified Kelsall model are proposed, enhancing the understanding and optimisation of PGE recovery in Platreef ore flotation.

Keywords: PGM species; PGE floatability; kinetic model; Platreef; recovery

check for updates

Citation: Doubra, P.; Carelse, C.; Chetty, D.; Manuel, M. Experimental and Modelling Study of Pt, Pd, and 2E+Au Flotation Kinetics for Platreef Ore by Exploring the Influence of Reagent Dosage Variations. *Minerals* **2023**, *13*, 1350. https://doi.org/10.3390/min13101350

Academic Editors: Fardis Nakhaei, Ahmad Hassanzadeh, Luis A. Cisternas and Hyunjung Kim

Received: 13 September 2023
Revised: 9 October 2023
Accepted: 18 October 2023
Published: 23 October 2023

1. Introduction

Platinum group elements (PGEs) comprise a group of six metallic elements, namely, platinum, palladium, rhodium, ruthenium, iridium, and osmium [1,2]. These elements display common physical and chemical properties, including high melting points, resistance to wear and tear, and exceptional catalytic activity [3]. Due to their unique properties, PGEs are considered rare and valuable and find numerous applications in various industries such as automotive [4], aerospace [5], electronics [6,7], production of fertilisers [8], plastics [9], and pharmaceuticals [10].

PGEs make up platinum-group minerals (PGMs), and the latter are mainly found in deposits that are rich in nickel, copper, and other metals. The primary sources of PGMs are deposits in South Africa, Russia, and Canada, with smaller deposits being found in Zimbabwe, Australia, and the United States [11]. The Bushveld Complex in South Africa holds approximately 75% of the world's Pt resources and 50% of its Pd resources, with these precious metals being predominantly found in the Merensky, UG2, and Platreef layers [12,13]. The Platreef is a layered mafic-ultramafic intrusion located in the northern limb of the Bushveld Complex. This is a complex assemblage of different rock types, including serpentinites, pyroxenites, and calc-silicates, hosting predominantly PGE tellurides, alloys, arsenides, and sulphides. While Pt and Pd tellurides are the major contributors to the PGM assemblage, the Platreef is characterised by a scarcity of Pt-Pd sulphides compared with other reefs in the Bushveld Complex. The high concentration of telluride minerals in the Platreef is primarily represented by merenskyite and moncheite, whereas the arsenides are mainly composed of sperrylite and palladoarsenide [14–16].

The recovery of PGEs from the Platreef is achieved through flotation, a selective separation process based on differences in hydrophobicity of minerals [17]. Platreef ore poses a challenge in flotation as the ore contains a larger proportion of fine-grained PGMs in association with gangue minerals compared with other reefs in the Bushveld Complex [18]. This disturbs the froth stability owing to the low mass and high surface area of fine particles [19], unlike Merensky ore, with a straightforward flotation process [18]. Froth stability in flotation is essential for effectively recovering valuable minerals and preventing the entrainment of gangue minerals during water recovery [20]. Consequently, the frother dosage is slightly higher (30–50 g/t) for Platreef ore due to froth instability and bubble challenges compared with other reefs. The overall floatability of valuable minerals in the ore results in varying collector requirements, where each collector performs a different action and therefore may target different sized particles [21].

The development of a kinetic model exclusively tailored through data-fit constants for Platreef ore flotation considering the reagents and test conditions stands as a significant advancement, distinct from UG2 and Merensky ores within the Bushveld Complex. This study aims at quantifying Pt, Pd, and 2E+Au (i.e., Pt+Pd+Au) behaviour through modelling the unique flotation response to variable reagent dosages for Platreef ore. By focusing solely on Platreef ore flotation, the model captures the intricacies and complexities inherent to this particular reef, allowing for the prediction of flotation behaviour.

The flotation behaviour of PGMs can be influenced by several factors, including the mineralogy (such as mineral association), particle size, reagent regime, and flotation conditions. The choice of flotation reagents and their concentrations can significantly affect the recovery of PGMs, and the effectiveness of these reagents may depend on the mineralogical attributes of the PGM species [22]. Flotation conditions, such as pH, temperature, and agitation rate, can also influence the recovery of PGMs [23–25].

In the early days of flotation, the initial flotation model developed by Gaudin [26] in the 1930s expressed flotation recovery as an exponential function of time. Since then, a range of flotation models has been proposed, published, and refined over time. These models often take into account various factors, including particle size, reagent chemistry, and process conditions, to predict flotation performance and optimise process efficiency [27]. They range from simple empirical equations to more complex and mechanistic formulations. The selection of an appropriate kinetic model for a given flotation experiment or plant depends on the specific conditions and objectives.

One key aspect of the flotation process is the rate at which the commodity of value is recovered. An example of a model to describe the rate is the first-order model, which assumes that the rate of mineral recovery decreases exponentially over time. Nonetheless, other models have also been developed based on the probability of particle–bubble collision, attachment, detachment, particle size distribution, and distribution of floatability [28–30]. In recent years, the growing knowledge of the sub-processes that occur in the flotation cell has led to the development of more efficient flotation models that are used for process analysis, simulation, and optimisation. These models can be used to predict the flotation performance of different materials and machine types, as well as to optimise the operating conditions for maximum recovery and grade. However, the application of these models in the plant is challenged by the uncertainties and complexities of industrial operations. Therefore, the development of accurate and reliable kinetic models is crucial for the optimisation of froth flotation in the mineral processing industry. Readers are referred to the review papers and theses cited here [27,31,32].

The prevailing scientific consensus holds that the flotation process is governed by a first-order kinetic model that is characterised by a dependence on particle concentration and a rate constant. This view has primarily been advanced in [33–35], among others [27]. To monitor and quantify the efficacy of this process, it is customary to evaluate the recovery of a specific component over time, known as R [12,36].

In practice, the most common approach for characterising each fraction involves using deterministic k-R_{max} pairs. This approach assumes that the flotation rate constants are deterministic and time-invariant for narrower fractions with respect to particle properties. However, such an approach may not be flexible enough to represent slow and sustained increasing recovery trends that are often observed in flotation responses with slow-floating components. In addition, over-fitting can occur when using a large number of discrete rate constants to represent a category, especially when the number of model parameters is comparable to the number of experimental data points. The use of more flexible models to represent such responses [37–39] has thus been suggested.

The attachment of mineral particles to bubbles is a complex process that involves several physical and chemical mechanisms, including adsorption, desorption, and chemical reactions. These mechanisms can be described mathematically using various models, which typically involve a set of differential equations that describe the concentration of particles and reagents in the flotation cell as a function of time [40,41]. The efficiency and selectivity of the flotation process are usually evaluated by measuring mineral recoveries and enrichment ratios at specific flotation or residence times. However, in many cases, it is necessary to investigate the performance of the process over time [42]. This is achieved by analysing the kinetic response of the process, which is essentially the change in mineral concentrations or cumulative recovery over time. The kinetic response has several important applications in mineral processing, including determining the maximum achievable recoveries, comparing different flotation types, investigating the effects of various operating conditions, scaling up metallurgical results, designing flotation circuits, and simulating flotation processes [32,43–45]. This study presents a novel and comprehensive investigation into the flotation kinetics of Pt, Pd, and 2E+Au based on experimental studies [46] using Platreef ore samples. The optimised parameters for the dosage of depressant, frother, and collector, as well as the monitoring of agitation rate and pH under different reagent conditions, allowed for a thorough and detailed assessment of the flotation performance. This paper is the first ever study to predict individual PGE flotation kinetics for the Platreef ore, achieving strong predictability despite the complex nature of the ore body as well as the reagent dosages. Notably, the most successful model demonstrated excellent capability for extrapolation.

2. Materials and Methods

2.1. Materials

A detailed description of the materials, methods, and experimental procedure is outlined in [46]. The chemicals employed in this investigation consisted of a frother (Senfroth522), depressant (Sendep30E), and collector (sodium isobutyl xanthate (SIBX) obtained from SENMIN. The Platreef ore feedstock was derived from a 100 kg bulk sample originating from the southern Platreef deposit.

2.2. Experimental Equipment and Procedure

The schematic of the experimental set-up is displayed in Figure 1. The experimental apparatus used in this study comprised a D12 Denver flotation machine with a cell capacity of 2.5 L. The cell was equipped with mechanical agitators to ensure uniform pulp mixing and an aeration system for controlled air bubble introduction with airflow measurements using a rotameter. This employs a linear scale, wherein 100% corresponds to an airflow rate of 21 NL/min at an absolute line pressure of 4.85 bar. This system is equipped with a pressure regulator to facilitate control. The incoming air is subjected to filtration through an in-line air cleaner, serving to inhibit the ingress of particulate matter and oil contaminants originating from the air compressor, thus safeguarding the integrity of the flotation cell [47]. Furthermore, a reagent addition system allowed precise dosing of chemicals. Within this configuration, the overflow from each test was designated as the rougher concentrate (RC), representing the fraction enriched with the desired minerals. Conversely, the collective effluent from the final experiment constituted the rougher tailing (RT), encompassing the particles that did not respond favourably to the flotation process. All products (concentrates and tails) were dried, and a sub-sample was split out for subsequent chemical analysis and mineralogical analysis. The samples were analysed using inductively coupled plasma atomic emission spectroscopy (ICP-OES), and the mineral liberation of PGMs was obtained using a mineral liberation analyser (MLA).

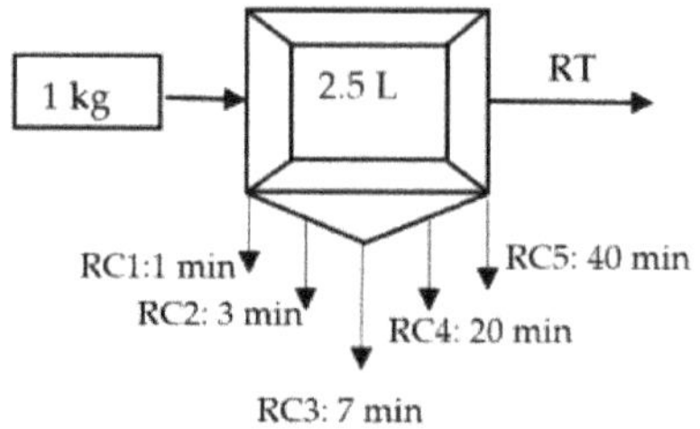

Figure 1. Schematic of the flotation set-up used in this work.

The behaviour of individual PGM species was investigated by liberating PGM grains from Platreef ore through fine grinding of the feed sample to 90% passing 75 µm. This fine grind was selected as the flotation response of liberated PGE species and was the focal point of our study. The Platreef feed sample contained PGE with a grade of approximately 2.8 g/t 2E (Au = 0.12 g/t, Pd = 1.61 g/t, and Pt = 1.17 g/t). The ore was subsequently crushed to 100% passing 1.7 mm using a jaw crusher and cone crusher, and 1 kg sub-samples were obtained using a rotary splitter for flotation test work.

Flotation test work was performed at a solids concentration of 35% and an impeller speed of 1200 rpm with a constant airflow. Tap water from the South African Rand Water supplier was used as the flotation medium. Table 1 presents the levels of cations, anions, conductivity, total dissolved solids (TDS), and pH identified in the tap water used for this study.

Table 1. Concentrations of selected ions in the sampled water [46].

Ions	Concentrations/ppm	Ions	Concentrations/Unit
Ag	<1	Si	<0.05 (ppm)
Al	<1	Li	<0.05 (ppm)
Ca	26.6	K	1.91 (ppm)
Cr	0.05	NO_3^-	5.14 (ppm)
Fe	0.51	Sulphide S	<0.05 (mg/L)
Mg	7.55	Conductivity	227 (uS/cm)
Pb	0.061	TDS	113.8 (mg/L)
V	0.06	pH	8.02

To investigate the response of individual liberated PGM species, reduced collector, depressant, and frother, dosages were tested. Figure 2 shows backscattered electron images of liberated PGM in the feed and concentrate.

The impact of the collector dosage on the recovery and flotation kinetics of PGEs was studied, while the frother's indirect effect was examined through froth phase effects. The reagents used and conditioning times are presented in Table 2. SIBX was introduced, and the slurry underwent a 2-min conditioning period. Sendep30E was added, and conditioning continued for an additional 3 min. Lastly, Senfroth522 was introduced, and the slurry underwent a final 1-min conditioning stage. The concentrate was manually collected by scraping once every 15 s using paddles.

Five timed RCs and an RT were produced from each test to ensure the reproducibility and statistical significance of the result. After each stage, an RC was collected, resulting in five rougher concentrates obtained at different cumulative flotation times of 1, 3, 7, 20, and 40 min, respectively (Table 2). Flotation progressed, resulting in the recovery of RC1 at 1 min, RC2 at 3 min after RC1 was taken, RC3 at 7 min after RC2, RC4 at 20 min after RC3, and RC5 at 40 min.

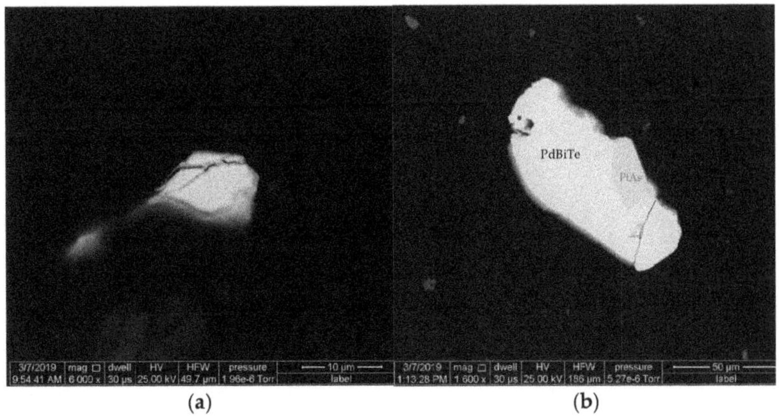

Figure 2. Backscattered electron image of (**a**) PtAs in the feed and (**b**) PdBiTe associated with PtAs in RC1.

Table 2. Platreef sample flotation test conditions.

Reagents & Dosages	Conditioning Time/min	Float Time/min
SIBX: 30 and 120 g/t	2	-
Sendep30E: 300 and 500 g/t	3	-
Senfroth522: 30 and 50 g/t	1	-
Rougher concentrate 1 (RC1)	-	1
Rougher concentrate 2 (RC2)	-	3
Rougher concentrate 3 (RC3)	-	7
Rougher concentrate 4 (RC4)	-	20
Rougher concentrate 5 (RC5)	-	40

The rougher flotation circuit is a vital component of the mineral processing circuit since it aims to achieve maximum recovery of valuable minerals from the feed material. Consequently, understanding the behaviour of PGM minerals during rougher flotation was crucial for optimising the overall recovery and efficiency of the mineral processing circuit.

2.3. Modelling

Various kinetic models have been studied to produce a comparative overview of their performance and to develop an accurate predictive tool. Flotation modelling has become increasingly sophisticated, with more advanced models being developed to incorporate additional complexities, such as the effect of particle size, the impact of froth zone turbulence, and the influence of inter-particle forces. However, the batch flotation models presented in this table remain highly relevant, especially in the context of the rougher stage of flotation processes.

Table 3 lists the flotation kinetic models parameterised in this study, including details on their equations, parameters, and associated references. The Classic, Klimpel, and second-order Klimpel models describe the mineral recovery over time in batch flotation processes, where R represents the recovery, k represents the flotation rate constant at which the minerals are recovered during the flotation process, t represents time, and R_{max} represents the species maximum recovery.

Table 3. Kinetic models for flotation separation.

Model Name	Equation	Kinetic Parameters	Ref.
Classical first-order model	$R = R_{max}(1 - e^{-Kt})$	R_{max} and k	[28]
Klimpel model	$R = R_{max}\left[1 - \frac{1}{k_{max}t}\left(1 - e^{-k_{max}t}\right)\right]$	R_{max} and k_{max}	[29]
Second-order	$R = \frac{R_{max}{}^2 k_{max}t}{1+R_{max}k_{max}t}$	$R_{max}\ and\ k_{max}$	[28]
Second-order Klimpel	$R = R_{max}\left[1 - \frac{1}{k_{max}t}ln(1 + k_{max}t)\right]$	$R_{max}\ and\ k_{max}$	[30]
Kelsall	$R = R_{fast}\left(1 - e^{-k_{fast}t}\right) + R_{slow}\left(1 - e^{-k_{slow}t}\right)$	$100\% = R_{fast} + R_{slow}$ $k_{slow}\ and\ k_{fast}$	[47]
Modified Kelsall	$R = R_{fast}\left(1 - e^{-k_{fast}t}\right) + R_{slow}\left(1 - e^{-k_{slow}t}\right)$	$R_{max} = R_{fast} + R_{slow}$ $k_{slow}\ and\ k_{fast}$	[33,34]

In the Kelsall and modified Kelsall models, R_{fast} and R_{slow} represent the fractions of species undergoing fast and slow flotation, respectively, while k_{fast} and k_{slow} denote the corresponding fast and slow flotation rate constants. In the Kelsall model, species are categorised as either fast or slow floaters, ensuring that the combined fractions of these species always amount to 100%, as demonstrated in Equation (1). However, in the modified Kelsall model, the sum of these fractions corresponds to R_{max}, signifying the maximum recoverable fraction of species, as depicted in Equation (2):

$$R_{fast}+R_{slow} = 100\%, \tag{1}$$

$$R_{fast}+R_{slow} = R_{max}. \tag{2}$$

The provided flotation models were assessed to gauge their accuracy in predicting the behaviour of PGEs, including Pt, Pd, and 2E+Au. To evaluate their performance, each model was fitted to experimental data to obtain the most appropriate kinetic parameters. The effectiveness of each model was subsequently determined by comparing its projected flotation kinetics with the actual kinetics observed during laboratory testing.

2.4. Statistical Analysis

In statistical evaluation, R-squared (R^2) and root mean squared error ($RMSE$) were used to assess the effectiveness of regression models. The formula used to calculate R^2 is given in Equation (3):

$$R^2 = 1 - \frac{RSS}{TSS}, \tag{3}$$

where RSS is the sum of the squared residuals (the difference between actual and predicted values), and TSS is the total sum of squares (the difference between actual values and the mean value). The formula used to calculate $RMSE$ is shown in Equation (4):

$$RMSE = \sqrt{\sum_{i=1}^{N}\left(z_{f_i} - z_{o_i}\right)^2/N}. \tag{4}$$

Here, z_{f_i} is the predicted value, z_{o_i} is the actual value, N is the number of observations in the dataset, and the sum is taken over all observations.

3. Results and Discussion

Regression analysis was used to determine the best data fit model among the models listed above that could provide a quantitative understanding of the underlying mechanisms that influence the flotation process. The residual study, a statistical method used to assess the accuracy of the models, was used to evaluate the goodness of fit of the models.

3.1. Experimental Results

Table 4 provides the recovery percentages of the three elements—Pt, Pd, and Au—at different stages of flotation using different combinations of collector, frother, and depressant. Table 4 focuses on the cumulative recovery percentage. For example, at 4 min (1+3 min), the recovery data are collected from both rougher concentrate 1 (collected after 1 min) and rougher concentrate 2 (collected after 3 min). Recovery is defined as the percentage of the valuable elements present in the ore that is recovered to the concentrate product. The outcomes indicate that the overall Pt recovery remains stable under diverse test conditions, whereas the recovery percentages for Pd are impacted by variations in reagent dosages. The optimal Pd recovery of 86.27% is achieved under the experimental parameters of 30 g/t collector, 30 g/t frother, and 300 g/t depressant.

Table 4. Impact of collector, frother, and depressant concentrations on the flotation performance of Platreef ore with cumulative times of 1, 4, 11, 31, and total of 40 min for the tests from RC1 to RC1+2+3+4+5, respectively.

Cumulative	Platinum %	Palladium %	2E + Au %
Collector 120 g/t, frother 50 g/t, and depressant 300 g/t			
RC1	55.44	52.28	52.13
RC1+RC2	72.63	68.64	68.37
RC1+RC2+RC3	78.45	75.48	74.56
RC1+RC2+RC3+RC4	83.57	80.99	79.74
RC1+RC2+RC3+RC4+RC5	86.96	84.64	83.16
Collector 30g/t, frother 50 g/t, and depressant 300 g/t			
RC1	56.69	53.57	56.01
RC1+RC2	72.76	69.70	71.39
RC1+RC2+RC3	78.96	76.54	77.68
RC1+RC2+RC3+RC4	84.59	82.05	83.06
RC1+RC2+RC3+RC4+RC5	86.84	84.62	85.45
Collector 120 g/t, frother 30 g/t, and depressant 300 g/t			
RC1	67.65	56.50	62.16
RC1+RC2	77.40	69.43	73.21
RC1+RC2+RC3	80.89	75.06	77.71
RC1+RC2+RC3+RC4	84.38	79.39	81.57
RC1+RC2+RC3+RC4+RC5	86.81	82.43	84.29
Collector 120 g/t, frother 50 g/t, and depressant 500 g/t			
RC1	60.87	46.23	53.21
RC1+RC2	74.45	63.27	68.35
RC1+RC2+RC3	79.55	70.57	74.48
RC1+RC2+RC3+RC4	83.88	77.53	80.10
RC1+RC2+RC3+RC4+RC5	86.72	81.92	83.71
Collector 30 g/t, frother 30 g/t, and depressant 300 g/t			
RC1	64.32	57.03	60.32
RC1+RC2	73.98	70.40	72.01
RC1+RC2+RC3	79.51	76.99	78.12
RC1+RC2+RC3+RC4	83.71	82.47	83.03
RC1+RC2+RC3+RC4+RC5	86.63	86.27	86.43

As expected, the results reveal that the selection of collector, frother, and depressant can significantly impact the recovery of the valuable elements. This is likely due to better selectivity achieved at the lower collector concentration. The higher dosage of SIBX is more applicable to UG2; compared with Platreef, UG2 ore may require a slightly higher dosage due to the limited floatability of valuable mineral species in combination with a co-collector such as dithiophosphate [19]. In contrast, the Platreef ore's complex composition may contain Pd minerals with a high affinity for collectors, allowing for effective flotation at lower dosages. To address the issue of higher gangue recovery, many operations have opted to utilise higher concentrations of depressants [48]. However, increasing the depressant concentration from 300 g/t to 500 g/t with the same collector and frother dosage resulted in a decrease in recovery for Pd, likely due to the excessive use of depressant, leading to hindered collector attachment and froth destabilisation. In this context, the disparities in Pt concentrations are insignificant.

3.2. Kinetic Model

Figure 3 presents recovery percentages of Pt and Pd at various flotation stages, using different combinations of collector, frother, and depressant. Appendix A contains extra model results for 2E+Au. The results reveal distinct variations in the floatability of these PGEs. Among the models employed, the modified Kelsall model exhibited remarkable performance, demonstrating high R_{max} values and yielding excellent R^2 and low RMSE values (Table 5). The recovery percentages on the graphs have been adjusted to enhance visibility within the range of 45–85%, and the graphs commence from an initial recovery of zero at time zero. This suggests that the modified Kelsall model accurately captured the flotation kinetics and proved effective in predicting the floatability of Pt and Pd. Notably, Pt demonstrated favourable floatability, as indicated by its high R_{max} value and the model's robust fit. Pd exhibited slightly lower R_{max} values but still demonstrated good flotation response according to the model.

Table 5. Modified Kelsall model for PGE recovery.

Modified Kelsall	R_{max}	R_{fast}	k_{fast} min^{-1}	k_{slow} min^{-1}	R^2	**RMSE**
	Collector 120 g/t, frother 50 g/, and depressant 300 g/t					
Pt	0.888	0.700	1.52	0.050	0.99822	0.006
Pd	0.856	0.646	1.561	0.060	0.99758	0.005
2E+Au	0.844	0.649	1.543	0.056	0.99787,	0.005
	Collector 30 g/t, frother 50 g/t, and depressant 300 g/t					
Pt	0.874	0.683	1.668	0.071	0.99935	0.002
Pd	0.849	0.646	1.6411	0.077	0.99886	0.002
2E+Au	0.859	0.668	1.706	0.073	0.99910	0.001
	Collector 120 g/t, frother 30 g/t, and depressant 300 g/t					
Pt	0.895	0.758	2.170	0.037	0.99900	0.001
Pd	0.829	0.659	1.841	0.063	0.99785	0.002
2E+Au	0.855	0.708	2.025	0.051	0.99842	0.001
	Collector 120 g/t, frother 50 g/t, and depressant 500 g/t					
Pt	0.877	0.715	1.827	0.056	0.99846	0.002
Pd	0.859	0.651	1.618	0.049	0.99741	0.011
2E+Au	0.859	0.651	1.618	0.049	0.99790	0.005
	Collector 30 g/t, frother 30 g/t, and depressant 300 g/t					
Pt	0.870	0.704	2.293	0.066	0.99796	0.000
Pd	0.876	0.666	1.826	0.056	0.99730	0.002
2E+Au	0.873	0.683	2.018	0.060	0.99758	0.001

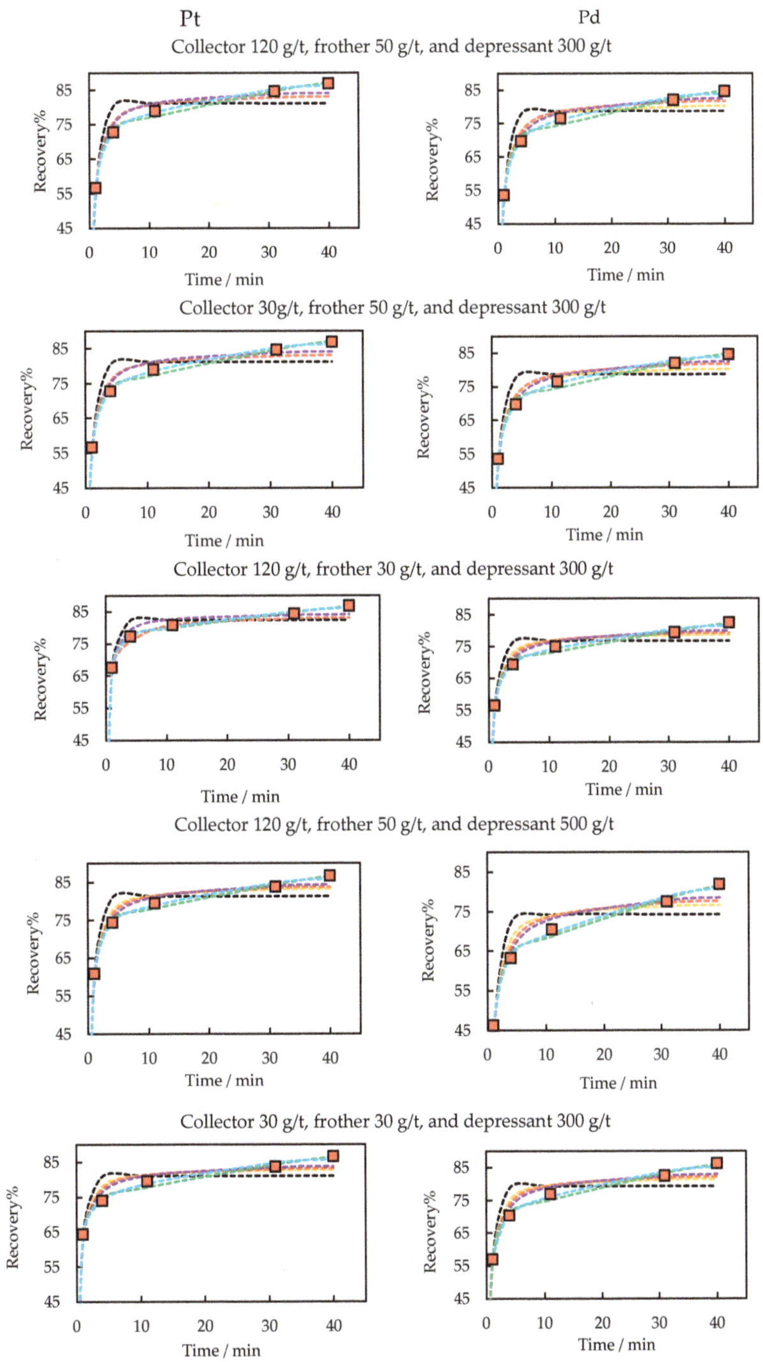

Figure 3. PGEs recovery data fit results for Pt and Pd, using different combinations of collector, frother, and depressant. ■ Experimental data; --- classic model; --- Klimpel model; --- second order, --- second-order Klimpel; --- Kelsall; and --- modified Kelsall.

3.3. Model Extrapolation Performance

Table 5 presents parameters and statistics for the modified Kelsall model under different reagent conditions. Additional model results are provided in Appendix A. The modified Kelsall model's high R_{max} values for Pt, Pd, and 2E+Au suggest efficient recovery due to favourable floatability. It effectively captures flotation kinetics, offering insights into particle attachment and detachment. The modified Kelsall model demonstrates accuracy in predicting Pt, Pd, and 2E+Au flotation in Platreef ore with strong correlations (R^2) and low $RMSE$ values. The R_{fast} for Pt exhibits the highest floatability, followed by 2E+Au and Pd. Pd-bearing minerals are found to be more oxidised than Pt-bearing minerals, which could be a contributing factor since they offer fewer active sites for collector adsorption [49].

Based on Ramlall, the flotation for 2E+Au in the UG2 deposit involved the use of SIBX as collector at 150 g/t, KU5 as frother at 30 g/t, and dowfroth 200 as depressant at 20 g/t [31,47]. The high value of R_{max} (88.26%) suggests efficient flotation recovery, indicating the successful separation of the valuable 2E+Au. In the flotation tests for Platreef ore, different reagents were utilised compared to UG2, including different combinations of collector, frother, and depressant. The highest achieved recovery (R_{max}) of 87.60% in Platreef ore was observed for 2E+Au under the optimal variable conditions, including collector at 30 g/t, frother at 30 g/t, and depressant at 300 g/t. The flotation testing results for 2E+Au in UG2 and Platreef indicate successful separation of the valuable minerals using various reagent combinations.

Irrespective of different reagents and dosages in the tests, for Platreef ore, R_{max} was determined as 0.873, comprising 0.68 R_{fast} and 0.19 R_{slow}. In contrast, UG2 ore demonstrated a higher R_{max} of 0.8826, divided into 0.62 R_{fast} and 0.26 R_{slow} [31,47]. UG2 ore demonstrated a slightly higher overall recovery compared with Platreef ore. Moreover, Platreef ore exhibited a higher proportion of fast-floating particles, while UG2 ore demonstrated a larger fraction of slow-floating species.

For the Platreef ore, the rate constant values for 2E+Au at the optimum condition of collector 30 g/t, frother 30 g/t and depressant 300 g/t are calculated as k_{fast} = 2.02 min^{-1} and k_{slow} = 0.06 min^{-1}. In contrast, the UG2 ore demonstrates higher rate constant values, with k_{fast} = 2.26 min^{-1} and k_{slow} = 0.13 min^{-1}. Platreef ore exhibits a comparatively lower k_{fast}, indicating a slower initial attachment of particles to bubbles. Conversely, UG2's higher k_{fast} points to a swifter initial particle–bubble interaction. The variance in k_{slow} values mirrors a similar trend: Platreef ore presents a relatively lower rate of slow flotation compared to UG2, which is affected by the differences in the mineralogy of the two ores as mentioned in the introduction.

The study on recovering Pt, Pd, and 2E+Au from Platreef ore has identified the modified Kelsall model as the most fitting model for the system, among others tested. The modified Kelsall model's superior performance is attributed to its ability to capture the nuances of the flotation process through its incorporation of two rate constants, k_{fast} and k_{slow}, which describe the flotation behaviour of two distinct fast and slow floating populations. The model's success is also due to its balance between accuracy and complexity. Despite its increased complexity compared to alternative models, the modified Kelsall model remains interpretable and practical, making it an ideal tool for both understanding the flotation of Platreef ore and optimising process conditions. Moreover, the model's capacity to handle changes in reagent conditions tested for Platreef ore more effectively than other models likely contributes to its enhanced accuracy in representing the flotation process. Generally, higher k values suggest faster flotation kinetics. In this case, the k values for the three elements are relatively consistent across the different tests, indicating that the flotation kinetics are not significantly affected by the changes in reagent dosages.

Table 6 lists three empirical correlations based on the modified Kelsall model to predict the flotation of individual PGEs/PGE groupings from Platreef ore for the frother, collector and depressant dosages in the ranges of 30–50, 30–120, and 300–500 g/t, respectively. Figure 4 depicts the recovery data fit results of modified Kelsall for Pt under all the conditions tested. In the case of the Platreef ore, a depressant dosage of 300 g/t of

Sendep suppresses gangue minerals with stronger flotation tendencies. An elevated dosage improves inhibition, thereby enhancing selectivity for valuable minerals.

Table 6. Empirical correlation based on modified Kelsall model to predict the flotation of individual PGEs/PGE groups from Platreef ore for the collector, frother, and depressant dosage ranges of 30–120, 30–50, and 300–500 g/t.

PGEs	Model
Pt	$R = 0.71\left(1 - e^{-1.86t}\right) + 0.17\left(1 - e^{-0.06t}\right)$
Pd	$R = 0.64\left(1 - e^{-1.65t}\right) + 0.21\left(1 - e^{-0.06t}\right)$
2E+Au	$R = 0.67\left(1 - e^{-1.77t}\right) + 0.18\left(1 - e^{-0.06t}\right)$

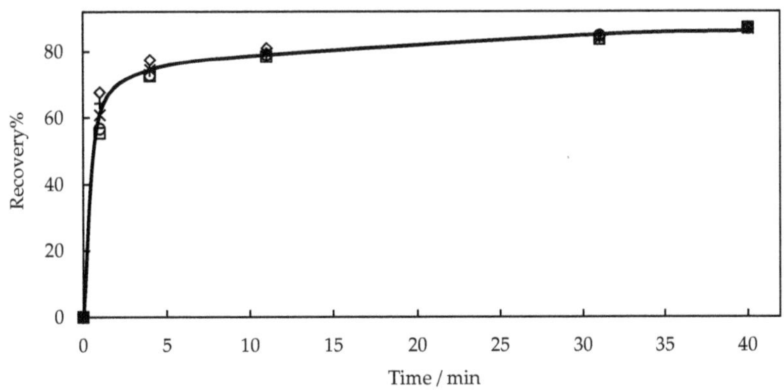

Figure 4. Recovery data fit results of Pt for all the conditions tested. ___ modified Kelsall; depressant, collector, and frother concentrations of □ 300, 120, and 50; ○ 300, 30, and 50; ♦ 300, 50, and 30; × 500, 120, and 50; and + 300, 120, and 50.

The ability of the modified Kelsall model to accurately describe the data as per findings here suggests that it is an appropriate model for predicting metal recovery from Platreef ore under different reagent dosage conditions. Moreover, the residual analysis of the data demonstrated that the modified Kelsall model fitted in this study was able to extrapolate results beyond the range of the data used to fit the model, while the precision of predictions requires further consideration. While the model's predictions consistently approximate the true value, an asymmetrical distribution of error values and deviation from the mean indicates a lack of precision in predictions.

Figure 5 depicts the scatter of residuals as red rings, which provide a one-sided representation of the deviation from the precision of the data fit model in relation to experimental results. The relatively small error figures suggest that a single correlated modified Kelsall model is highly accurate overall. However, this finding implies that the model may not be able to make precise predictions for extended flotation tests, which can be a critical aspect of mineral processing optimisation.

Figures 6–9 showcase the recovery and residuals for Pd and 2E+Au species. Despite the observed deviations for Pt (see Figure 3), the modified Kelsall model has exhibited superior performance in these instances, with the residuals scattering randomly around both sides of the coordinates.

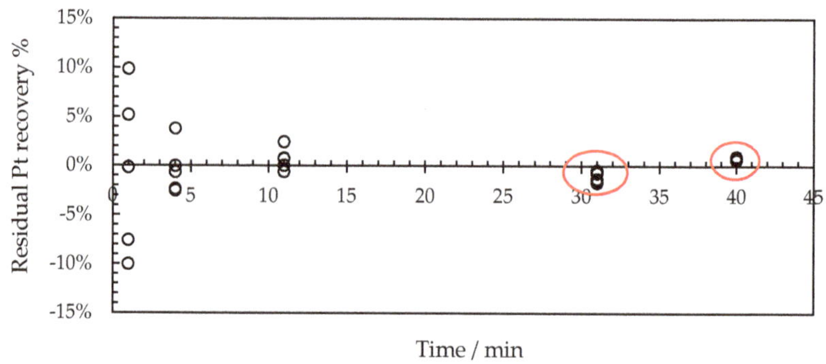

Figure 5. Residual study of modified Kelsall model for Pt recovery in five different tests. ○ Discrepancies associated with each data point when compared to experimental measurements.

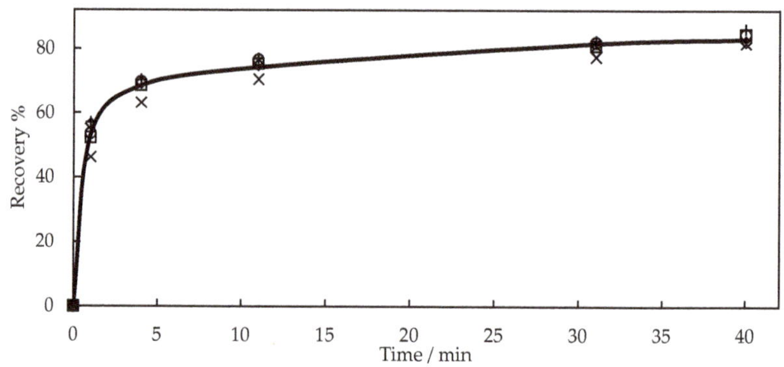

Figure 6. Recovery data fit results of Pd for all the tested conditions: __ modified Kelsall; depressant, collector, and frother concentrations of □ 300, 120, and 50; ○ 300, 30, and 50; ♦ 300, 50, and 30; × 500, 120, and 50; and + 300, 120, and 50.

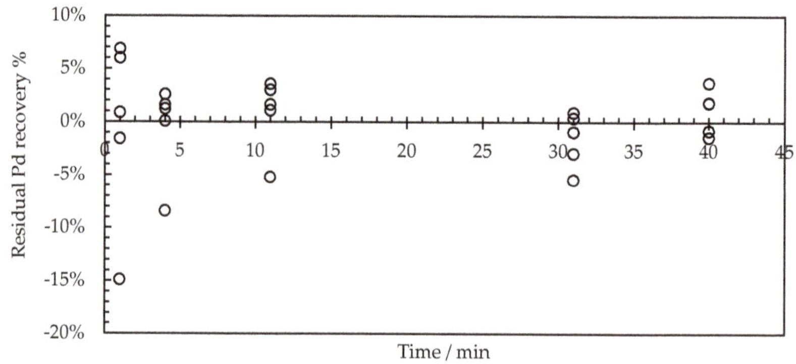

Figure 7. Residual study of modified Kelsall model Pd recovery in five different tests. ○ Discrepancies associated with each data point when compared to experimental measurements.

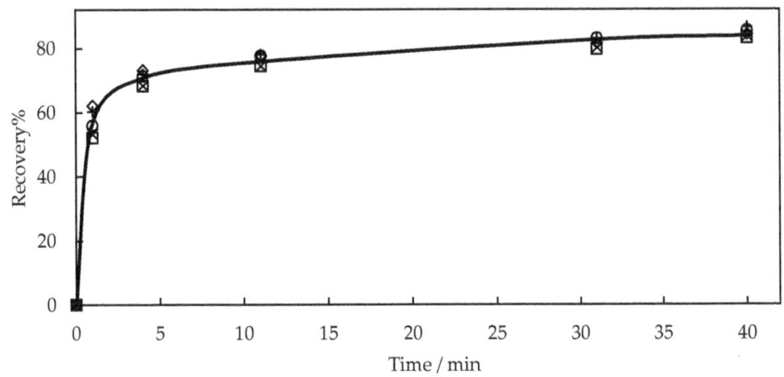

Figure 8. Recovery data fit results of 2E+Au for all the tested conditions: __ modified Kelsall; depressant, collector, and frother concentrations of □ 300, 120, and 50; ○ 300, 30, and 50; ⧫ 300, 50, and 30; × 500, 120, and 50; and + 300, 120, and 50.

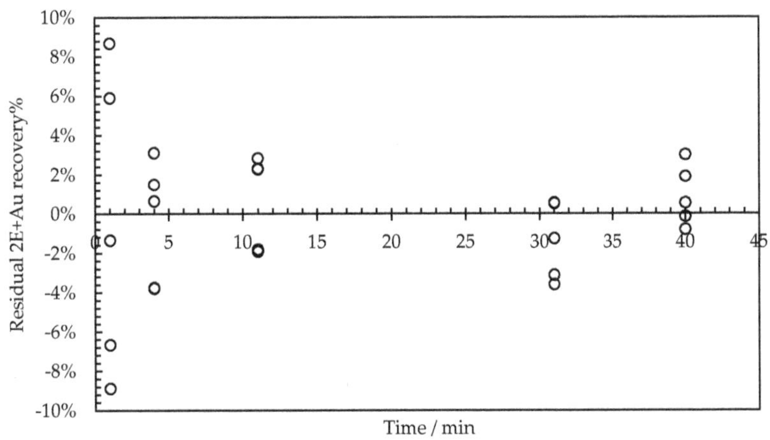

Figure 9. Residual study of modified Kelsall model 2E+Au recovery in five different tests. ○ Discrepancies associated with each data point when compared to experimental measurements.

4. Conclusions

Experimental data were generated for the flotation of Platreef ore using various depressant, frother, and collector dosages at a controlled agitation rate, and the recoveries of individual PGE (Pt, Pd, 2E+Au) have been analysed using six different kinetic models. The modified Kelsall model was found to be the most suitable model for accurately describing the kinetics of the flotation process and predicting metal recovery under different reagent dosage conditions. The model's ability to account for the distinct flotation behaviours of two distinct PGE, viz., Pt and Pd, as well as 2E+Au, through the incorporation of two rate constants, $k_{fas}t$ and k_{slow}, is a significant advantage in modelling the complex Platreef ore flotation system. The findings suggest that Pt has the fastest floatability, followed by Pd and 2E+Au. Pd minerals, with higher oxidation levels, may have fewer active adsorption sites, possibly accounting for the collector adsorption difference compared to Pt minerals.

The modified Kelsall model is a highly effective method among models studied for predicting the recovery of Pt, Pd, and 2E+Au from Platreef ore through flotation. The residual analysis approach enabled the refinement of the model and increased confidence in its ability to accurately predict data outside the range of the training data. Three empirical correlations following the modified Kelsall model were proposed for Pt, Pd, and 2E+Au recoveries, making this a novel advancement for Platreef ore flotation performance prediction.

Author Contributions: Conceptualization, all authors.; methodology, all authors.; software, P.D.; validation, all authors.; formal analysis, P.D.; investigation, P.D. and C.C.; data curation, P.D. and C.C.; writing—original draft preparation, P.D.; writing—review and editing, all authors.; visualization, P.D.; supervision, D.C. and M.M.; project administration, C.C.; funding acquisition, C.C. All authors have read and agreed to the published version of the manuscript.

Funding: Funding for this work was provided via Mintek's Science Vote Grant.

Data Availability Statement: All data have been provided in the manuscript and Appendix A.

Acknowledgments: The authors would like to thank Mintek for providing financial support to this project. Colleagues in the Minerals Processing Division are acknowledged for conducting the flotation tests.

Conflicts of Interest: The authors declare no conflict of interest.

Appendix A

The test conditions are summarised in Table A1 and represented in Figures A1–A5.

Table A1. Experimental flotation tests on the Platreef samples with different reagent dosing.

Test No.	Depressant [1]	Collector [1]	Frother [1]	Mixer Rate [2]	pH	Representation
1	300	120	50	1200 (all tests)	9 (all tests)	Figure A1
2	300	30	50			Figure A2
3	300	120	30			Figure A3
4	500	30	50			Figure A4
5	500	30	30			Figure A5

[1] Reagent concentration in g/t. [2] Mixer rate in rpm.

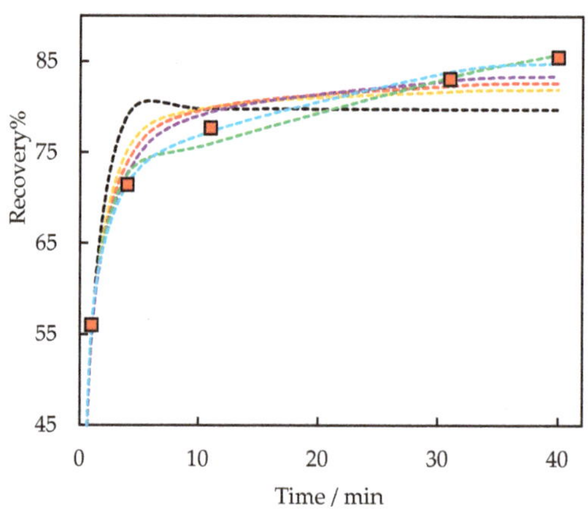

Figure A1. PGE recovery data fit results 2E+Au with depressant, collector, and frother concentrations of 300, 120, and 50 g/t, respectively. ■ Experimental data; --- classic model; --- Klimpel model; --- second order, --- second-order Klimpel; --- Kelsall; and --- modified Kelsall.

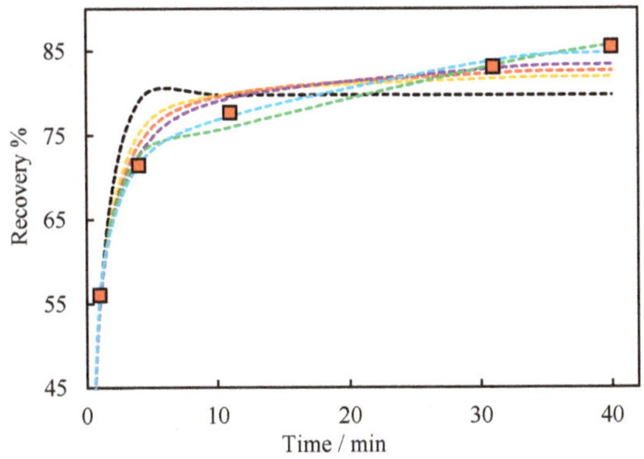

Figure A2. Recovery data fit results 2E+Au with depressant, collector, and frother concentrations of 300, 50, and 30 g/t, respectively. ■ Experimental data; --- classic model; --- Klimpel model; --- second order, --- second-order Klimpel; --- Kelsall; and --- modified Kelsall.

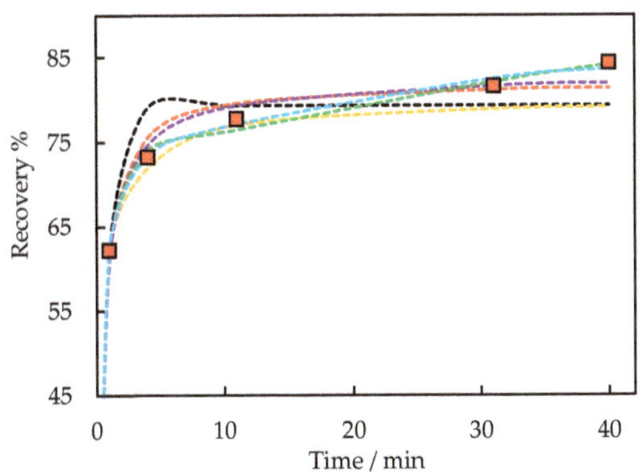

Figure A3. Recovery data fit results 2E+Au with depressant, collector, and frother concentrations of 300, 120, and 30 g/t, respectively. ■ Experimental data; --- classic model; --- Klimpel model; --- second order, --- second-order Klimpel; --- Kelsall; and --- modified Kelsall.

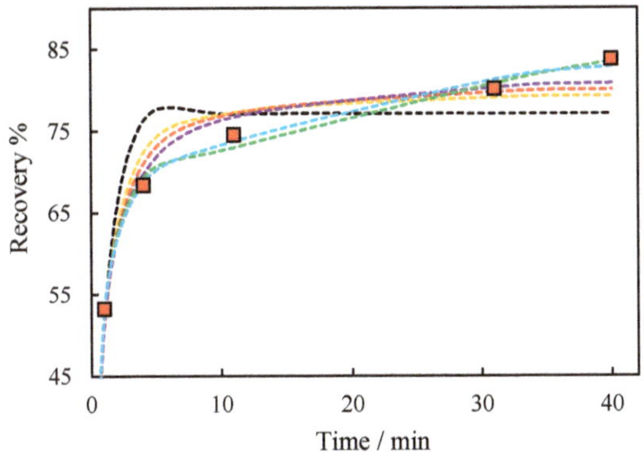

Figure A4. Recovery data fit results 2E+Au with depressant, collector, and frother concentrations of 500, 120, and 50 g/t, respectively. ■ Experimental data; --- classic model; --- Klimpel model; --- second order, --- second-order Klimpel; --- Kelsall; and --- modified Kelsall.

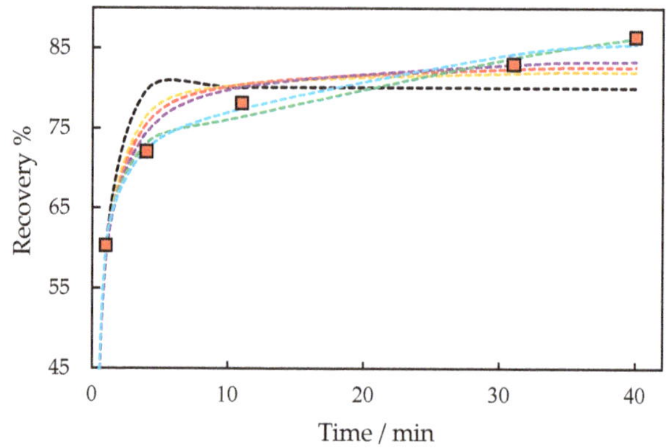

Figure A5. Recovery data fit results 2E+Au with depressant, collector, and frother concentrations of 300, 30, and 30 g/t, respectively. ■ Experimental data; --- classic model; --- Klimpel model; --- second order, --- second-order Klimpel; --- Kelsall; and --- modified Kelsall.

Table A2 presents PGE recovery data fit results of Pt, Pd, and 2E+Au with depressant, collector, and frother concentrations of 300, 120, and 50 g/t, respectively. The analysis of various flotation models for Pt, Pd, and 2E+Au in Platreef ore for this experimental condition reveals that the modified Kelsall model performs exceptionally well in predicting the flotation behaviour of these PGEs. The model exhibits high R_{max} values for Pt, Pd, and 2E+Au, indicating their favourable floatability and potential for efficient recovery. The model's ability to capture the kinetics of flotation, including fast and slow flotation rate constants, provides valuable insights into the attachment and detachment processes of PGM particles during flotation. With high correlation coefficients (R2) and low root mean square error (RMSE) values, the modified Kelsall model (parameterised for this work) demonstrates its accuracy and reliability in predicting the flotation performance of Pt, Pd, and 2E+Au in Platreef ore. Overall, Pt demonstrates the highest floatability, followed by Pd and 2E+Au.

Table A3 presents recovery data fit results of Pt, Pd, and 2E+Au with depressant, collector, and frother concentrations of 30, 50, and 300 g/t, respectively. The results reveal distinct variations in the floatability of these PGEs. Among the models employed, the modified Kelsall model exhibited remarkable performance, demonstrating high R_{max} values and yielding excellent R^2 and low RMSE values. This suggests that the modified Kelsall model accurately captured the flotation kinetics and proved effective in predicting the floatability of Pt, Pd, and 2E+Au. Notably, Pt demonstrated favourable floatability, as indicated by its high R_{max} value and the model's robust fit. Pd exhibited slightly lower R_{max} values but still demonstrated good flotation response according to the model. On the other hand, 2E+Au displayed lower R_{max} values compared to Pt and Pd, indicating relatively lower floatability for this PGE group.

Table A4 presents PGE recovery data fit results of Pt, Pd, and 2E+Au with depressant, collector, and frother concentrations of 120, 30, and 300 g/t, respectively. Among the models employed, the modified Kelsall model exhibited excellent performance, with high R_{max} values, strong R^2, and low RMSE values. This indicates the model's ability to accurately describe the flotation kinetics of Pt, Pd, and 2E+Au under condition 3. Notably, Pt displayed favourable floatability, as evidenced by its high R_{max} value and the robust fit obtained from the model. Pd exhibited slightly lower R_{max} values but still demonstrated good flotation response according to the model. In contrast, 2E+Au exhibited relatively lower R_{max} values compared to Pt and Pd, indicating reduced floatability for this PGE grouping under condition 3.

Table A2. PGE recovery data fit results (a) Pt, (b) Pd, and (c) 2E+Au with collector, frother, and depressant concentrations of 120, 50, and 300g/t, respectively.

	Pt
Classic	R_{max} = 0.808, k = 1.119 min^{-1}, R^2 = 0.92580, RMSE = 0.414
Klimpel	R_{max} = 0.839, k = 2.619 min^{-1}, R^2 = 0.97588, RMSE = 0.509
Second order	R_{max} = 0.850, k = 2.077 min^{-1}, R^2 = 0.98560, RMSE = 0.473
Second-order Klimpel	R_{max} = 0.870, k = 4.686 min^{-1}, R^2 = 0.99281, RMSE = 0.275
Kelsall	R_{fast} = 0.717, k_{fast} = 1.448 min^{-1}, k_{slow} = 0.019 min^{-1}, R^2 = 0.99758, RMSE = 0.020
Modified Kelsall	R_{max} = 0.888, R_{fast} = 0.70, k_{fast} = 1.52 min^{-1}, k_{slow} = 0.05 min^{-1}, R^2 = 0.99822, RMSE = 0.006
	Pd
Classic	R_{max} = 0.780, k = 1.058 min^{-1}, R^2 = 0.90703, RMSE = 0.560
Klimpel	R_{max} = 0.811, k = 2.427 min^{-1}, R^2 = 0.96647, RMSE = 0.664
Second order	R_{max} = 0.824, k = 1.939 min^{-1}, R^2 = 0.97979, RMSE = 0.650
Second-order Klimpel	R_{max} = 0.846, k = 4.153 min^{-1}, R^2 = 0.98972, RMSE = 0.452
Kelsall	R_{fast} = 0.678, k_{fast} = 1.434 min^{-1}, k_{slow} = 0.018 min^{-1}, R^2 = 0.99614, RMSE = 0.027
Modified Kelsall	R_{max} = 0.856, R_{fast} = 0.646, k_{fast} = 1.561 min^{-1}, k_{slow} = 0.060 min^{-1}, R^2 = 0.99758, RMSE = 0.005
	2E+Au
Classic	R_{max} = 0.769, k = 1.086 min^{-1}, R^2 = 0.91573, RMSE = 0.474
Klimpel	R_{max} = 0.799, k = 2.514 min^{-1}, R^2 = 0.97094, RMSE = 0.574
Second order	R_{max} = 0.811, k = 2.063 min^{-1}, R^2 = 0.98258, RMSE = 0.552
Second-order Klimpel	R_{max} = 0.832, k = 4.389 min^{-1}, R^2 = 0.99127, RMSE = 0.360
Kelsall	R_{fast} = 0.677, k_{fast} = 1.434 min^{-1}, k_{slow} = 0.016 min^{-1}, R^2 = 0.99669, RMSE = 0.025
Modified Kelsall	R_{max} = 0.844, R_{fast} = 0.649, k_{fast} = 1.543 min^{-1}, k_{slow} = 0.056 min^{-1}, R^2 = 0.99787, RMSE = 0.005

Table A3. PGE recovery data fit results (a) Pt, (b) Pd, and (c) 2E+Au with collector, frother, and depressant concentrations of 30, 50, and 300 g/t, respectively.

	Pt
Classic	R_{max} = 0.812, k = 1.162 min^{-1}, R^2 = 0.92237, RMSE = 0.382
Klimpel	R_{max} = 0.859, k = 2.757 min^{-1}, R^2 = 0.97779, RMSE = 0.00
Second order	R_{max} = 0.853, k = 2.171 min^{-1}, R^2 = 0.98509, RMSE = 0.543
Second-order Klimpel	R_{max} = 0.872, k = 4.94 5 min^{-1}, R^2 = 0.99336, RMSE = 0.366
Kelsall	R_{fast} = 0.719, k_{fast} = 1.514 min^{-1}, k_{slow} = 0.0197 min^{-1}, R^2 = 0.99719, RMSE = 0.020
Modified Kelsall	R_{max} = 0.874, R_{fast} = 0.683, k_{fast} = 1.668 min^{-1}, k_{slow} = 0.071 min^{-1}, R^2 = 0.99935, RMSE = 0.002
	Pd
Classic	R_{max} = 0.787, k = 1.095 min^{-1}, R^2 = 0.91489, RMSE = 0.4490
Klimpel	R_{max} = 0.819, k = 2.70 min^{-1}, R^2 = 0.97530, RMSE = 0.00
Second order	R_{max} = 0.830, k = 2.033 min^{-1}, R^2 = 0.98381, RMSE = 0.601
Second-order Klimpel	R_{max} = 0.851, k = 4.431 min^{-1}, R^2 = 0.99285, RMSE = 0.408
Kelsall	R_{fast} = 0.691, k_{fast} = 1.453 min^{-1}, k_{slow} = 0.018 min^{-1}, R^2 = 0.99591, RMSE = 0.028
Modified Kelsall	R_{max} = 0.849, R_{fast} = 0.646, k_{fast} = 1.6411 min^{-1}, k_{slow} = 0.077 min^{-1}, R^2 = 0.99886, RMSE = 0.002
	2E+Au
Classic	R_{max} = 0.798, k = 1.174 min^{-1}, R^2 = 0.91997, RMSE = 0.368
Klimpel	R_{max} = 0.827, k = 2.767 min^{-1}, R^2 = 0.97316, RMSE = 0.552
Second order	R_{max} = 0.838, k = 2.238 min^{-1}, R^2 = 0.98393, RMSE = 0.561
Second-order Klimpel	R_{max} = 0.857, k = 5.016 min^{-1}, R^2 = 0.99264, RMSE = 0.394
Kelsall	R_{fast} = 0.707, k_{fast} = 1.532 min^{-1}, k_{slow} = 0.018 min^{-1}, R^2 = 0.99664, RMSE = 0.020
Modified Kelsall	R_{max} = 0.859, R_{fast} = 0.668, k_{fast} = 1.706 min^{-1}, k_{slow} = 0.073 min^{-1}, R^2 = 0.99910, RMSE = 0.001

Table A4. PGE recovery data fit results (a) Pt, (b) Pd, and (c) 2E+Au with collector, frother, and depressant concentrations of 120, 30, and 300 g/t, respectively.

Model	Parameters
	Pt
Classic	R_{max} = 0.824, k = 1.712 min^{-1}, R^2 = 0.95475, RMSE = 0.047
Klimpel	R_{max} = 0.990, k = 3.00 min^{-1}, R^2 = 0.98522, RMSE = 0.00
Second order	R_{max} = 0.847, k = 4.403 min^{-1}, R^2 = 0.98621, RMSE = 0.342
Second-order Klimpel	R_{max} = 0.858, k = 4.686 min^{-1}, R^2 = 0.99161, RMSE =0.296
Kelsall	R_{fast} = 0.767, k_{fast} = 2.101 min^{-1}, k_{slow} = 0.014 min^{-1}, R^2 = 0.99879, RMSE = 0.002
Modified Kelsall	R_{max} = 0.895, R_{fast} = 0.758, k_{fast} = 2.170 min^{-1}, k_{slow} = 0.037 min^{-1}, R^2 = 0.99900, RMSE = 0.001
	Pd
Classic	R_{max} = 0.768, k = 1.304 min^{-1}, R^2 = 0.92176, RMSE = 0.227
Klimpel	R_{max} = 0.794, k = 3.161 min^{-1}, R^2 = 0.97113, RMSE = 0.485
Second order	R_{max} = 0.802, k = 2.752 min^{-1}, R^2 = 0.98062, RMSE = 0.528
Second-order Klimpel	R_{max} = 0.819, k = 6.074 min^{-1}, R^2 = 0.98958, RMSE = 0.413
Kelsall	R_{fast} = 0.689, k_{fast} = 1.677 min^{-1}, k_{slow} = 0.014 min^{-1}, R^2 = 0.99624, RMSE = 0.012
Modified Kelsall	R_{max} = 0.829, R_{fast} = 0.659, k_{fast} = 1.841 min^{-1}, k_{slow} = 0.063 min^{-1}, R^2 = 0.99785, RMSE = 0.002
	2E+Au
Classic	R_{max} = 0.793, k = 1.518 min^{-1}, R^2 = 0.93867, RMSE = 0.101
Klimpel	R_{max} = 0.941, k = 2.759 min^{-1}, R^2 = 0.97315, RMSE = 0.006
Second order	R_{max} = 0.821, k = 3.548 min^{-1}, R^2 = 0.98561, RMSE = 0.431
Second-order Klimpel	R_{max} = 0.834, k = 8.474 min^{-1}, R^2 = 0.99007, RMSE = 0.361
Kelsall	R_{fast} = 0.726, k_{fast} = 1.902 min^{-1}, k_{slow} = 0.014 min^{-1}, R^2 = 0.99768, RMSE = 0.005
Modified Kelsall	R_{max} = 0.855, R_{fast} = 0.708, k_{fast} = 2.025 min^{-1}, k_{slow} = 0.051 min^{-1}, R^2 = 0.99842, RMSE = 0.001

Table A5 presents recovery data fit results of Pt, Pd, and 2E+Au with depressant, collector, and frother concentrations of 120, 50, and 500 g/t, respectively. Among the models applied, the modified Kelsall model demonstrated superior performance, exhibiting high R_{max} values, strong R^2, and low RMSE values. This indicates the model's effectiveness in describing the flotation kinetics of Pt, Pd, and 2E+Au under this test condition. Notably, Pt displayed favourable floatability, as evidenced by its high R_{max} value and the excellent fit obtained from the model. Pd exhibited slightly lower R_{max} values but still exhibited satisfactory flotation response according to the model. In contrast, 2E+Au exhibited relatively lower R_{max} values compared to Pt and Pd, suggesting reduced floatability for this PGE grouping.

Table A6 presents recovery data fit results of Pt, Pd, and 2E+Au with depressant, collector, and frother concentrations of 120, 50, and 500 g/t, respectively. Among the models applied, the modified Kelsall model exhibited excellent performance, demonstrating high R_{max} values, strong R^2, and low RMSE values. This indicates the effectiveness of the model in describing the flotation kinetics of Pt, Pd, and 2E+Au. Pt displayed favourable floatability, as evidenced by its high R_{max} value and the excellent fit obtained from the model. Pd exhibited slightly lower R_{max} values but still demonstrated satisfactory flotation response according to the model. 2E+Au exhibited relatively lower R_{max} values compared to Pt and Pd, indicating its reduced floatability.

Table A5. PGE recovery data fit results of (a) Pt, (b) Pd, and (c) 2E+Au with collector, frother, and depressant concentrations of 120, 50, and 500 g/t, respectively.

	Pt
Classic	R_{max} = 0.813, k = 1.359 min^{-1}, R^2 = 0.93434, RMSE = 0.181
Klimpel	R_{max} = 0.839, k = 3.369 min^{-1}, R^2 = 0.97721, RMSE = 0.421
Second order	R_{max} = 0.847, k = 2.834 min^{-1}, R^2 = 0.98469, RMSE = 0.450
Second-order Klimpel	R_{max} = 0.863, k = 6.731 min^{-1}, R^2 = 0.99200, RMSE = 0.333
Kelsall	R_{fast} = 0.737, k_{fast} = 1.712 min^{-1}, k_{slow} = 0.017 min^{-1}, R^2 = 0.99756, RMSE = 0.009
Modified Kelsall	R_{max} = 0.877, R_{fast} = 0.715, k_{fast} = 1.827 min^{-1}, k_{slow} = 0.056 min^{-1}, R^2 = 0.99846, RMSE = 0.002
	Pd
Classic	R_{max} = 0.771, k = 1.129 min^{-1}, R^2 = 0.88364, RMSE = 1.010
Klimpel	R_{max} = 0.801, k = 2.614 min^{-1}, R^2 = 0.95269, RMSE = 0.912
Second order	R_{max} = 0.812, k = 2.148 min^{-1}, R^2 = 0.97116, RMSE = 0.859
Second-order Klimpel	R_{max} = 0.832, k = 4.592 min^{-1}, R^2 = 0.98391, RMSE = 0.618
Kelsall	R_{fast} = 0.673, k_{fast} = 1.520 min^{-1}, k_{slow} = 0.017 min^{-1}, R^2 = 0.99664, RMSE = 0.035
Modified Kelsall	R_{max} = 0.859, R_{fast} = 0.651, k_{fast} = 1.618 min^{-1}, k_{slow} = 0.049 min^{-1}, R^2 = 0.99741, RMSE = 0.011
	2E+Au
Classic	R_{max} = 0.771, k = 1.129 min^{-1}, R^2 = 0.90524, RMSE = 0.436
Klimpel	R_{max} = 0.801, k = 2.614 min^{-1}, R^2 = 0.96364, RMSE = 0.625
Second order	R_{max} = 0.812, k = 2.148 min^{-1}, R^2 = 0.97652, RMSE = 0.646
Second-order Klimpel	R_{max} = 0.832, k = 4.592 min^{-1}, R^2 = 0.98711, RMSE = 0.483
Kelsall	R_{fast} = 0.673, k_{fast} = 1.520 min^{-1}, k_{slow} = 0.017 min^{-1}, R^2 = 0.99702, RMSE = 0.017
Modified Kelsall	R_{max} = 0.859, R_{fast} = 0.651, k_{fast} = 1.618 min^{-1}, k_{slow} = 0.049 min^{-1}, R^2 = 0.99790, RMSE = 0.005

Table A6. PGE recovery data fit results of (a) Pt, (b) Pd, and (c) 2E+Au with collector, frother, and depressant concentrations of 30, 30, and 300 g/t, respectively.

	Pt
Classic	R_{max} = 0.810, k = 1.563 min^{-1}, R^2 = 0.92304, RMSE = 0.104
Klimpel	R_{max} = 0.833, k = 4.058 min^{-1}, R^2 = 0.96640, RMSE = 0.487
Second order	R_{max} = 0.839, k = 3.560 min^{-1}, R^2 = 0.97401, RMSE = 0.583
Second-order Klimpel	R_{max} = 0.853, k = 8.671 min^{-1}, R^2 = 0.98363, RMSE = 0.546
Kelsall	R_{fast} = 0.733, k_{fast} = 2.051 min^{-1}, k_{slow} = 0.017 min^{-1}, R^2 = 0.99641, RMSE = 0.004
Modified Kelsall	R_{max} = 0.870, R_{fast} = 0.704, k_{fast} = 2.293 min^{-1}, k_{slow} = 0.066 min^{-1}, R^2 = 0.99796, RMSE = 0.00
	Pd
Classic	R_{max} = 0.793, k = 1.232 min^{-1}, R^2 = 0.89751, RMSE = 0.339
Klimpel	R_{max} = 0.822, k = 2.895 min^{-1}, R^2 = 0.95712, RMSE = 0.662
Second order	R_{max} = 0.833, k = 2.358 min^{-1}, R^2 = 0.97028, RMSE = 0.741
Second-order Klimpel	R_{max} = 0.852, k = 6.527 min^{-1}, R^2 = 0.98247, RMSE = 0.623
Kelsall	R_{fast} = 0.692, k_{fast} = 1.684 min^{-1}, k_{slow} = 0.020 min^{-1}, R^2 = 0.99615, RMSE = 0.012
Modified Kelsall	R_{max} = 0.876, R_{fast} = 0.666, k_{fast} = 1.826 min^{-1}, k_{slow} = 0.056 min^{-1}, R^2 = 0.99730, RMSE = 0.002
	2E+Au
Classic	R_{max} = 0.801, k = 1.374 min^{-1}, R^2 = 0.90802, RMSE = 0.205
Klimpel	R_{max} = 0.827, k = 3.355 min^{-1}, R^2 = 0.96085, RMSE = 0.576
Second order	R_{max} = 0.836, k = 2.821 min^{-1}, R^2 = 0.97148, RMSE = 0.670
Second-order Klimpel	R_{max} = 0.852, k = 6.526 min^{-1}, R^2 = 0.98271, RMSE = 0.593
Kelsall	R_{fast} = 0.711, k_{fast} = 1.839 min^{-1}, k_{slow} = 0.019 min^{-1}, R^2 = 0.99623, RMSE = 0.007
Modified Kelsall	R_{max} = 0.873, R_{fast} = 0.683, k_{fast} = 2.018 min^{-1}, k_{slow} = 0.060 min^{-1}, R^2 = 0.99758, RMSE = 0.001

References

1. Dubiella-Jackowska, A.; Polkowska, Ż.; Namieśnik, J. Platinum group elements in the environment: Emissions and exposure. *Rev. Environ. Contam. Toxicol.* **2008**, *199*, 1–25.
2. Cabri, L.J.; Harris, D.C.; Weiser, T.W. Mineralogy and distribution of platinum-group mineral (PGM) placer deposits of the world. *Explor. Min. Geol.* **1996**, *2*, 73–167.
3. Kettler, P.B. Platinum group metals in catalysis: Fabrication of catalysts and catalyst precursors. *Org. Process Res. Dev.* **2003**, *7*, 342–354. [CrossRef]
4. Wei, X.; Liu, C.; Cao, H.; Ning, P.; Jin, W.; Yang, Z.; Wang, H.; Sun, Z. Understanding the features of PGMs in spent ternary automobile catalysts for development of cleaner recovery technology. *J. Clean. Prod.* **2019**, *239*, 118031. [CrossRef]
5. Merker, J.; Lupton, D.; Töpfer, M.; Knake, H. High temperature mechanical properties of the platinum group metals. *Platin. Met. Rev.* **2001**, *45*, 74–82.
6. Zientek, M.L.; Loferski, P.J. *Platinum-Group Elements–So Many Excellent Properties*; US Geological Survey Fact Sheet; US Geological Survey: Reston, VA, USA, 2014; p. 3064.
7. Zientek, M.L.; Loferski, P.J.; Parks, H.L.; Schulte, R.F.; Seal, R.R., II. *Platinum-Group Elements*; N1411339916; US Geological Survey: Reston, VA, USA, 2017.
8. González, J.D.; Yang, Y.; Haynes, B.S.; Montoya, A. Surface coverage effect on ammonia oxidation over Pt (211). *Mol. Catal.* **2023**, *540*, 113048. [CrossRef]
9. Chen, Q.; Huang, Y.; Hu, X.; Hu, B.; Liu, M.; Bi, J.; Liu, L.; Li, N. A novel ion-solvating polymer electrolyte based on imidazole-containing polymers for alkaline water electrolysis. *J. Membr. Sci.* **2023**, *668*, 121186. [CrossRef]
10. Golding, T.M.; Omondi, R.O.; Biswas, S.; Chakraborty, S.; Prince, S.; Smith, G.S. Quinolyl-based PGM metallarectangles: Antiproliferative activity, DNA and BSA protein interactions and a molecular docking perspective. *ChemBioChem* **2023**, *24*, e202300271. [CrossRef]
11. Hughes, A.E.; Haque, N.; Northey, S.A.; Giddey, S. Platinum group metals: A review of resources, production and usage with a focus on catalysts. *Resources* **2021**, *10*, 93. [CrossRef]
12. Cawthorn, R. The platinum and palladium resources of the Bushveld Complex. *S. Afr. J. Sci.* **1999**, *95*, 481–489.
13. Cawthorn, R.G. Seventy-fifth anniversary of the discovery of the platiniferous Merensky Reef. *Platin. Met. Rev.* **1999**, *43*, 146–148.
14. McDonald, I.; Holwell, D.A.; Armitage, P.E. Geochemistry and mineralogy of the Platreef and "Critical Zone" of the northern lobe of the Bushveld Complex, South Africa: Implications for Bushveld stratigraphy and the development of PGE mineralisation. *Miner. Depos.* **2005**, *40*, 526–549. [CrossRef]
15. van der Merwe, F.; Viljoen, F.; Knoper, M. The mineralogy and mineral associations of platinum group elements and gold in the Platreef at Zwartfontein, Akanani Project, Northern Bushveld Complex, South Africa. *Mineral. Petrol.* **2012**, *106*, 25–38. [CrossRef]
16. Holwell, D.; McDonald, I.; Armitage, P. Platinum-group mineral assemblages in the Platreef at the Sandsloot Mine, northern Bushveld Complex, South Africa. *Mineral. Mag.* **2006**, *70*, 83–101. [CrossRef]
17. Miller, J.; Li, J.; Davidtz, J.; Vos, F. A review of pyrrhotite flotation chemistry in the processing of PGM ores. *Miner. Eng.* **2005**, *18*, 855–865. [CrossRef]
18. Rule, C. Energy considerations in the current PGM processing flowsheet utilizing new technologies. *J. South. Afr. Inst. Min. Metall.* **2009**, *109*, 39–46.
19. Corin, K.C.; McFadzean, B.J.; Shackleton, N.J.; O'Connor, C.T. Challenges related to the processing of fines in the recovery of platinum group minerals (PGMs). *Minerals* **2021**, *11*, 533. [CrossRef]
20. Neethling, S.; Cilliers, J. The entrainment of gangue into a flotation froth. *Int. J. Miner. Process.* **2002**, *64*, 123–134. [CrossRef]
21. Corin, K.; Bezuidenhout, J.; O'Connor, C. The role of dithiophosphate as a co-collector in the flotation of a platinum group mineral ore. *Miner. Eng.* **2012**, *36*, 100–104. [CrossRef]
22. Taguta, J.; Ross, V. The Application of a Nitrile-based Collector in the Flotation of a Platreef PGM Ore. *Miner. Process. Extr. Metall. Rev.* **2022**, *43*, 716–719. [CrossRef]
23. Muzenda, E.; Afolabi, A.S.; Abdulkareem, A.S.; Ntuli, F. In Effect of pH on the Recovery and grade of base metal sulphides (PGMs) by flotation. In Proceedings of the World Congress on Engineering and Computer Science, San Francisco, CA, USA, 19–21 October 2011; pp. 19–21.
24. Mokadze, A.M.; Ndlovu, S.; Shemi, A.; Dworzanowski, M. The Reduction of Chrome in UG-2 Flotation Concentrate by Hydrometallurgical Means. *Int. J. Miner. Process. Extr. Metall.* **2021**, *6*, 41–52. [CrossRef]
25. Uys, N. Flotation of a UG2 Ore in a Novel Pneumo-Mechanical Laboratory Cell. Doctoral Thesis, Stellenbosch University, Stellenbosch, South Africa, 2018.
26. Ralston, O.C. Principles of Flotation. By Ian W. Wark. *J. Phys. Chem.* **1939**, *43*, 816–818. [CrossRef]
27. Gharai, M.; Venugopal, R. Modeling of flotation process—An overview of different approaches. *Miner. Process. Extr. Metall. Rev.* **2016**, *37*, 120–133. [CrossRef]
28. Fichera, M.; Chudacek, M. Batch cell flotation models—A review. *Miner. Eng.* **1992**, *5*, 41–55. [CrossRef]
29. Zuniga, H.G. Flotation recovery is an exponential function of its rate. *Boln. Soc. Nac. Min.* **1935**, *47*, 83–86.
30. Morris, T. Measurement and evaluation of the rate of flotation as a function of particle size. *Trans. Am. Inst. Min. Metall. Pet. Eng.* **1952**, *193*, 794–798.

31. Ramlall, N.; Loveday, B. A comparison of models for the recovery of minerals in a UG2 platinum ore by batch flotation. *J. South. Afr. Inst. Min. Metall.* **2015**, *115*, 221–228. [CrossRef]
32. McCall, M.-J. Mineralogical and Geochemical Variations in the UG2 Reef at BOOYSENDAL and Zondereinde Mines, with Implications for Beneficiation of PGM. Doctoral Thesis, Stellenbosch University, Stellenbosch, South Africa, 2016.
33. Kelsall, D. Application of probability assessment of flotation systems. *Trans. Inst. Min. Metall.* **1961**, *70*, 191–204.
34. Arbiter, N.; Harris, C. *Froth Flotation*; Fuerstenau, D.W., Ed.; AIME: New York, NY, USA, 1962; p. 215.
35. Mao, L.; Yoon, R.-H. Predicting flotation rates using a rate equation derived from first principles. *Int. J. Miner. Process.* **1997**, *51*, 171–181. [CrossRef]
36. Hay, M.P.; Roy, R. A case study of optimising UG2 flotation performance. Part 1: Bench, pilot and plant scale factors which influence Cr_2O_3 entrainment in UG2 flotation. *Miner. Eng.* **2010**, *23*, 855–867. [CrossRef]
37. Cnudde, V.; Boone, M.N. High-resolution X-ray computed tomography in geosciences: A review of the current technology and applications. *Earth-Sci. Rev.* **2013**, *123*, 1–17. [CrossRef]
38. Godel, B.; Barnes, S.-J.; Maier, W.D. 3-D distribution of sulphide minerals in the Merensky Reef (Bushveld Complex, South Africa) and the JM Reef (Stillwater Complex, USA) and their relationship to microstructures using X-ray computed tomography. *J. Petrol.* **2006**, *47*, 1853–1872. [CrossRef]
39. Ketcham, R.A.; Carlson, W.D. Acquisition, optimization and interpretation of X-ray computed tomographic imagery: Applications to the geosciences. *Comput. Geosci.* **2001**, *27*, 381–400. [CrossRef]
40. Nguyen, A.; Schulze, H.J. *Colloidal Science of Flotation*; CRC Press: Boca Raton, FL, USA, 2003; Volume 118.
41. Chen, J.; Chimonyo, W.; Peng, Y. Flotation behaviour in reflux flotation cell–A critical review. *Miner. Eng.* **2022**, *181*, 107519. [CrossRef]
42. Von Gruenewaldt, G.; Merkle, R. Platinum group element proportions in chromitites of the Bushveld complex: Implications for fractionation and magma mixing models. *J. Afr. Earth Sci.* **1995**, *21*, 615–632. [CrossRef]
43. Cilek, E. Estimation of flotation kinetic parameters by considering interactions of the operating variables. *Miner. Eng.* **2004**, *17*, 81–85. [CrossRef]
44. Webb, S.J.; Cawthorn, R.G.; Nguuri, T.; James, D. Gravity modelling of Bushveld Complex connectivity supported by Southern African seismic experiment results. *S. Afr. J. Geol.* **2004**, *107*, 207–218. [CrossRef]
45. Boudreau, A.E. Modeling the Merensky Reef, Bushveld Complex, Republic of South Africa. *Contrib. Mineral. Petrol.* **2008**, *156*, 431–437. [CrossRef]
46. Carelse, C.; Manuel, M.; Chetty, D.; Taguta, J.; Safari, M.; Youlton, K. The flotation behaviour of liberated Platinum Group minerals in Platreef ore under reduced reagent conditions. *Miner. Eng.* **2022**, *190*, 107913. [CrossRef]
47. Ramlall, N.V. An Investigation into the Effects of UG2 Ore Variability on Froth Flotation. Doctoral Thesis, University of KwaZulu-Natal, Durban, South Africa, 2013.
48. Corin, K.C.; Wiese, J.G. Investigating froth stability: A comparative study of ionic strength and frother dosage. *Miner. Eng.* **2014**, *66–68*, 130–134. [CrossRef]
49. Becker, M.; Wiese, J.; Ramonotsi, M. Investigation into the mineralogy and flotation performance of oxidised PGM ore. *Miner. Eng.* **2014**, *65*, 24–32. [CrossRef]

minerals

MDPI

Article

A Model of Froth Flotation with Drainage: Simulations and Comparison with Experiments

Fernando Betancourt [1], Raimund Bürger [2,*], Stefan Diehl [3], Leopoldo Gutiérrez [4], M. Carmen Martí [5] and Yolanda Vásquez [2]

1 Centro de Investigación en Ingeniería Matemática, Departamento de Ingeniería Metalúrgica, Facultad de Ingeniería, Universidad de Concepción, Casilla 160-C, Concepción 4030000, Chile
2 Centro de Investigación en Ingeniería Matemática, Departamento de Ingeniería Matemática, Facultad de Ciencias Físicas y Matemáticas, Universidad de Concepción, Casilla 160-C, Concepción 4030000, Chilel
3 Centre for Mathematical Sciences, Lund University, P.O. Box 118, S-221 00 Lund, Sweden
4 Departamento de Ingeniería Metalúrgica, Facultad de Ingeniería, Universidad de Concepción, Casilla 160-C, Concepción 4030000, Chile
5 Departament de Matemàtiques, Universitat de València, Avda. Vicent Andrés Estellés s/n, E-46100 Burjassot, Spain
* Correspondence: rburger@ing-mat.udec.cl; Tel. +56-41-2661317

Abstract: The operation of a froth flotation column can be described by a nonlinear convection–diffusion partial differential equation that incorporates the solids–flux and drift–flux theories as well as a model of foam drainage. The resulting model predicts the bubble and (gangue) particle volume fractions as functions of height and time. The steady-state (time-independent) version of the model defines so-called operating charts that map conditions on the gas and pulp feed rates that allow for operation with a stationary froth layer. Operating charts for a suitably adapted version of the model are compared with experimental results obtained with a laboratory flotation column. Experiments were conducted with a two-phase liquid–bubble flow. The results indicate good agreement between the predicted and measured conditions for steady states. Numerical simulations for transient operation, in part for the addition of solid particles, are presented.

Keywords: froth flotation; drainage; drift flux; mathematical model; partial differential equation; steady state; numerical simulation

Citation: Betancourt, F.; Bürger, R.; Diehl, S.; Gutiérrez, L.; Martí, M.C.; Vásquez, Y. A Model of Froth Flotation with Drainage: Simulations and Comparison with Experiments. *Minerals* **2023**, *13*, 344. https://doi.org/10.3390/min13030344

Academic Editor: Saeed Chehreh Chelgani

Received: 31 January 2023
Revised: 22 February 2023
Accepted: 25 February 2023
Published: 28 February 2023

1. Introduction

Froth flotation is the most important concentration operations in mineral processing and is widely used for the recovery of valuable minerals from low-grade ores (cf. [1], ([2], Chapter 12) or ([3], Part 7)). This unit operation is an important stage particularly for copper mining in Chile. The flotation process selectively separates hydrophobic materials (that are repelled by water) from hydrophilic (that would be attracted to water), where both are suspended in a viscous fluid. It is well known that a flotation column works as follows: gas is introduced close to the bottom and generates bubbles that rise through the continuously injected pulp that contains the solid particles.

The hydrophobic particles (the valuable mineral particles) attach to the rising bubbles, forming froth that is removed through a launder. The hydrophilic particles (slimes or gangue) do not attach to bubbles but settle to the bottom (unless they are trapped in the bulk upflow) and are removed continuously as flotation tailings. Close to the top, additional wash water can be injected to assist with the rejection of entrained impurities and to increase the froth stability [1,4,5]. This unit operation is particularly suitable for processing low-grade ores, such as copper ores in Chilean deposits; however, this requires huge amounts of process water.

Since water is a scarce resource for most economic activity in Chile—in particular in the desert areas where most mines are located—the improvement of the scientific understanding of flotation processes and the development of suitable tools for the design, simulation and control of flotation devices is of critical economical, ecological and societal importance. This situation has motivated collaborative research between applied mathematicians and metallurgical engineers at Universidad de Concepción.

Modelling flotation and developing strategies to control this process are research areas that have generated many contributions [6–15]. The development of control strategies requires dynamic models along with a classification of steady-state (stationary) solutions of such models. These models should focus on the separation process aligned with gravity and are, therefore, spatially one-dimensional. (In fact, we wish to avoid the additional computational effort associated with spatially two- or three-dimensional models, mostly based on computational fluid dynamics (CFD) that also involve the solution of additional equations for the motion of the mixture; we refer to [16] for a review on CFD-based models of flotation).

The sought unknowns are the volume fractions of gas (bubbles), liquid, and possibly solid particles as functions of both time and spatial position, so that the resulting governing equations are partial differential equations (PDEs). With the aim of developing controllers, some authors [9–12] used hyperbolic systems of PDEs for the froth or pulp regions coupled to ordinary differential equations (ODEs) for the lower part of the column. These include the attachment and detachment processes; however, with their approaches, the phases seem to have constant velocities, which is not in agreement with the established drift–flux theory [17] that establishes that the velocity of a unit of the disperse phase (droplet, bubble or particle) is a function of the local volume fraction (or concentration). Nonlinear dependence of the phase velocities on the volume fractions gives rise to discontinuities in the concentration profiles, which was confirmed experimentally [6].

Narsimhan [18] showed realistic conceptual transient solutions of an initially homogeneous bubble–liquid suspension. The rising bubbles form a layer of foam at the top, which can undergo compressibility due to gravity and capillarity. Separate equations are derived for the foam region, and boundary assumptions between regions have to be imposed. The purpose of our previous contribution [19] (see also [20]), which is utilized herein in a slightly modified form, is to let one single equation govern the bubble–liquid behaviour in the whole column under any dynamic situation without any imposed boundary conditions.

Such are automatically assumed by the PDE solutions, which satisfy the so-called entropy conditions by definitions of the PDE coefficients that are discontinuous across the feed, discharge and overflow levels of height. When solids are also fed into the column, an additional equation modelling the settling of solids outside the bubbles is needed—still without any imposed boundary conditions.

Phenomenological models for two-phase systems with bubbles rising (or, analogously, particles settling) in a liquid, are derived from the physical laws of conservation of mass and momentum [21,22]. Under certain simplifying assumptions on the stress tensor and partial pressure of the bubbles/solids, one can obtain first- or second-order PDEs involving one or two constitutive (material specific) functions, respectively. The resulting first-order PDE modelling such a separation process in a one-dimensional column of rising bubbles is a scalar conservation law

$$\partial_t\phi + \partial_z j_{\mathrm{b}}(\phi) = 0, \tag{1}$$

where t is time, z is a spatial position (height), $\phi = \phi(z,t)$ is the sought volume fraction of bubbles, and

$$j_{\mathrm{b}}(\phi) := \phi\tilde{v}(\phi) \tag{2}$$

is the bubble (aggregate) batch flux density function, where $\tilde{v}(\phi)$ is a given drift–flux velocity function. The formulations (1) and (2) are in agreement with the drift–flux theory [17]. With additional bulk flows due to the inlets and outlets of the column, that theory has mostly been used for investigations of steady states of flotation columns [5,23–26]. Models of and numerical schemes for column froth flotation with the drift–flux assumption and possibly simultaneous sedimentation have been presented in [27–30].

The analogy of the drift–flux theory for sedimentation is the established solids–flux theory [17,31–33]. With an additional constitutive assumption on sediment compressibility, the corresponding model becomes a second-order degenerate parabolic PDE [21]. Sedimentation in a clarifier–thickener unit is mathematically similar to column flotation. A full PDE model of such a vessel necessarily contains source terms and spatial discontinuities at both inlets and outlets. Steady-state analyses, numerical schemes, dynamic simulations and the control of such models are reported in [32,34–37].

The first-order PDE of the flotation process advanced in [28] does not include capillarity in the foam. Such effects have been studied intensively [38–40]; see also [13]. Solids motion in froth was investigated in [41]. A generalized model PDE that captures both the rising bubbles of low concentrations and the formation and drainage of froth at high concentration was recently presented in [19]. That article contains a generalized model when settling particles are also present in a specific flotation column with a common feed inlet for both pulp and gas.

In this contribution, we adjust the generalized drainage model to an experimental laboratory flotation column with separate inlets for gas and mixture (Figure 1), derive the so-called desired steady state (which has a foam layer in the upper part and no bubbles leaving at the underflow) and find a single set of parameters in the model that qualitatively captures several steady state experiments. The numerical simulations presented here are made with the numerical method in [19] and adapted to the setup of the pilot column (Figure 1).

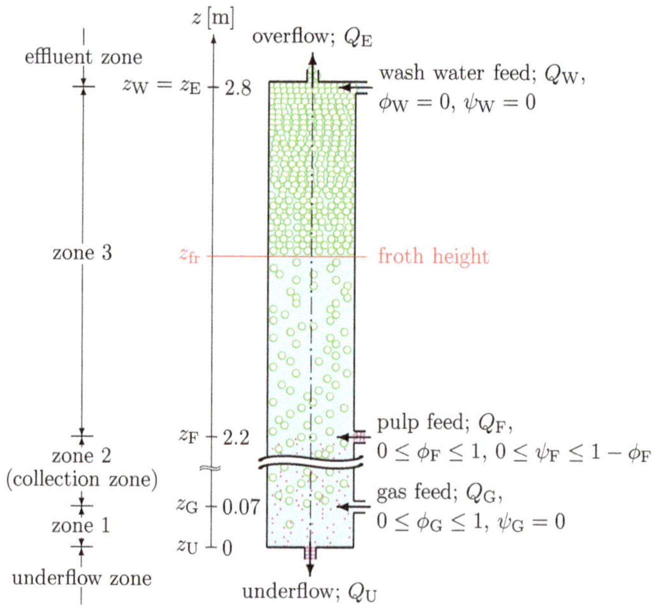

Figure 1. Conceptual drawing of the model of the pilot flotation column: (**left**) denomination of zones, (**middle**) height axis (z-axis) showing the locations of the feed and discharge levels and (**right**) schematic of the column. The green open circles and solid magenta dots represent bubbles and hydrophilic particles (bubbles and particles are not drawn to scale), respectively. The information to the right indicates the overflow rate, volume feed rates and concentrations and the underflow volume rate along with limitations that the feed concentrations must satisfy. The denomination of zone 2 as a "collection zone" is common in mineral processing, although the process of collection (the adhesion of hydrophobic particles to bubbles) is not part of the model.

2. Materials and Methods

2.1. Pilot Flotation Column and Experimental Setup

The experiments were conducted at the laboratory of the Department of Metallurgical Engineering of Universidad de Concepción with a laboratory-scale flotation column; see Figure 2a. This column was made of acrylic to visualize the internal phenomena that occur in both the collection and cleaning areas. The column had a volume of 54.7 L and an interior diameter of 6 inches so that its cross-sectional area was $A = (\pi/4)(0.1524\,\mathrm{m})^2 \approx 0.0182415\,\mathrm{m}^2$ and was 2.8 m high (see Table 1).

The air was injected from the lower central part of the column through a sparger whose pores were 1 mm in diameter. The locations of the inlets and outlets are detailed in Figure 1. The column was instrumented as shown in Figure 2b. For the tests, the use of solids was not considered, the only reagent to be used was a mix (1:1) of MIBC (methyl isobutyl carbinol, an organic chemical compound used primarily as a frother in mineral flotation) and polyglycol as a frother at a dosage of 100 g/L of water.

Table 1. Dimensions of the pilot flotation column.

Symbol	Significance	Value
z_U	underflow level	0 m
z_G	gas feed level	0.07 m
z_F	pulp feed level	2.20 m
z_W	wash water feed level	2.80 m
z_E	overflow level	2.80 m
A	interior cross-sectional area	0.018241 m^2

(a) (b)

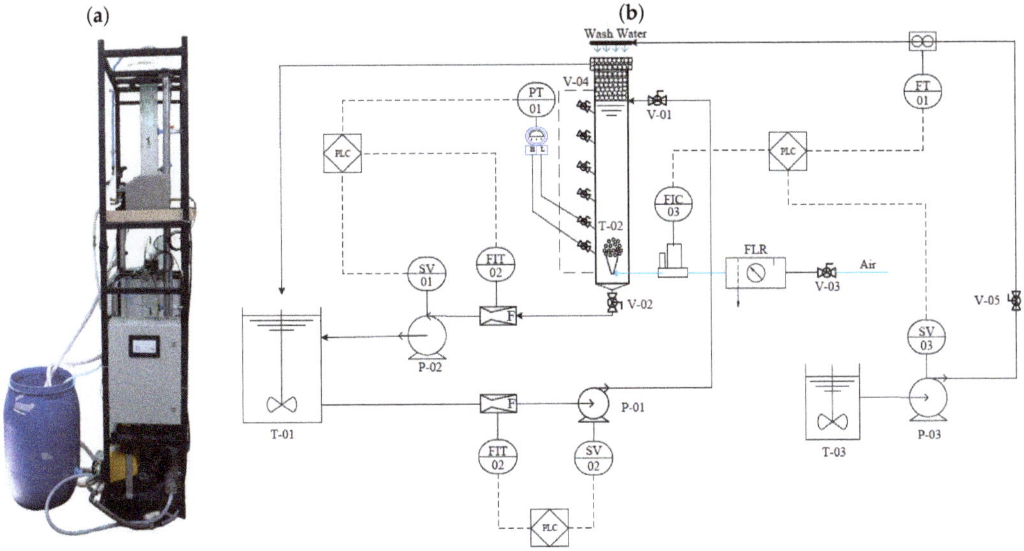

Figure 2. (**a**) Photograph of the laboratory column and (**b**) schematic of the piping and instrumentation devices (P & ID). See Table 2 for explanation.

Table 2. Legend of the P & ID schematic (Figure 2, right).

Instrument	Tag	Quantity Measured	Connected to PLC?
Mass flowmeter transmitter	FIT-01/02	feed/discharge flowrate	yes
Mass flowmeter controller	FIC-03	air flowrate	yes
Magnetic flowmeter	FT-01	wash water flowrate	yes
Variable frequency drive	SV-01/02/03	pump velocity	yes
Differential pressure transmitter	PT-01	holdup	yes
Feed manual valve	V-01	——	no
Discharge manual valve	V-02	——	no
Air manual valve	V-03	——	no
Pressure taps valve	V-04	——	no
Wash water manual valve	V-05	——	no
Equipment	**Tag**	**Type**	**Range/Dimensions and Unit**
Feed pump	P-01	centrifuge	20–110 L/min
Discharge pump	P-02	peristaltic	0–18 L/min
Wash water pump	P-03	peristaltic	0–12 L/min
Regulator filter with water decanter	FLR	manual	0–16 bar
Pulp tank	T-01	plastic cylinder	200 L
Flotation column	T-02	acrylic tube	55 L
Wash water tank	T-03	plastic cylinder	200 L

2.2. Experimental Determination of Stability Regions

We performed five steady-state experiments; see Table 3. There are 36 data points in the interval $[0, 1.630]$ of the underflow velocity $q_U = Q_U/A$. The resulting upper and lower limits of q_F within which a pulp–froth interface is present in zone 3 are presented in Figures 3 and 4. The enclosed region between the two curves may be called the *stability region* since, for values of (q_U, q_F) therein, a stable pulp–froth interface in zone 3 was observed experimentally. On the other hand, for choices of (q_U, q_F), outside that region, unstable operation was observed, which means that either no froth layer was produced at all or that the froth layer reached into zone 2 and possibly that bubbles left through the underflow.

A stable pulp–froth interface in zone 2 is a valid stationary solution (corresponding to a mode of operation in which the pulp feed acts as a submerged feed source), but such steady states are very difficult to control, and we, therefore, address them as "unstable operation". This procedure is consistent with the theoretical steady-state analysis of [19] (see also Section 3.4) and in particular the construction of operating charts in which any theoretical situation in which the froth level cannot be accommodated within zone 3 is deemed "unstable," independently of whether the parameters permit a froth layer within zone 2 or not (i.e., bubbles leave through the underflow).

Table 3. Overview of the five steady-state experiments.

Experiment No.	1	2	3	4	5
ϕ_G [−]	1	1	1	1	1
ϕ_F [−]	0	0	0	0	0
q_G [cm/s]	1.3	1.8	2.3	1.8	1.8
q_W [cm/s]	0	0	0	0.3	0.5
q_U [cm/s]	$[0, 1.630]$	$[0, 1.630]$	$[0, 1.630]$	$[0, 1.630]$	$[0, 1.630]$
q_F [cm/s]	see Figure 3	see Figure 3	see Figure 3	see Figure 4	see Figure 4

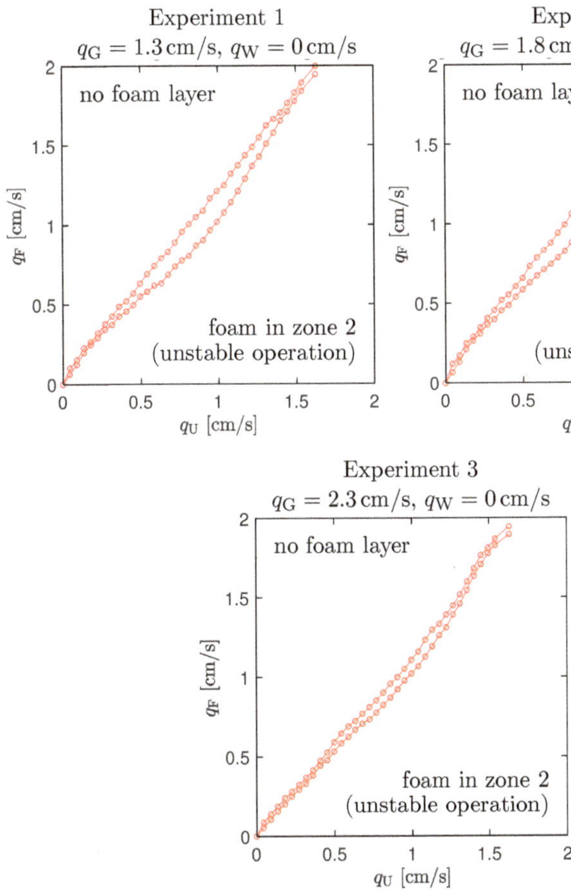

Figure 3. Experiments 1–3 (no wash water added): stability regions enclosed between two curves formed by points in the (q_U, q_F) plane.

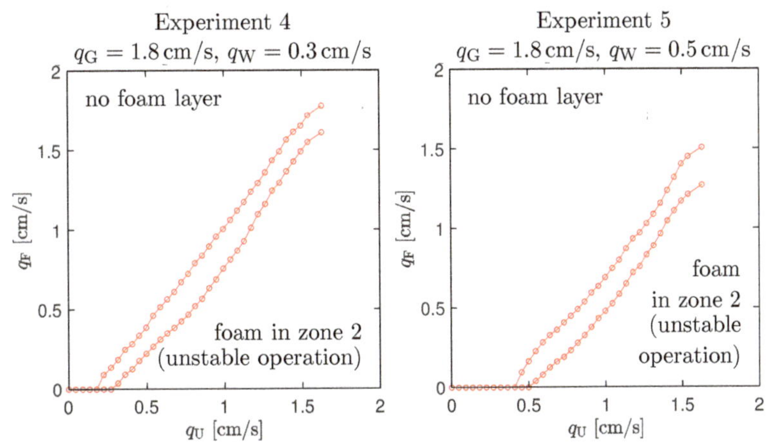

Figure 4. Experiments 4 and 5 (with wash water added): stability regions enclosed between two curves formed by points in the (q_U, q_F) plane.

In order to ensure the reproducibility of the experiments, the tests were performed in duplicate and randomly, and we report only the average values. The standard deviation of the tests was less than 5%.

3. Theory

3.1. Mathematical Model

The governing model combines the setup with separated inlets for gas and the pulp formulated in [27] with the approach of describing foam drainage developed in [19]. The model is formulated as a three-phase model formed of the gas bubbles and (hydrophilic gangue) solid particles as primary and secondary disperse phases that move in the fluid that forms the continuous phase.

The present experimental support refers to a two-phase flow model only, but for illustrative purposes, we present one numerical simulation corresponding to a (hypothetical) solids feed. The three phases and their (dimensionless) volume fractions are the fluid $\phi_\mathrm{f} = \phi_\mathrm{f}(z,t)$, the solids $\psi = \psi(z,t)$ and the bubbles (aggregates) $\phi = \phi(z,t)$, where $\phi_\mathrm{f} + \psi + \phi = 1$. A mixture of fluid and solid particles is addressed as a suspension. The volume fraction of solids within the suspension that fills the interstices between bubbles φ is defined by

$$\varphi := \frac{\psi}{\psi + \phi_\mathrm{f}} = \frac{\psi}{1-\phi}.$$

The system of PDEs that governs the evolution of ϕ and ψ can be formulated as

$$\begin{aligned} &A(z)\partial_t \begin{pmatrix} \phi \\ \psi \end{pmatrix} + \partial_z \left(A(z) \begin{pmatrix} J(\phi,z,t) \\ -\tilde{F}(\psi,\phi,z,t) \end{pmatrix} \right) \\ &= \partial_z \left(A(z)\gamma(z) \begin{pmatrix} 1 \\ -\psi/(1-\phi) \end{pmatrix} \partial_z D(\phi) \right) + \sum_{S\in\{\mathrm{G,F,W}\}} Q_\mathrm{S}(t) \begin{pmatrix} \phi_\mathrm{S}(t) \\ \psi_\mathrm{S}(t) \end{pmatrix} \delta(z - z_\mathrm{S}). \end{aligned} \tag{3}$$

Apart from the variables introduced in the context of (1) and above, here, $A = A(z)$ is the cross-sectional area of the tank, and $J = J(\phi,z,t)$ and $\tilde{F} = \tilde{F}(\psi,\phi,z,t)$ are convective flux functions that depend discontinuously on z at the locations of the gas inlet ($z = z_\mathrm{G}$), the pulp feed inlet ($z = z_\mathrm{F}$), the wash water inlet ($z = z_\mathrm{W}$), the underflow outlet ($z = z_\mathrm{U}$) at the bottom and the overflow outlet ($z = z_\mathrm{E}$) at the top; see Figure 1.

The system (3) is valid for $t > 0$ and all z, $-\infty < z < \infty$. The characteristic function γ indicates the interior of the tank:

$$\gamma(z) := \begin{cases} 1 & \text{inside the tank, i.e., if } z_\mathrm{U} \le z \le z_\mathrm{E}, \\ 0 & \text{outside the tank, i.e., if } z < z_\mathrm{U} \text{ or } z > z_\mathrm{E}. \end{cases}$$

The nonlinear function D, see Figure 5, models the capillarity present when bubbles are in contact and is given by

$$D(\phi) := \int_0^\phi d(s)\,\mathrm{d}s, \tag{4}$$

where the function d (introduced in [19] and specified later in this work) is assumed to satisfy

$$d(\phi) = D'(\phi) \begin{cases} = 0 & \text{for } 0 \le \phi \le \phi_\mathrm{c}, \\ > 0 & \text{for } \phi_\mathrm{c} < \phi \le 1. \end{cases} \tag{5}$$

Consequently, at each point (z,t) where $\phi(z,t) \le \phi_\mathrm{c}$, there holds $D(\phi(z,t)) = 0$, and therefore (3) degenerates at such points into a first-order system of conservation laws of hyperbolic type. (Precise algebraic definitions of J, $\tilde{F}$ and d are provided further below).

The last term on the right-hand side describes three singular sources located at the level $z = z_S$, $S \in \{G, F, W\}$, where $Q_S(t)$ is the corresponding volume feed rate (as a given function of time) and $\phi_S(t)$ and $\psi_S(t)$ are the respective bubble and solids feed concentrations. Of course, under normal circumstances the solids feed concentration at the gas inlet should be zero ($\psi_G \equiv 0$), the bubble feed concentrations at the pulp feed inlet should be zero ($\phi_F \equiv 0$), and the concentrations of both disperse phases at the wash water inlet should be zero ($\phi_W \equiv 0$, $\psi_W \equiv 0$). Outside the tank, the mixture is assumed to follow the outlet streams. Consequently, boundary conditions are not needed; the conservation of mass determines the outlet volume fractions in a natural way.

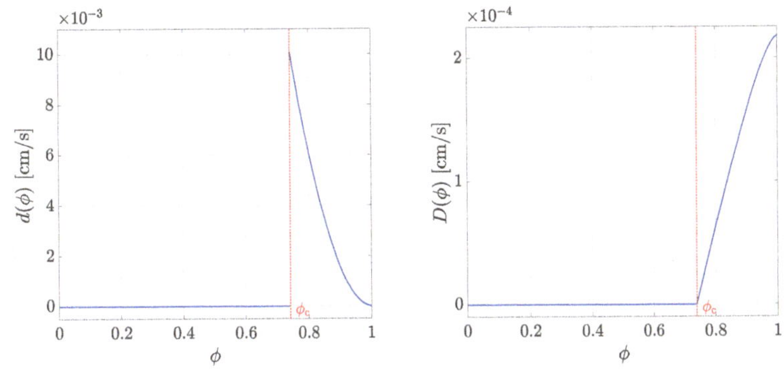

Figure 5. Functions $d(\phi)$ (**left**) and $D(\phi)$ (**right**) modelling the capillarity. In both figures, we set $\phi_c = 0.74$.

Since the wash water is located at the top of the column, we have $z_W = z_E$. Thus, the interior of the flotation column can be subdivided into three zones. In what follows, for the ease of discussion, we refer to the z-subintervals $z \geq z_E$ as "the effluent zone," $z_F \leq z < z_E$ as "zone 3," $z_G \leq z < z_F$ as "zone 2," $z_U \leq z < z_G$ as "zone 1" and $z < z_U$ as "the underflow zone."

Applying the conservation of mass to each of the three phases, introducing the volume-average velocity, or bulk velocity, of the mixture q and the relative velocities of both the aggregate suspension and the solid–fluid, Bürger et al. [28] derived a PDE model similar to (3) without the capillarity function $D(\phi)$ and assuming that there is one joint inlet for both the pulp and the gas. Extending this approach to the present setup, we obtain the flow rates (velocities) in and out of the flotation column

$$q(z,t) := \begin{cases} q_E := (-Q_U + Q_G + Q_F + Q_W)/A & \text{in the effluent zone,} \\ q_3 := (-Q_U + Q_G + Q_F)/A & \text{in zone 3,} \\ q_2 := (-Q_U + Q_G)/A & \text{in zone 2,} \\ q_1 = q_U := -Q_U/A & \text{in zone 1 and the underflow.} \end{cases} \tag{6}$$

The drift–flux and solids–flux theories utilize constitutive functions for the aggregate upward batch flux $j_b(\phi)$ and the solids batch sedimentation flux

$$f_b(\varphi) := \varphi v_{hs}(\varphi),$$

where $v_{hs}(\varphi)$ is the hindered–settling function. For simplicity, we employ the Richardson–Zaki expression [42]

$$v_{hs}(\varphi) = v_\infty (1-\varphi)^{n_{RZ}}, \quad \text{where } n_{RZ} > 1, \tag{7}$$

and v_∞ is the velocity of a single particle. In the underflow and effluent zones, all phases are assumed to have the same velocity, i.e., they follow the bulk flow. Then, the total convective fluxes for ϕ and φ are given by

$$J(\phi,z,t) = \begin{cases} j_{\mathrm{E}}(\phi,t) := q_{\mathrm{E}}(t)\phi & \text{in the effluent zone,} \\ j_3(\phi,t) := q_3(t)\phi + j_{\mathrm{b}}(\phi) & \text{in zone 3,} \\ j_2(\phi,t) := q_2(t)\phi + j_{\mathrm{b}}(\phi) & \text{in zone 2,} \\ j_1(\phi,t) := q_1(t)\phi + j_{\mathrm{b}}(\phi) & \text{in zone 1,} \\ j_{\mathrm{U}}(\phi,t) := q_1(t)\phi & \text{in the underflow zone,} \end{cases} \tag{8}$$

$$F(\varphi,\phi,z,t) = \begin{cases} f_{\mathrm{E}}(\varphi,\phi,t) := -(1-\phi)q_{\mathrm{E}}(t)\varphi & \text{in the effluent zone,} \\ f_3(\varphi,\phi,t) & \text{in zone 3,} \\ f_2(\varphi,\phi,t) & \text{in zone 2,} \\ f_1(\varphi,\phi,t) & \text{in zone 1,} \\ f_{\mathrm{U}}(\varphi,\phi,t) := -(1-\phi)q_1(t)\varphi & \text{in the underflow zone,} \end{cases} \tag{9}$$

with the zone-settling flux functions (positive in the direction of sedimentation, that is, decreasing z)

$$\begin{aligned} f_k(\varphi,\phi,t) &:= (1-\phi)f_{\mathrm{b}}(\varphi) + \big(j_{\mathrm{b}}(\phi) - (1-\phi)q_k(t)\big)\varphi \\ &= (1-\phi)f_{\mathrm{b}}(\varphi) + \big(j_k(\phi,t) - q_k(t)\big)\varphi, \quad k = 1,2,3. \end{aligned}$$

Here, the batch drift–flux function $j_{\mathrm{b}} = j_{\mathrm{b}}(\phi)$ is given by (2), where $\tilde{v}(\phi)$ is given by

$$\tilde{v}(\phi) := \begin{cases} v_{\mathrm{term}}(1-\phi)^{n_{\mathrm{b}}} & \text{for } 0 \le \phi \le \phi_{\mathrm{c}}, \\ v_{\mathrm{term}}\dfrac{(1-\phi)^{2n_{\mathrm{S}}+1}}{(1-\phi_{\mathrm{c}})^{2n_{\mathrm{S}}+1-n_{\mathrm{b}}}} & \text{for } \phi_{\mathrm{c}} < \phi \le 1. \end{cases} \tag{10}$$

Here, v_{term} is the constant velocity of a single bubble in liquid, and n_{b} is a dimensionless constant. The expression in the first case of (10) is valid as long as the bubbles are not all in contact with each other. This contact is assumed to occur whenever ϕ exceeds the critical concentration ϕ_{c}. The expression in the second case of (10) is derived from a compatibility condition that makes it possible to express the drainage velocity of the liquid in the froth relative to the bubbles with respect to gravity and dissipation in terms of v_{term} and the dimensionless constant n_{S}.

The latter emerges from empirical connections between the radius of the Plateau borders in the foam, the radius of the bubbles and the volume fraction of the liquid in the foam $1-\phi$; see [19] for all details. The function $d(\phi)$ arising in (5), and which describes capillarity, is given by

$$d(\phi) := \begin{cases} 0 & \text{for } 0 \le \phi \le \phi_{\mathrm{c}}, \\ v_{\mathrm{term}}d_{\mathrm{cap}}\dfrac{\phi(1-\phi)^{n_{\mathrm{S}}}}{(1-\phi_{\mathrm{c}})^{2n_{\mathrm{S}}+1-n_{\mathrm{b}}}} & \text{for } \phi_{\mathrm{c}} < \phi \le 1, \end{cases} \tag{11}$$

where d_{cap} is a capillarity-to-gravity constant present in the froth when $\phi > \phi_{\mathrm{c}}$ and involving, among others, the surface tension of water; see [19]. In light of (11), we obtain

$$D(\phi) = \begin{cases} 0 & \text{for } 0 \le \phi \le \phi_{\mathrm{c}}, \\ v_{\mathrm{term}}d_{\mathrm{cap}}\dfrac{\omega(\phi_{\mathrm{c}}) - \omega(\phi)}{(1-\phi_{\mathrm{c}})^{2n_{\mathrm{S}}+1-n_{\mathrm{b}}}(n_{\mathrm{S}}+1)(n_{\mathrm{S}}+2)} & \text{for } \phi_{\mathrm{c}} < \phi \le 1, \end{cases} \tag{12}$$

where $\omega(\phi) := (1-\phi)^{n_S+1}((n_S+1)\phi+1)$, and we reconfirm the property (5). Finally, we define the total convective flux for the solids appearing in the governing system (3) by

$$\tilde{F}(\psi,\phi,z,t) := \begin{cases} F\left(\dfrac{\psi}{1-\phi},\phi,z,t\right) & \text{if } 0 \le \phi < 1, \\ 0 & \text{if } \phi = 1. \end{cases}$$

3.2. Reduced Model for Two-Phase Flow of Bubbles in Liquid

The description provided so far refers to the full model that involves a feed of both gas bubbles and solids in the water. The currently available experimental data, however, are, for the moment, limited to a two-phase gas–liquid system; measurements of the system behaviour with solids feed are currently being made and will be presented in forthcoming work. We here focus on the dynamics of the gas–liquid system considering the effects of froth drainage. Furthermore, since the pilot column is cylindrical, we assume that A is constant (we utilize $A = 0.0182415\,\mathrm{m}^2$ as indicated in Section 2.1).

Under this assumption (that is, the presence of particles is neglected) the model reduces to the scalar PDE

$$\partial_t\phi + \partial_z J(\phi,z,t) = \partial_z\big(\gamma(z)\partial_z D(\phi)\big) + \sum_{S\in\{\mathrm{G,F,W}\}} q_S(t)\phi_S(t)\delta(z-z_S), \tag{13}$$

where we define the velocities

$$q_S(t) := Q_S(t)/A, \quad S \in \{\mathrm{G,F,W}\},$$

and the definitions $J(\phi,z,t)$, (8), $D(\phi)$, (12) and $\gamma(z)$ remain in effect. These and other variables have the range of values given in Table 4.

Table 4. Range of parameters according to the literature (typical values employed in [43–47]) and used in the present work.

Parameter	Symbol	Working Range (Literature)	Range in Present Work
Froth height	[m]	0.5–2.0	0.5–1.5
Bubble diameter	[mm]	0.5–2.0	0.5–1.3
Hold-up in zone 2	ϕ_2 [−]	0.05–0.30	0.09–0.20
Gas feed rate	q_G [cm/s]	0.5–3.0	1.3–2.3
Pulp feed rate	q_F [cm/s]	0.2–2.0	0.8–1.5
Discharge rate	q_U [cm/s]	0.2–2.0	1.0–1.4
Wash water rate	q_W [cm/s]	0.2–1.0	0.3–0.5

3.3. Numerical Method

The numerical method used for the solution of the complete model is outlined, and in part analysed, in [19]. It is based on subdividing the computational domain, corresponding to a z-interval that encloses the tank (that is, the interval $[z_\mathrm{U}, z_\mathrm{E}]$ into a number N of layers (subintervals) of equal height Δz, and time is discretized through time points $t_n = n\Delta t$, $n = 0,1,2,\dots$. Without entering into any details, assume that the unknowns of the scheme are ϕ_j^n and ψ_j^n, where these quantities are approximate values of ϕ and ψ in cell j at time t_n, respectively. The general scheme can then be written in the form

$$\begin{aligned} \phi_j^{n+1} &= \mathcal{H}(\phi_{j-1}^n, \phi_j^n, \phi_{j+1}^n, j, n), \\ \psi_j^{n+1} &= \mathcal{K}(\phi_{j-1}^n, \phi_j^n, \phi_{j+1}^n, \psi_{j-1}^n, \psi_j^n, \psi_{j+1}^n, j, n), \quad j = 1,\dots,N; \quad n = 0,1,2,\dots, \end{aligned} \tag{14}$$

We refer to [19] for all details regarding the precise algebraic forms of the functions $\mathcal{H}$ and $\mathcal{K}$, which are chosen in such a way that (14) represents a consistent finite difference approximation of the system (3) and all ingredients outlined in Section 3.1. While any specific information is omitted here (for brevity), the general formulation (14) is useful to indicate some particular properties of the numerical scheme of [19]: first, the scheme is *explicit*, that is, from the given initial values ϕ_j^0 and ψ_j^0, $j = 1, \dots, N$, one successively calculates ϕ_j^1 and ψ_j^1, $j = 1, \dots, N$, then ϕ_j^2 and ψ_j^2, $j = 1, \dots, N$ and so on for $n = 3, 4, \dots$.

Furthermore, the system (3) is *triangular*, which means that the first equation, the PDE for the update of ϕ, contains, apart from $\partial_t \phi$, only terms that depend on known functions and ϕ and its z-derivatives. On the contrary, the second PDE, for the update of ψ, contains, apart from $\partial_t \psi$, terms that depend on both ϕ and ψ. Thus, the bubble volume fraction ϕ can be updated independently from the solids volume fraction ψ, which is also reflected in the different arguments of $\mathcal{H}$ and $\mathcal{K}$ in (14).

Consequently, the first update formula of (14) is a valid numerical scheme for the one-equation reduced model outlined in Section 3.2. Furthermore, the functions $\mathcal{H}$ and $\mathcal{K}$ are based on particular numerical fluxes that satisfy the so-called monotonicity property, which ensures that, if the initial values are physically relevant, i.e.,

$$\phi_j^n \geq 0, \quad \psi_j^n \geq 0, \quad \phi_j^n + \psi_j^n \leq 1 \quad \text{for all } j$$

is in effect for $n = 0$, then the same property is valid for all $n = 1, 2, \dots$. The latter property makes the approach of [19] interesting for practical applications. That said, for a given layer thickness Δz, one needs to choose the time step Δt in such a way that the Courant–Friedrichs–Lewy (CFL) condition is satisfied. Such a condition also ensures that the numerical approximations converge (as $\Delta z, \Delta t \to 0$) to an exact solution of the model as is outlined in [48].

3.4. Desired Steady States for the Two-Phase System

There are many different steady-state solutions of (13) depending on the values of all the feed velocities in and out of the column and volume fractions of the inlets. We are interested in the *desired* steady states, which means that no bubbles leave through the underflow, and there is a froth level. This is the interface in zone 3 (above the feed inlet $z = z_F$) above which the froth is located. Similar desired steady states were presented in [19] for the general model (3) but for the special case when the gas and slurry feed inlets coincide.

We here follow the description in [19], which, in turn, refers to [28,49], and we leave out the mathematical details. The latter involves uniqueness issues and entropy conditions for discontinuities of the solution $\phi = \phi(z,t)$. Such discontinuities arise in three different situations, namely: (1) in regions where $\phi(z,t) < \phi_c$ and the PDE is hyperbolic; (2) across the discontinuity from a lower concentration $\phi < \phi_c$ up to the critical concentration ϕ_c, beyond which the PDE is parabolic; and (3) across the z-positions of inlet and outlets, where the total flux function $J = J(\phi, z, t)$ of (13) is discontinuous (with respect to z).

We directly let $\phi_F = 0$, since there is no gas in that feed inlet. If we write the delta symbol on the right-hand side of (13) formally as $\delta(z - z_S) = H'(z - z_S)$, where

$$H(x) = \begin{cases} 0 & \text{if } x \leq 0, \\ 1 & \text{if } x > 0 \end{cases}$$

is the Heaviside function, then time-independent solutions $\phi = \phi(z)$ of (13) satisfy the second-order ODE

$$\frac{\mathrm{d}}{\mathrm{d}z}\left(J(\phi, z) - \gamma(z) \frac{\mathrm{d}D(\phi)}{\mathrm{d}z} - q_G \phi_G H(z - z_G) - q_W \phi_W H(z - z_W) \right) = 0.$$

This ODE can be integrated with respect to z to yield

$$J(\phi, z) - \gamma(z) d(\phi) \frac{\mathrm{d}\phi(z)}{\mathrm{d}z} - q_G \phi_G H(z - z_G) - q_W \phi_W H(z - z_W) = M \quad \text{for all } z, \tag{15}$$

where the constant mass flux per area unit M can be determined by considering (15) outside the tank—that is, for setting either $z < z_U$ or $z > z_E$, such that $\gamma(z) = 0$, and hence

$$M = j_U(\phi_U) = -q_U\phi_U,$$
$$M = j_E(\phi_E) - q_G\phi_G = q_E\phi_E - q_G\phi_G, \tag{16}$$

Since, in a desired steady state, ϕ_U and $M = 0$, (16) implies that the effluent concentration is given by

$$\phi_E = \frac{q_G\phi_G}{q_E} = \frac{q_G\phi_G}{q_W + q_F + q_G - q_U}. \tag{17}$$

In zone 2, we want a solution that satisfies $\phi_2(z) < \phi_c$; hence, $d(\phi_2) = 0$, and choosing $z \in (z_G, z_F)$ in (15), we obtain

$$j_2(\phi; q_2) - q_G\phi_G = 0, \tag{18}$$

where writing out the dependence of j_2 on q_2 is convenient when investigating the dependence of steady-state solutions on the bulk velocities. We denote by $\bar{\phi}_2$ the smallest solution of (18), which is thus the constant solution $\phi_2(z) = \bar{\phi}_2$ in zone 2. The conditions for this solution are

$$q_G\phi_G \leq j_2(\phi_2^M(q_2); q_2), \tag{19}$$
$$\bar{\phi}_2 \leq \phi_{1Z}(q_1), \tag{20}$$

where ϕ_2^M is the maximum point of j_2 for given q_2 and $\phi_{1Z}(q_1)$ the positive zero of $j_1(\phi; q_1)$; see [19] for exact definitions. The analogous conditions can be found in [19].

In Figure 6 (left), we can see a possible steady-state value for zone 2, with $\bar{\phi}_2 \leq \phi_2^M$ and satisfying both conditions (19) and (20). Another steady state with high volume fractions greater than ϕ_2^M in the entire zone 2 could, in some cases, be possible theoretically; however, this is not a desirable steady state. Here, and in the rest of the text, $\phi_{2M}(q_2)$ will denote the local minimum point greater than the inflection point of j_2, analogously for $\phi_{3M}(q_3)$; see [19] for details.

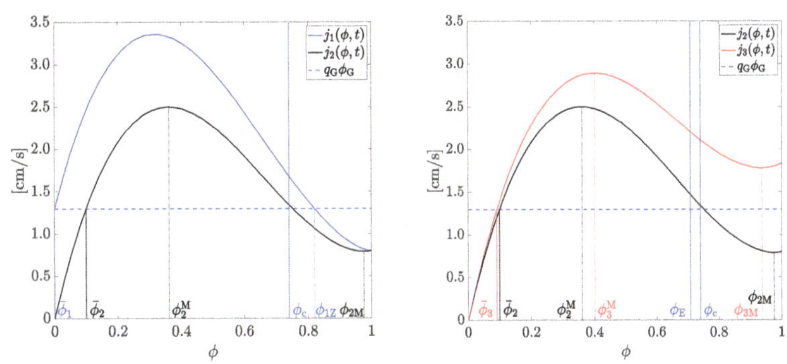

Figure 6. Possible steady-state values $\bar{\phi}_2$ satisfying condition (18) for zone 2 (**left**) and $\bar{\phi}_3$ satisfying (23) for zone 3 (**right**). Note that, in zone 1, we suppose $\bar{\phi}_1 = 0$, since we suppose that that no bubbles leave through the underflow. These graphics were obtained using $\phi_G = 1$, $\phi_F = 0$, $q_U = 0.5\,\text{cm/s}$, $q_G = 1.3\,\text{cm/s}$, $q_F = 1.035\,\text{cm/s}$ and $q_W = 0\,\text{cm/s}$.

With z in zone 3, Equation (15) yields

$$j_3(\phi; q_3) - d(\phi)\frac{\mathrm{d}\phi(z)}{\mathrm{d}z} = q_G\phi_G, \quad z_F < z < z_E. \tag{21}$$

As in [19], we construct a solution in zone 3, which is

$$\phi_3(z) = \begin{cases} \bar{\phi}_3, & z_F < z < z_{fr}, \\ \phi_{3par}(z), & z_{fr} < z \leq z_E, \end{cases} \tag{22}$$

where $\bar{\phi}_3$ is a constant volume fraction above $z = z_F$ and below the pulp–froth interface located at $z = z_{fr}$. Flux continuity and an entropy condition not detailed here yield that $\bar{\phi}_3$ is the smallest solution of

$$j_3(\phi; q_3) = j_2(\bar{\phi}_2; q_2), \tag{23}$$

which means that $\bar{\phi}_3 < \bar{\phi}_2$ when $q_F > 0$. Furthermore, in (22), $\phi_{3par}(z)$ is the strictly increasing solution of the ODE

$$\frac{d\phi(z)}{dz} = \frac{j_3(\phi; q_3) - q_G\phi_G}{d(\phi)}; \quad \phi(z_{fr}) = \phi_c, \quad \phi(z_E) = \phi_E \tag{24}$$

(see (21)), where z_{fr} is the unknown location of the pulp–froth interface $\phi = \phi_c$. Thus, problem (24) defines a function Z_{fr} from all input variables to the pulp–froth interface z_{fr} via (recall that ϕ_E is given by (17))

$$z_{fr} = Z_{fr}(\phi_F, q_G, q_F, q_U, q_W).$$

Necessary conditions for the existence of a steady-state solution profile $\phi_3 = \phi_3(z)$ given by (22) are

$$q_G\left(1 - \frac{\phi_G}{\phi_c}\right) < q_U - q_F - q_W \leq q_G(1 - \phi_G), \tag{25}$$

$$z_F < Z_{fr}(\phi_G, q_G, q_F, q_U, q_W), \tag{26}$$

$$q_G\phi_G \begin{cases} < j_3(\phi_{3M}(q_3); q_3) & \text{if } \phi_{3M}(q_3) < \phi_E, \\ \leq j_3(\phi_E; q_3) & \text{if } \phi_{3M}(q_3) \geq \phi_E. \end{cases} \tag{27}$$

In Figure 6 (right), we can see a possible steady-state value for zone 3, with $\bar{\phi}_3 < \phi_M^3$ satisfying (23). Note that, in this case, there does not exist a steady-state value with $\bar{\phi}_3 > \phi_3^M$, since the straight line given by $q_G\phi_G$ does not intersect with the flux function $j_3(\phi, t)$ for values of $\phi > \phi_3^M$. Moreover, since condition (27) is not satisfied for the values of the volumetric flows chosen, the solution in zone 3 will be constant and equal to $\bar{\phi}_3$, i.e., the solution $\phi_{3par}(z)$ of the ODE (24) does not exist, and hence a froth layer is not possible in this scenario.

In conclusion, the desired steady state for the two-phase gas–liquid system is, thus,

$$\phi_{SS}(z) = \begin{cases} 0 & \text{in the underflow zone and zone 1,} \\ \bar{\phi}_2 & \text{in zone 2,} \\ \bar{\phi}_3 & \text{in zone 3 for } z_F < z < z_{fr}, \\ \phi_{3par}(z) & \text{in zone 3 for } z_{fr} < z < z_E, \\ \phi_E & \text{in the effluent zone.} \end{cases} \tag{28}$$

In Figure 7, some desired steady states of this type are shown. This solution can only be obtained if the conditions (19), (20) and (25)–(27) are satisfied. We visualize these conditions in two-dimensional operating charts in the (q_U, q_F)-plane for given q_G, ϕ_G and q_W. To exemplify, we illustrate these conditions in Figure 8 for the first steady-state experiment.

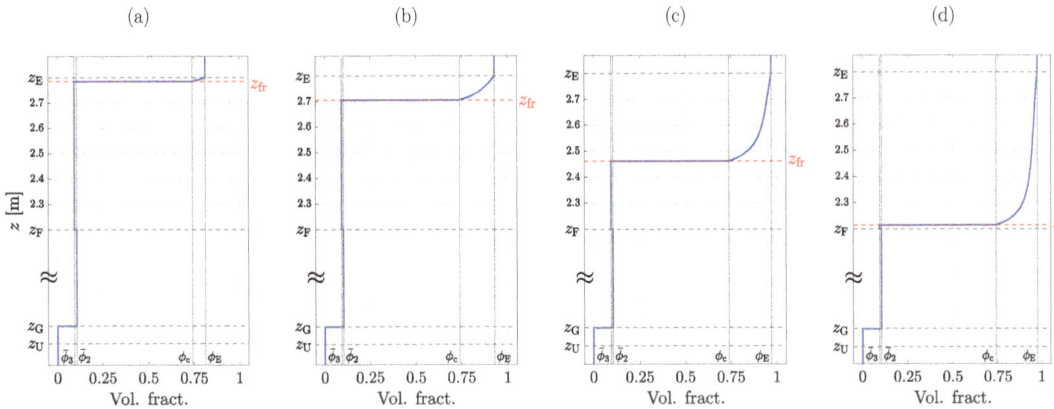

Figure 7. Examples of desired steady states for the gas phase, given by (28). To obtain these figures, we used fixed values of $\phi_G = 1$, $\phi_F = 0$, $q_U = 1\,\mathrm{cm/s}$, $q_G = 1.3\,\mathrm{cm/s}$ and $q_W = 0\,\mathrm{cm/s}$, and we varied q_F, choosing: (**a**) $q_F = 1.3\,\mathrm{cm/s}$, (**b**) $q_F = 1.1\,\mathrm{cm/s}$, (**c**) $q_F = 1.04\,\mathrm{cm/s}$ and (**d**) $q_F = 1.032\,\mathrm{cm/s}$. With these values for the volumetric flows, we obtained a pulp–froth interface located at (**a**) $z_{fr} = 2.7856\,\mathrm{m}$, (**b**) $z_{fr} = 2.7062\,\mathrm{m}$, (**c**) $z_{fr} = 2.4628\,\mathrm{m}$ and (**d**) $z_{fr} = 2.2155\,\mathrm{m}$, respectively, and an effluent volumetric fraction of (**a**) $\phi_E = 0.8125$, (**b**) $\phi_E = 0.9286$, (**c**) $\phi_E = 0.9701$ and (**d**) $\phi_E = 0.9760$.

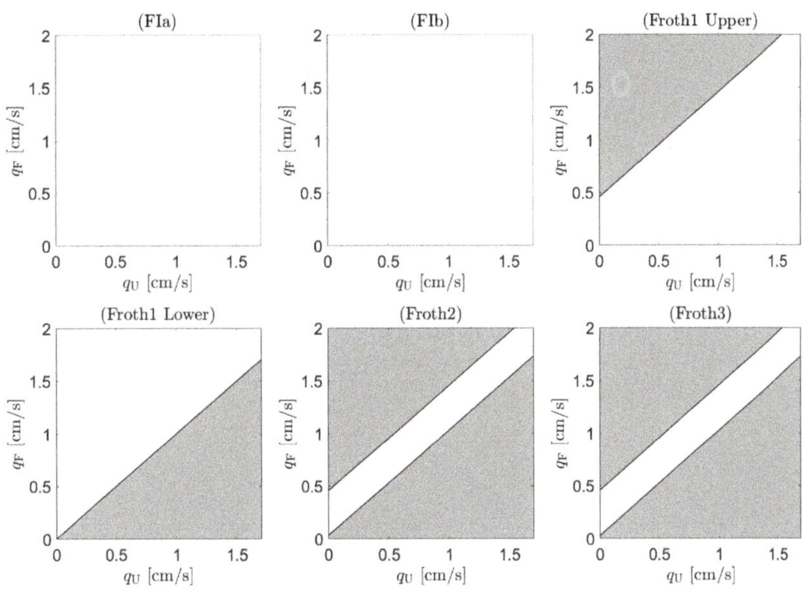

Figure 8. Operating charts showing the theoretical conditions for a steady-state with a froth present in the upper part of the flotation column for the first steady-state experiment. A white region means that the condition is satisfied. The superposition of all these charts results in a white strip, which can be seen in the (q_U, q_F)-plane in Figure 9 (Experiment 1). The labels (FIa), (FIb), (Froth1), (Froth2), and (Froth3) correspond the notation in [19] and to the respective inequalities (19), (20), (25), (26), and (27) in the present work.

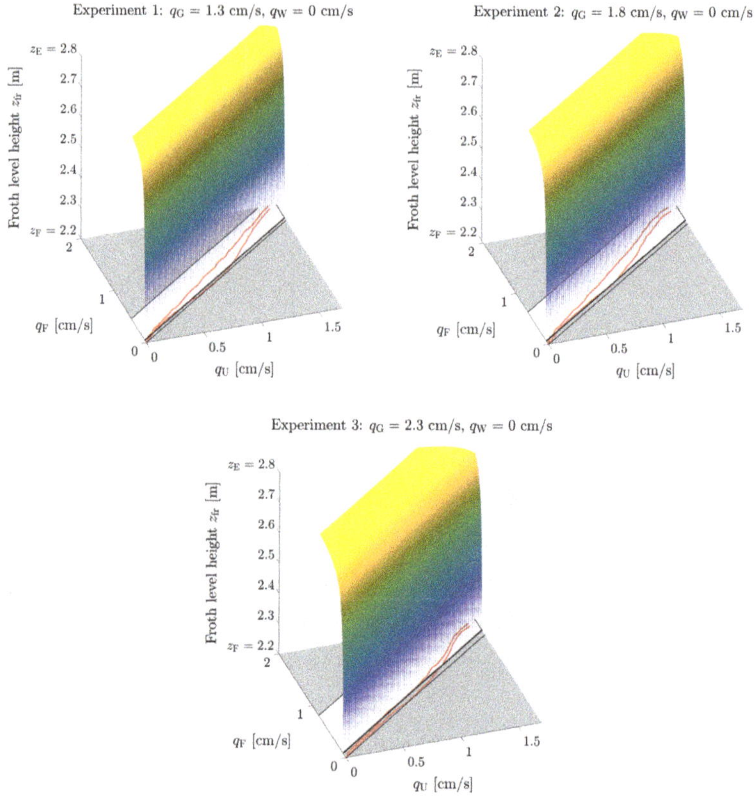

Figure 9. Experiments 1–3 (no wash water): Comparisons between the model with stationary conditions. Here, and in Figure 10, each (q_U, q_F)-plane shows the operating chart in which the white region shows the theoretical conditions for a pulp–froth level above the feed level $z = z_F$ (cf. Figure 8). The red lines in that plane (see also Figures 3 and 4) show the experimental lower and upper values of q_F for each given q_U, in between which, a pulp–froth level was observed. The yellow surface is the graph of the function $(q_U, q_F) \mapsto Z_{fr}(\phi_F, q_G, q_F, q_U, q_W)$, i.e., the estimated height of the pulp–froth interface by the model.

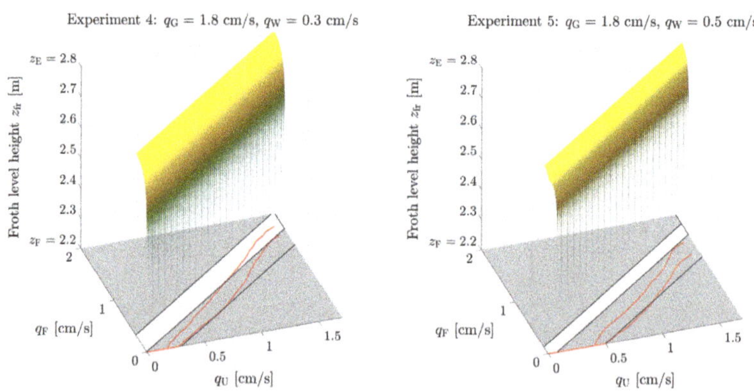

Figure 10. Experiments 4 and 5 (with wash water added): comparisons between the model and experiments with stationary conditions.

4. Results

4.1. Choice of Parameters

The model contains several parameters. We fixed $\phi_c = 0.74$ and $n_S = 0.46$ with the arguments given in [19] for a general froth and let n_b, v_{term} and d_{cap} be those that should be adjusted to reproduce the experiments—at least qualitatively. We thus aimed to find the same fixed values for all experiments. As stated in [28], there exist a number of methods to calculate v_{term}. The generalized correlation by Wallis [50] is recommended; see [23], Appendix A for details. This correlation involves additional quantities, such as the equilibrium surface tension and the viscosity of the fluid. Its discussion is beyond our focus.

Values for n_b normally range from 2 to 3.2. For given values of q_U and q_G, the constant volume fraction $\bar{\phi}_2$ in zone 2 could be estimated experimentally. Equation (18) could then be used to estimate

$$v_{term} = \frac{q_G\phi_G + (q_U - q_G)\bar{\phi}_2}{\bar{\phi}_2(1-\bar{\phi}_2)^{n_b}}, \tag{29}$$

where $\phi_G = 1$ in our experiments. Choosing $n_b = 2$, we obtained the value $v_{term} = 15\,\text{cm/s}$, which gave qualitatively similar operating charts by the model as from the experiments. As for the capillary-to-gravity parameter d_{cap} involved in the modelling of the capillary effect in the froth, we found that the single value $d_{cap} = 2$ cm could be used for a qualitative description of all five steady-state experiments.

All the numerical results were obtained with a spatial discretization of $N = 800$ computational cells, which means a spatial step size of $\Delta z = 3.50 \times 10^{-3}$ m and a time step $\Delta t = 4.43 \times 10^{-3}$ s satisfying the CFL condition.

4.2. Comparison between the Model and Experimental Stationary Data

Comparisons between the model and the five steady-state experiments can be seen in Figures 9 and 10. The goal is that the region between the two experimental lines coincides with the white region of the theoretical operating chart, i.e., the (q_U, q_F)-plane. The qualitative agreement between the experimental data and the model must be considered very good, considering that the theoretical model contains several idealized assumptions and several parameters whose values were taken from the literature for a general sludge. We emphasize that the same parameter values were used in the model for all experiments. The model prediction of the froth level z_{fr} is given by the yellow surface in each subplot of Figures 9 and 10.

In Figure 11, we explain the use of the operating charts of Figures 9 and 10 for the particular case of Experiment 1 (Figure 9a). For a fixed value of $q_U = 1.02\,\text{cm/s}$, we chose four values of q_F on the line $(1.02, q_F)$—shown in dark green in the 3D plot (Figure 11a) inside the white region of the operating chart. For each point chosen, using the correspondence $(q_U, q_F) \mapsto Z_{fr}(\phi_F, q_G, q_F, q_U, q_W)$, we know the estimated height of the pulp–froth interface given by the model; see Figure 11b. In Figure 11c–f, we show the graphs of the steady states for the gas phase recovered with the values of (q_U, q_F) chosen, which are in accordance with the results shown in Figure 11b.

Figure 12 shows a dynamic simulation of the model from a column initially filled with only water and with the parameters of Experiment 1. As can be seen, a steady-state solution arises with a froth layer at the top and with constant volume fractions of bubbles in each zone otherwise. In particular, the steady-state concentration in zone 2, $\bar{\phi}_2$, is slightly larger than the volume fraction $\bar{\phi}_3$ in the lower part of zone 3 as the theoretical steady states predict.

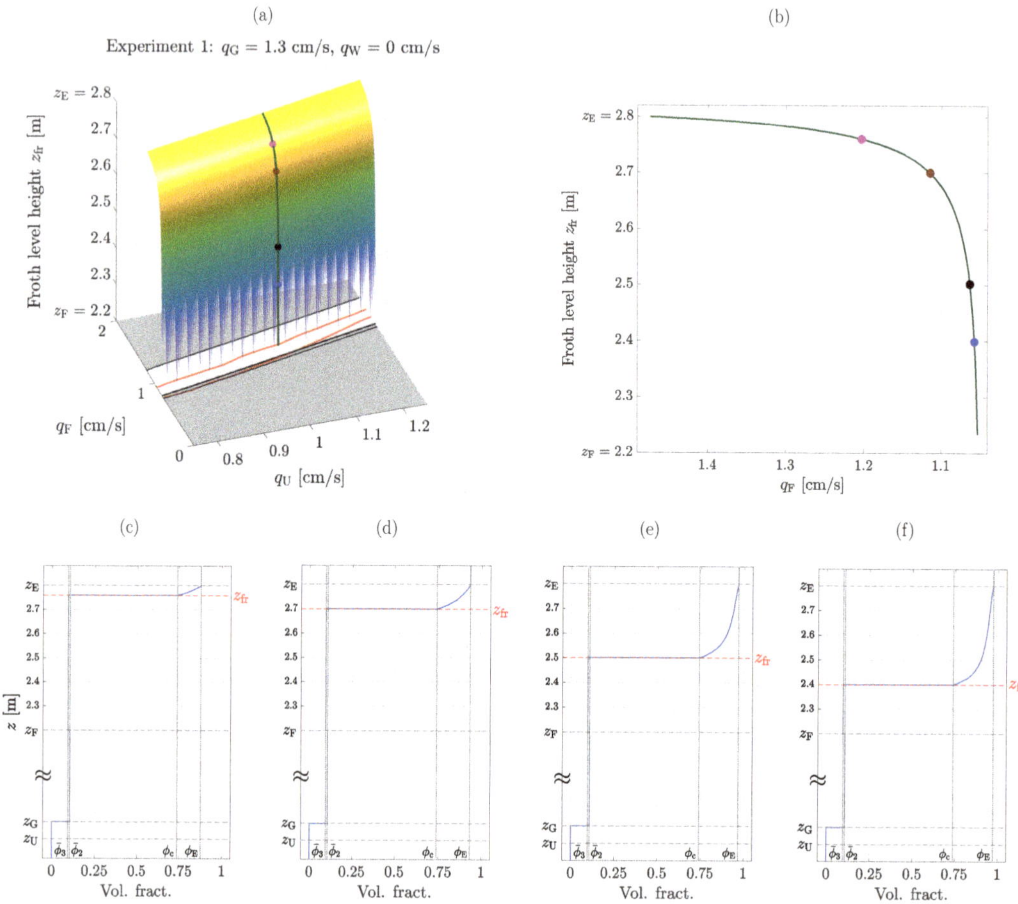

Figure 11. Experiment 1: Example of the use of the operating charts in Figures 9 and 10. (**a**) Enlarged view of Figure 9a. For a fixed value of $q_U = 1.02\,\text{cm/s}$, the line (q_U, q_F) is marked in dark green, and four points with q_F between 1.0568 and 1.2039 cm/s inside the white region of the operating chart were chosen. (**b**) Cross section of the surface $Z_{fr}(\phi_F, q_G, q_F, q_U, q_W)$ for $q_U = 1.02\,\text{cm/s}$ (the q_F-axis is oriented in decreasing order for ease of comparison with plot (**a**)). (**c**–**f**) Steady states for the gas phase obtained with the points (q_U, q_F) marked in plots (**a**,**b**). The values used in each figure are (**c**) $q_F = 1.2039\,\text{cm/s}$, (**d**) $q_F = 1.1150\,\text{cm/s}$, (**e**) $q_F = 1.0628\,\text{cm/s}$ and (**f**) $q_F = 1.0568\,\text{cm/s}$.

The qualitative agreement between the model and experiments lies in the fact that both the model and the experiments confirm that it is only for a small region in the four-dimensional space of (q_U, q_G, q_F, q_W)-values that a froth layer exists. Both the model and the experiments verify that, once a steady state has been found with a froth level, a small change in any of the bulk velocities will either make the froth layer be flushed out upwards or the entire column filled with bubbles, which also leave through the underflow.

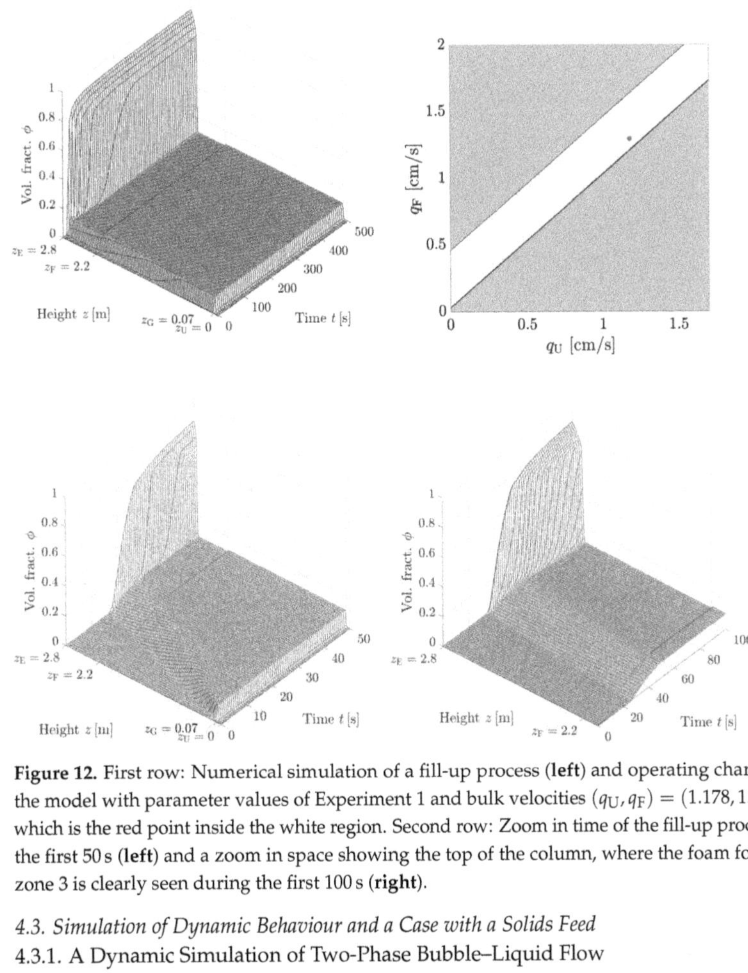

Figure 12. First row: Numerical simulation of a fill-up process (**left**) and operating chart (**right**) of the model with parameter values of Experiment 1 and bulk velocities $(q_U, q_F) = (1.178, 1.292)$ cm/s, which is the red point inside the white region. Second row: Zoom in time of the fill-up process during the first 50 s (**left**) and a zoom in space showing the top of the column, where the foam formation in zone 3 is clearly seen during the first 100 s (**right**).

4.3. *Simulation of Dynamic Behaviour and a Case with a Solids Feed*

4.3.1. A Dynamic Simulation of Two-Phase Bubble–Liquid Flow

We start with the tank full of only fluid at time $t = 0$ s ($\phi(z,0) = \psi(z,0) = 0$ for all z) when we start pumping gas and fluid, with $\phi_F = 0$ and $\psi_F = 0$ at the feed inlet and $\phi_G = 1$ and $\psi_G = 0$ at the gas inlet. We choose the flow rates

$$(q_U, q_F, q_G, q_W) = (1.132, 1.292, 1.8, 0)\,\text{cm/s}$$

in the white region of the theoretical operating chart in Figure 13—marked in red colour. With these parameters, we obtain a desired steady state with a froth layer at the top of the column and no bubbles leaving through the underflow after about $t = 600$ s; see Figure 13 (left).

Once the system is in a steady state at $t = 600$ s, we make a step change from $q_F = 1.292$ cm/s to $q_F = 0.9$ cm/s. The new point chosen lies in the grey region of the theoretical operating chart—marked in blue in Figure 13 (right). As one expects after this change, the froth layer increases, and bubbles at high volume fractions fill the entire column as can be seen in Figure 13 (left). This illustrates how our model is capable of predicting changes in the system and thus is able to take the appropriate control actions. In this case, the control action leads to an undesired steady state since there are gas bubbles leaving through the underflow.

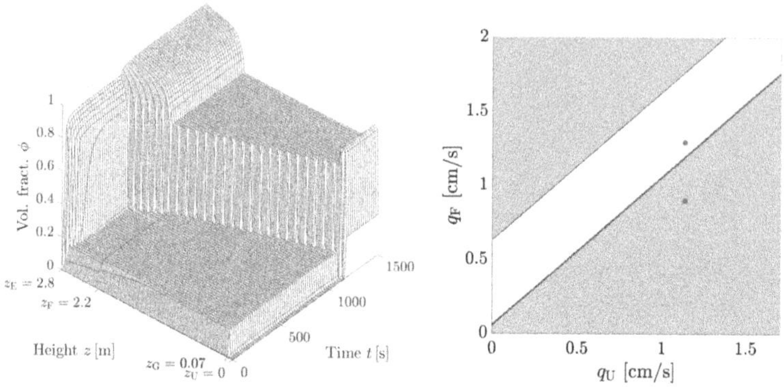

Figure 13. Dynamic simulation of a bubble–liquid flow. (**Left**) Time evolution of the volumetric fraction profile of loaded gas bubbles ϕ from $t = 0\,\mathrm{s}$ to $t = 1500\,\mathrm{s}$. (**Right**) Theoretical operating chart for the simulation in Section 4.3.1. The point marked in red corresponds to $(q_U, q_F, q_G, q_W) = (1.132, 1.292, 1.8, 0)\,\mathrm{cm/s}$ and the one in blue to $(q_U, q_F, q_G, q_W) = (1.132, 0.9, 1.8, 0)\,\mathrm{cm/s}$.

4.3.2. A Dynamic Simulation of Three-Phase Bubble–Solids–Liquid Flow

As in the previous example, the column is initially full of only fluid when we start pumping bubbles and solids with the volume fractions $\phi_G = 1$, $\psi_G = 0$, $\phi_F = 0$ and $\psi_F = 0.1$ along with fluid and wash water. We choose the flow rates

$$(q_U, q_G, q_F, q_W) = (1.087, 1.8, 1.120, 0.3)\,\mathrm{cm/s}$$

Figure 14 shows the time evolution of the volume fractions ϕ and ψ. It can be seen that a desired steady state arises with a layer of froth at the top, while the solids are present only below the feed level z_F where they settle to the bottom.

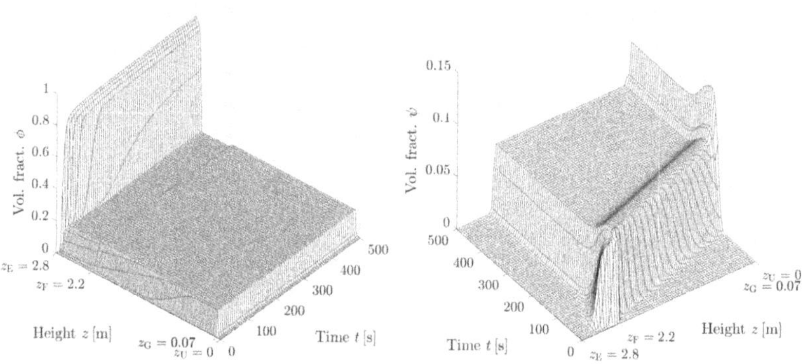

Figure 14. Dynamic simulation of bubble–solids–liquid flow. Time evolution from $t = 0\,\mathrm{s}$ to $t = 500\,\mathrm{s}$ of the volume fraction profiles of bubbles ϕ (**left**) and solids ψ (**right**).

5. Discussion

The theoretically derived PDE model automatically captures several different phenomena (hindered bubble rise, hindered settling of particles and the formation of foam) without any boundary conditions. The model predicts that a desired steady-state solution with a froth layer above the feed level is only possible in a thin region in (q_U, q_G, q_F, q_W)-space. The same set of model parameters was used for all comparisons with the five experiments. The model involves nonlinearities in both the convective and diffusive parts and is strongly degenerate in the diffusive part and, therefore, gives rise to discontinuities in the concentration profiles, which was confirmed experimentally [6] (as emphasized in [19]; see also ([3], p. 915, Figure 3)).

This property is inherent to drift–flux analyses (that disregard a diffusive term that, in our context, models capillarity) (see, for instance, [5,23–25,27]). However, this contrasts with the approaches by Azhin et al. [11,12] and Tian et al. [9,10] that are based on linearized models and where continuity conditions and boundary conditions are imposed explicitly, and therefore continuous steady-state profiles are obtained. The qualitative agreement between our model and the experiments is interesting and shows the possibilities for further investigations and model calibration. A reason for the discrepancies between the model output and the experimentally determined regions in the operating charts, when a stationary froth level in zone 3 is possible (see Figure 9), is the following.

Above the white region in each theoretical operating chart, i.e., the (q_U, q_F)-plane, the combination of the bulk velocities are such that no layer of foam can exist according to the PDE model. In decreasing the value of q_F for fixed q_U, q_G and q_W such that the point (q_U, q_F) lies in the white region, then a froth level z_{fr} is possible in zone 3 ($z_F < z < z_E$). In the upper strip of the white region, the model predicts a froth level z_{fr} (the yellow surface) close to the effluent level z_E, which means a very thin layer of froth. It may well be that such thin froth layers were not registered as valid in the experiments. It is therefore natural that the upper experimental red line lies some distance below the upper line of the white region.

The fact that the lower red line, at least when wash water is present, lies further down in the grey region, means that a froth level is observed close to, but above, the feed level z_F, i.e., almost the entire zone 3, is filled with froth. The theoretical model does not fully capture this behaviour and predicts that, for points (q_U, q_F) below the white region, the entire zone 3 is filled with foam, and there are possibly bubbles dragged down to the underflow. This discrepancy between the model and experiments near the location of the pulp–froth interface when wash water is applied should be further investigated.

In the model development in [19], several reasonable assumptions (partially verified by reported experiments) for the drainage in the froth were assumed to hold for volume fractions close to but above the critical concentration ϕ_c in order to obtain a unified model. It appears that further modelling is needed for the behaviour near the pulp–froth interface. That said, we suggest that that the foam model and the description of that interface by a critical concentration is consistent with the approach by Neethling and Cilliers [39] (which is further elaborated, e.g., in [40]).

6. Conclusions

The conclusions arising from the specific findings of our theory, simulations and experiments are outlined in Section 5. In summary, we can say that the model (3), plus constitutive equations and specifications of control functions, is based on several existing theories (the drift–flux and solids–flux models as well as the model of foam drainage) and lays the ground for a complete simulator of a flotation column in one space dimension without the need to impose boundary conditions or track, for instance, the pulp–froth interface. We once again refer to [19] for an exposition of all technical details. Let us emphasize here the following aspect.

Our previous work [19,28,29], as well as the transient simulations presented herein, have demonstrated that the (theoretical) steady-state theory is consistent with (numerical) simulations of transient scenarios in the sense that steady states are assumed precisely when the various parameters are chosen such that the point (q_U, q_F) lies in the "white region" of the corresponding operating chart. In addition, the model does not catastrophically "break down" when the desired steady-state conditions are violated but makes precise predictions of the "unwanted" behaviour (e.g., bubbles leave through the underflow as seen in Figure 13). This consistency is internal to the mathematical model; however, the comparison with experimental results conducted in this work are the first results that indicate that the model is also consistent with experimental observations. That said, further experiments and comparisons with simulations should be conducted with a focus on transient behaviour and involving solids.

With respect to the potential use of the model in real flotation practice (for instance, to optimize the flotation performance) we mention that the model presented is an advancement of the phenomenological models currently reported in relation to the description of the foam level within the column as well as the gas hold-up. Although it does not yet consider the attachment and detachment mechanism of particles to bubbles, the model is reasonably accurate in determining the stable operating zones, which would allow its use as a complement to current control systems [7,8,10,12–15,45]. Examples of control systems based on the involved phenomenological models (coupled PDEs) have been reported and used in other unit operations [32,51–56].

The present approach captures the multiphase hydrodynamics of aggregates (bubbles) and gangue particles in the column but does not model the aggregation process itself (that is, the attachment of hydrophobic (valuable) particles to gas bubbles). That process usually takes place in the collection zone (zone 2 in Figure 1). To add realism and to explore the interdependence of velocities and reaction kinetics, the flotation model (3) should be extended to include the process of attachment of hydrophobic (valuable) particles. One option consists of considering the valuable and gangue particles as two independent disperse solid phases (while the present approach only includes the gangue) and adding another field variable that describes the local state of aggregation. This procedure leads to two additional PDEs for the two new variables and likely involves spatial variants of known kinetic models for the adhesion of particles (as reviewed, for instance, in [16]).

Author Contributions: Conceptualization, F.B., R.B. and Y.V.; methodology, R.B. and S.D.; software, M.C.M. and Y.V.; validation, F.B. and L.G.; investigation, R.B., S.D., M.C.M. and Y.V.; writing—original draft preparation, R.B.; writing—review and editing, F.B., R.B., M.C.M. and S.D.; visualization, M.C.M. and Y.V. All authors have read and agreed to the published version of the manuscript.

Funding: This research was funded by ANID (Chile) grant numbers ANID/Fondecyt/1210610 and ANID/Fondecyt/1211705; ANID/FSQE210002; Anillo ANID/ACT210030; Centro de Modelamiento Matemático (CMM), projects ACE210010 and FB210005 of BASAL funds for Centers of Excellence; and CRHIAM, project ANID/FONDAP/15130015. It was also funded by the Swedish Research Council (Vetenskapsrådet, 2019-04601), grant PID2020-117211GB-I00 funded by Ministerio de Ciencia e Investigación MCIN/AEI/10.13039/501100011033, by Conselleria de Innovación, Universidades, Ciencia y Sociedad Digital through project CIAICO/2021/ 227 and by IFARHU-SENACYT (Panama).

Data Availability Statement: No new data were created or analyzed in this study apart from those shown in tables and figures. Data sharing is not applicable to this article.

Conflicts of Interest: The authors declare no conflict of interest.

Glossary

List of Symbols

The following symbols are used in this manuscript:

Symbol	Significance and Unit
A	interior cross-sectional area of column $[\mathrm{m}^2]$
$D(\phi)$	integrated capillarity function $[\mathrm{m}^2/\mathrm{s}]$
$J(\phi, z, t)$	convective flux function of bubbles $[\mathrm{m}/\mathrm{s}]$
$F(\varphi, \phi, z, t)$	convective flux function of solids $[\mathrm{m}/\mathrm{s}]$
$\tilde{F}(\psi, \phi, z, t)$	convective flux function of solids $[\mathrm{m}/\mathrm{s}]$
$H(z)$	Heaviside step function $[-]$
N	no. of numerical subintervals for numerical method $[-]$
Q	volumetric flow $[\mathrm{m}^3/\mathrm{s}]$
$Z_{\mathrm{fr}}(\phi_{\mathrm{G}}, q_{\mathrm{G}}, q_{\mathrm{F}}, q_{\mathrm{U}}, q_{\mathrm{W}})$	function giving height of pulp–froth interface $[\mathrm{m}]$
$d(\phi)$	capillarity function $[\mathrm{m}^2/\mathrm{s}]$

d_{cap}	capillarity constant [m]
j_{b}	bubble batch flux function [m/s]
f_{b}	solids batch sedimentation flux function [m/s]
n_{b}	constant exponent in bubble batch flux [−]
n_{S}	constant exponent related to Plateau borders in foam [−]
n_{RZ}	Richardson–Zaki exponent [−]
q	bulk velocity, flow rate [m/s]
t	time [s]
$\tilde{v}(\phi)$	drift–flux velocity function [m/s]
$v_{\mathrm{hs}}(\phi)$	hindered–settling velocity function [m/s]
v_{term}	terminal velocity of single bubble [m/s]
v_{∞}	terminal velocity of single particle [m/s]
z	height [m]
z_{fr}	height of pulp–froth interface [m]
$\partial_t\Box$	$= \partial\Box/\partial t$ [s^{-1}]
δ	Dirac delta distribution [m^{-1}]
Δt	temporal step size of numerical method [s]
Δz	spatial step size of numerical method [m]
$\gamma(z)$	characteristic function; = 1 inside column; = 0 outside [−]
ϕ	volume fraction of bubbles (aggregates) [−]
$\phi_{\mathrm{SS}}(z)$	steady-state solution [−]
ϕ_{c}	critical volume fraction [−]
ϕ_{f}	volume fraction of fluid [−]
ϕ_j^n	volume fraction of bubbles of numerical method [−]
φ	volume fraction of solids in suspension outside bubbles [−]
ψ	volume fraction of solids [−]

Subscripts and Superscript

The following sub- and superscripts are used in this manuscript:

Sub-/Superscript	**Significance**
$\Box_1, \Box_2, \Box_3$	zone 1, zone 2, zone 3
$\Box_{\mathrm{E}}$	effluent
$\Box_{\mathrm{F}}$	feed
$\Box_{\mathrm{G}}$	gas
$\Box_{\mathrm{M}}$	(local) minimum point
$\Box_{\mathrm{SS}}$	steady state
$\Box_{\mathrm{U}}$	underflow
$\Box_{\mathrm{W}}$	wash water
$\Box_{\mathrm{Z}}$	zero (of a function)
$\Box_{\mathrm{c}}$	critical
$\Box_{\mathrm{b}}$	batch
$\Box_{\mathrm{f}}$	fluid
$\Box_{\mathrm{fr}}$	froth
$\Box_{\mathrm{par}}$	parabolic
$\Box^{\mathrm{M}}$	(local) maximum point

Abbreviations

The following abbreviations are used in this manuscript:

CFD	computational fluid dynamics
CFL	Courant–Friedrichs–Lewy (condition)
MIBC	methyl isobutyl carbinol
ODE	ordinary differential equation
PDE	partial differential equation

References

1. Finch, J.A.; Dobby, G.S. *Column Flotation*; Pergamon Press: London, UK, 1990.
2. Wills, B.A.; Napier-Munn, T.J. *Wills' Mineral Processing Technology*, 7th ed.; Butterworth-Heinemann: Oxford, UK, 2006.
3. Dunne, R.C.; Kawatra, S.K.; Young, C.A. (Eds.) *SME Mineral Processing & Extractive Metallurgy Handbook*; Society for Mining, Metallurgy, and Exploration: Englewood, CO, USA, 2019.
4. Pal, R.; Masliyah, J.H. Flow characterization of a flotation column. *Can. J. Chem. Eng.* **1989**, *67*, 916–923. [CrossRef]

5. Vandenberghe, J.; Chung, J.; Xu, Z.; Masliyah, J. Drift flux modelling for a two-phase system in a flotation column. *Can. J. Chem. Eng.* **2005**, *83*, 169–176. [CrossRef]
6. Cruz, E.B. A Comprehensive Dynamic Model of the Column Flotation Unit Operation. Ph.D. Thesis, Virginia Tech, Blacksburg, VA, USA, 1997.
7. Maldonado, M.; Desbiens, A.; del Villar, R. Potential use of model predictive control for optimizing the column flotation process. *Int. J. Miner. Process.* **2009**, *93*, 26–33. [CrossRef]
8. Bergh, L.G.; Yianatos, J.B. The long way to multivariate predictive control of flotation processes. *J. Process Control* **2022**, *21*, 226–234. [CrossRef]
9. Tian, Y.; Azhin, M.; Luan, X.; Liu, F.; Dubljevic, S. Three-phases dynamic modelling of column flotation process. *IFAC-PapersOnLine* **2018**, *51*, 99–104. [CrossRef]
10. Tian, Y.; Luan, X.; Liu, F.; Dubljevic, S. Model predictive control of mineral column flotation process. *Mathematics* **2018**, *6*, 100. [CrossRef]
11. Azhin, M.; Popli, K.; Afacan, A.; Liu, Q.; Prasad, V. A dynamic framework for a three phase hybrid flotation column. *Miner. Eng.* **2021**, *170*, 107028. [CrossRef]
12. Azhin, M.; Popli, K.; Prasad, V. Modelling and boundary optimal control design of hybrid column flotation. *Can. J. Chem. Eng.* **2021**, *99* (Suppl. 1), S369–S388. [CrossRef]
13. Quintanilla, P.; Neethling, S.J.; Brito-Parada, P.R. Modelling for froth flotation control: A review. *Miner. Eng.* **2021**, *162*, 106718. [CrossRef]
14. Quintanilla, P.; Neethling, S.J.; Navia, D.; Brito-Parada, P.R. A dynamic flotation model for predictive control incorporating froth physics. Part I: Model development. *Miner. Eng.* **2021**, *173*, 107192. [CrossRef]
15. Quintanilla, P.; Neethling, S.J.; Mesa, D.; Navia, D.; Brito-Parada, P.R. A dynamic flotation model for predictive control incorporating froth physics. Part II: Model calibration and validation. *Miner. Eng.* **2021**, *173*, 107190. [CrossRef]
16. Wang, G.; Ge, L.; Mitra, S.; Evans, G.M.; Joshi, J.B.; Chen, S. A review of CFD modelling studies on the flotation process. *Miner. Eng.* **2018**, *127*, 153–177. [CrossRef]
17. Wallis, G.B. *One-Dimensional Two-Phase Flow*; McGraw-Hill: New York, NY, USA, 1969.
18. Narsimhan, G. Analysis of creaming and formation of foam layer in aerated liquid. *J. Colloid Interface Sci.* **2010**, *345*, 566–572. [CrossRef]
19. Bürger, R.; Diehl, S.; Martí, M.C.; Vásquez, Y. A degenerating convection-diffusion system modelling froth flotation with drainage. *IMA J. Appl. Math.* **2022**, *87*, 1151–1190. [CrossRef]
20. Vásquez, Y. Conservation Laws with Discontinuous Flux Modeling Flotation Columns. Doctoral Thesis, Universidad de Concepción, Concepción, Chile, 2022.
21. Bürger, R.; Wendland, W.L.; Concha, F. Model equations for gravitational sedimentation-consolidation processes. *Z. Angew. Math. Mech.* **2000**, *80*, 79–92. [CrossRef]
22. Bascur, O.A. A unified solid/liquid separation framework. *Fluid/Part. Sep. J.* **1991**, *4*, 117–122.
23. Stevenson, P.; Fennell, P.S.; Galvin, K.P. On the drift-flux analysis of flotation and foam fractionation processes. *Can. J. Chem. Eng.* **2008**, *86*, 635–642. [CrossRef]
24. Dickinson, J.E.; Galvin, K.P. Fluidized bed desliming in fine particle flotation, Part I. *Chem. Eng. Sci.* **2014**, *108*, 283–298. [CrossRef]
25. Galvin, K.P.; Dickinson, J.E. Fluidized bed desliming in fine particle flotation Part II: Flotation of a model feed. *Chem. Eng. Sci.* **2014**, *108*, 299–309. [CrossRef]
26. Galvin, K.P.; Harvey, N.G.; Dickinson, J.E. Fluidized bed desliming in fine particle flotation – Part III flotation of difficult to clean coal. *Miner. Eng.* **2014**, *66–68*, 94–101. [CrossRef]
27. Bürger, R.; Diehl, S.; Martí, M.C. A conservation law with multiply discontinuous flux modelling a flotation column. *Netw. Heterog. Media* **2018**, *13*, 339–371. [CrossRef]
28. Bürger, R.; Diehl, S.; Martí, M.C. A system of conservation laws with discontinuous flux modelling flotation with sedimentation. *IMA J. Appl. Math.* **2019**, *84*, 930–973. [CrossRef]
29. Bürger, R.; Diehl, S.; Martí, M.C.; Vásquez, Y. Flotation with sedimentation: Steady states and numerical simulation of transient operation. *Miner. Eng.* **2020**, *157*, 106419. [CrossRef]
30. Bürger, R.; Diehl, S.; Martí, M.C.; Vásquez, Y. Simulation and control of dissolved air flotation and column froth flotation with simultaneous sedimentation. *Water Sci. Technol.* **2020**, *81*, 1723–1732. [CrossRef] [PubMed]
31. Kynch, G.J. A theory of sedimentation. *Trans. Faraday Soc.* **1952**, *48*, 166–176. [CrossRef]
32. Diehl, S. Operating charts for continuous sedimentation I: Control of steady states. *J. Eng. Math.* **2001**, *41*, 117–144. [CrossRef]
33. Diehl, S. The solids-flux theory—Confirmation and extension by using partial differential equations. *Water Res.* **2008**, *42*, 4976–4988. [CrossRef]
34. Bürger, R.; Karlsen, K.H.; Risebro, N.H.; Towers, J.D. Well-posedness in BV_t and convergence of a difference scheme for continuous sedimentation in ideal clarifier-thickener units. *Numer. Math.* **2004**, *97*, 25–65. [CrossRef]
35. Bürger, R.; Karlsen, K.H.; Towers, J.D. A model of continuous sedimentation of flocculated suspensions in clarifier-thickener units. *SIAM J. Appl. Math.* **2005**, *65*, 882–940. [CrossRef]
36. Diehl, S. On scalar conservation laws with point source and discontinuous flux function. *SIAM J. Math. Anal.* **1995**, *26*, 1425–1451. [CrossRef]

37. Diehl, S. A conservation law with point source and discontinuous flux function modelling continuous sedimentation. *SIAM J. Appl. Math.* **1996**, *56*, 388–419. [CrossRef]
38. Neethling, S.J.; Lee, H.T.; Cilliers, J.J. A foam drainage equation generalized for all liquid contents. *J. Phys. Condens. Matter* **2002**, *14*, 331–342. [CrossRef]
39. Neethling, S.J.; Cilliers, J.J. Modelling flotation froths. *Int. J. Miner. Process.* **2003**, *72*, 267–287. [CrossRef]
40. Neethling, S.J.; Brito-Parada, P.R. Predicting flotation behaviour – The interaction between froth stability and performance. *Miner. Eng.* **2018**, *120*, 60–65. [CrossRef]
41. Neethling, S.J.; Cilliers, J.J. Solids motion in flowing froths. *Chem. Eng. Sci.* **2002**, *57*, 607–615. [CrossRef]
42. Richardson, J.F.; Zaki, W.N. Sedimentation and fluidisation: Part I. *Trans. Inst. Chem. Eng.* **1954**, *32*, 35–53. [CrossRef]
43. Xu, M.; Finch, J.A.; Uribe-Salas, A. Maximum gas and bubble surface rates in flotation columns. *Int. J. Miner. Process.* **1991**, *32*, 233–250. [CrossRef]
44. Bergh, L.G.; Yianatos, J.B. Experimental studies on flotation column dynamics. *Miner. Eng.* **1994**, *7*, 345–355 [CrossRef]
45. Bergh, L.G.; Yianatos, J.B. Flotation column automation: State of the art. *Control Eng. Pract.* **2003**, *11*, 67–72. [CrossRef]
46. Yianatos, J.B.; Bucarey, R.; Larenas, J.; Henríquez, F.; Torres, L. Collection zone kinetic model for industrial flotation columns. *Miner. Eng.* **2005**, *18*, 1373–1377. [CrossRef]
47. Yianatos, J.B.; Henríquez, F.H.; Oroz, A.G. Characterization of large size flotation cells. *Miner. Eng.* **2006**, *19*, 531–538. [CrossRef]
48. Bürger, R.; Diehl, S.; Martí, M.C.; Vásquez, Y. A difference scheme for a triangular system of conservation laws with discontinuous flux modeling three-phase flows. *Netw. Heterog. Media* **2023**, *18*, 140–190. [CrossRef]
49. Diehl, S. A uniqueness condition for nonlinear convection-diffusion equations with discontinuous coefficients. *J. Hyperbolic Differ. Equations* **2009**, *18*, 127–159. [CrossRef]
50. Wallis, G.B. The terminal speed of single drops or bubbles in an infinite medium. *Int. J. Multiph. Flow* **1974**, *1*, 491–511. [CrossRef]
51. Diehl, S. Operating charts for continuous sedimentation III: Control of step inputs. *J. Eng. Math.* **2006**, *54*, 225–259. [CrossRef]
52. Diehl, S. Operating charts for continuous sedimentation IV: Limitations for control of dynamic behaviour. *J. Eng. Math.* **2008**, *60*, 249–264. [CrossRef]
53. Diehl, S. A regulator for continuous sedimentation in ideal clarifier-thickener units. *J. Eng. Math.* **2008**, *60*, 265–291. [CrossRef]
54. Diehl, S.; Farås, S. Control of an ideal activated sludge process in wastewater treatment via an ODE-PDE model. *J. Process Control* **2013**, *23*, 359–381. [CrossRef]
55. Betancourt, F.; Bürger, R.; Diehl, S.; Farås, S. Modelling and controlling clarifier-thickeners fed by suspensions with time-dependent properties. *Miner. Eng.* **2014**, *62*, 91–101. [CrossRef]
56. Torfs, E.; Maere, T.; Bürger, R.; Diehl, S.; Nopens, I. Impact on sludge inventory and control strategies using the benchmark simulation model no. 1 with the Bürger-Diehl settler model. *Water Sci. Technol.* **2015**, *71*, 1524–1535. [CrossRef]

minerals

MDPI

Article

The Correlation between Macroscopic Image and Object Properties with Bubble Size in Flotation

Luis Vinnett [1,2,*], Iván Cornejo [1], Juan Yianatos [1,2], Claudio Acuña [1], Benjamín Urriola [1], Camila Guajardo [1] and Alex Esteban [1]

[1] Department of Chemical and Environmental Engineering, Universidad Técnica Federico Santa María, Valparaíso 2390123, Chile

[2] Automation and Supervision Centre for Mining Industry, CASIM, Universidad Técnica Federico Santa María, Valparaíso 2390123, Chile

* Correspondence: luis.vinnett@usm.cl

Abstract: This paper studies the correlation between different macroscopic features of image regions and object properties with the Sauter diameter (D_{32}) of bubble size in flotation. Bubbles were sampled from the collection zone of a two-dimensional flotation cell using a McGill Bubble Size Analyzer, and photographed bubbles were processed using image analysis. The Sauter mean diameters were obtained under different experimental conditions using a semiautomated methodology, in which non-identifiable bubbles were manually characterized to estimate the bubble size distribution. For the same processed images, different image properties from their binary representation were studied in terms of their correlation with D_{32}. The median and variability of the shadow percentage, aspect ratio, power spectral density, perimeter, equivalent diameters, solidity, and circularity, among other image or object properties, were studied. These properties were then related to the measured D_{32} values, from which four predictors were chosen to obtain a multivariable model that adequately described the Sauter diameter. After removing abnormal gas dispersion conditions, the multivariable linear model was able to represent D_{32} values (99 datasets) for superficial gas rates in the range of 0.4–2.5 cm/s, for four types of frothers and surfactant concentrations ranging from 0 to 32 ppm. The model was tested with 72 independent datasets, showing the generalizability of the results. Thus, the approach proved to be applicable at the laboratory scale for D_{32} = 1.3–6.7 mm.

Keywords: gas dispersion; flotation; bubble size; Sauter diameter

check for updates

Citation: Vinnett, L.; Cornejo, I.; Yianatos, J.; Acuña, C.; Urriola, B.; Guajardo, C.; Esteban, A. The Correlation between Macroscopic Image and Object Properties with Bubble Size in Flotation. *Minerals* **2022**, *12*, 1528. https://doi.org/10.3390/min12121528

Academic Editors: Fardis Nakhaei, Ahmad Hassanzadeh and Luis A. Cisternas

Received: 6 November 2022

Accepted: 27 November 2022

Published: 29 November 2022

1. Introduction

Flotation rate and efficiency critically depend upon the relationship between particle size and bubble size; both parameters play a significant role in successful collection and froth transport processes [1–5]. Since the development of bubble size analyzers, a better understanding of the impact of gas dispersion on flotation performance has been achieved. The most used analyzers consist of a bubble viewer for sampling along with an image processing tool to characterize bubbles [6–8]. These devices proved to have a good trade-off between the number of identified bubbles and their applicability at different flotation scales. More basic image processing algorithms identify bubbles by a single or a variety of shape factors, such as circularity, solidity, and others [9–11], removing irregular objects and overlapped bubbles from the analysis. This approach has been widely used in the flotation literature [10–12]; however, significant biases have been observed in bubble size estimations, especially in the presence of large bubbles [13–15]. To avoid these biases, some applications also incorporate segmentation algorithms to separate or identify bubbles in clusters (i.e., Watershed and Hough transforms) [7,13,16–19]. In any case, most current algorithms focus on the individualization of bubbles, removing objects that are not identifiable (e.g., complex clusters or cap-shaped bubbles) from a predefined performance criterion or threshold. This strategy has proven to be effective only under specific gas dispersion regimes [20]. Thus, robust bubble size characterizations have been limited to spherical and spherical-ellipsoidal regimes, which hinders the generalization of experimental results to non-ideal conditions.

Some alternative or indirect estimations of bubble size have been proposed in the literature. These approaches take advantage of either signals that are generated from bubble motion or correlations between bubble size with other measurable variables. For example, Steinemann and Buchholz [21] used conductivity measurements to estimate bubble size and bubble velocity. A two-point probe was proposed, in which terminals were disturbed by the rising bubbles. These disturbances generated pulse trains associated with bubble properties. Geometrical and physical relationships allowed for the bubble size estimations, for bubbles larger than 1.0 mm. Meernik and Yuen [22] presented an optical method based on disturbances to laser beams to estimate bubble size. Optical fiber was used at the injection and detection terminals, and a photodiode was employed as a transmitter. The measurement system was limited by the optical beam length to characterize single bubbles. Kracht et al. [23] proposed a stochastic approach based on the covariance function of the image backgrounds to determine bubble size distributions (BSDs). That methodology was successfully tested under a spherical regime from 10 images generated at the laboratory scale. Image simulations also supported this study. Kracht and Moraga [24] estimated the Sauter mean diameter of bubble populations $D_{32} = \sum d_i^3 / \sum d_i^2$ from acoustic measurements. Bubbles were exposed to an acoustic disturbance, whose responses (demodulated signals) were related to bubble size. An approximated linear trend between the signal intensities and the Sauter diameter was observed for $D_{32} \approx 0.8$–2.7 mm. The reference D_{32} values (ground truth) were estimated from image analysis. Vinnett and Alvarez-Silva [25] related the shadow percent from binary bubble images with D_{32} at different superficial gas rates, J_G. A linear model was proposed, which incorporated J_G and the shadow percent as predictors. This model presented acceptable results from laboratory and industrial datasets. However, the trends were rather noisy. Vinnett et al. [26] reported a technique to estimate D_{32} from the power spectral density of pulses generated by bubbles in binary images. The spatial bandwidth proved to be non-linearly correlated with D_{32}. A piecewise algorithm based on conventional image analysis in a spherical regime and a bandwidth correlation of $D_{32} > 2.0$ mm was proposed for industrial measurements [20]. Ilonen et al. [27] used the two-dimensional discrete Fourier transform to estimate BSDs in pulp delignification. The results provided by different circle detection techniques were compared with the estimations obtained from the 2D power spectral density. Principal component analysis was employed to reduce dimensionality. In addition, multivariable linear regression was used to obtain the bubble counts in ten size classes from the power spectral density. The technique showed adequate performance in spherical regimes. Bu et al. [28] correlated gas dispersion parameters with the variability of differential pressure measurements in a flotation column. A linear model was proposed for the bubble size, using the standard deviation of the differential pressure as a predictor. This model led to a coefficient of correlation of 0.77. High variability was obtained with this methodology.

The information provided in the previous paragraph shows that bubble size can be determined by several techniques, which involve parameters that can be measured or are influenced by the characteristics and behavior of bubbles in a swarm. Some techniques directly use bubble viewers along with image analysis, whereas other physical parameters may also be correlated with photographed bubble populations (e.g., differential pressure measurements and gas hold-up variability). This paper studies correlations between different image and object properties in binary representation and the Sauter mean diameter of bubble size distributions. These properties and their variability can be automatically determined, with negligible bias. A semiautomated algorithm that allowed all bubbles to be processed was used to obtain the Sauter diameters employed as ground truth. A multivariable linear model is proposed to estimate D_{32} from image and object features, which does not require the individualization of every single bubble in the photographed populations.

2. Materials and Methods

2.1. Experimental Procedure

Bubble size measurements were conducted in the laboratory-scale flotation cell depicted in Figure 1. This 2D cell emulated a slice of an industrial machine with a 140 × 140 cm cross-section and a width of 15 cm. The forced air was controlled and fed from 24 porous spargers. A McGill bubble size analyzer (MBSA) [6] was used for bubble sampling and image recording. This device was initially filled with conditioned water at the same surfactant concentration as in the flotation cell. The rising bubbles were then photographed in a 2D plane with a digital video camera (version, Teledyne Dalsa, Waterloo, ON, Canada), at a sampling rate of one frame per second. All measurements were conducted for 3 minutes at a resolution of 0.056 mm/pxl.

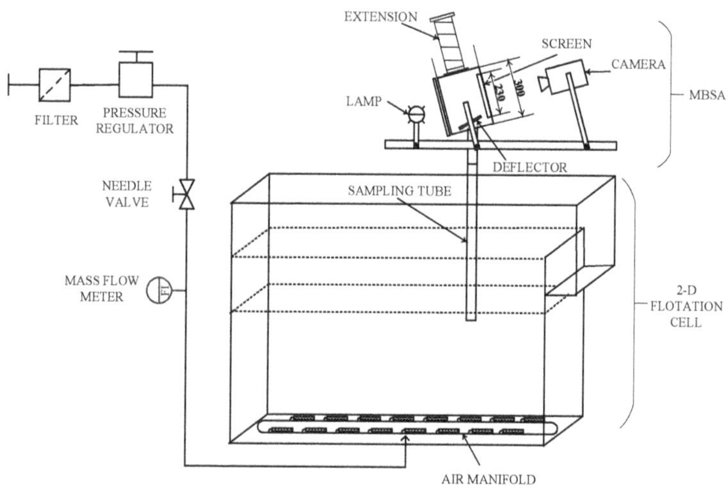

Figure 1. Two-dimensional flotation cell and installation of the McGill bubble size analyzer [29].

Four types of frothers were studied: methyl isobutyl carbinol (MIBC), AeroFroth® 70 (Cytec, Woodland Park, NJ, USA), OrePrep® F-507 (Cytec, Woodland Park, NJ, USA), and Flotanol® 9946 (Clariant Mining Solutions, Louisville, KY, USA). AeroFroth® 70 contains MIBC and diisobutyl ketone [30], OrePrep® F-507 contains glycol and other non-hazardous components [30], and Flotanol® 9946 corresponds to a 2-ethyl hexanol distillation bottom [31]. The experimental data were divided into training and testing datasets. Tables 1 and 2 present all the evaluated experimental conditions, including the distribution of the training and testing datasets. Frother concentrations of 0, 2, 4, 8, and 16 ppm were evaluated for all types of frothers, whereas 32 ppm was also assessed for AeroFroth® 70, OrePrep® F-507, and Flotanol® 9946. The superficial gas rates were set to 0.5, 1.0, 1.5, 2.0, and 2.5 cm/s for MIBC, and to 0.4, 0.8, 1.2, 1.6, and 2.0 cm/s for the rest of the frothers. Conditions with high J_G and low concentrations of MIBC favored the transition toward a churn-turbulent regime from an ellipsoidal regime, which was detected in the analysis. All tests were conducted at two locations in the flotation cell. From Tables 1 and 2, 104 experimental conditions were used in the training procedure and 72 for testing.

Table 1. Flotation tests for training.

Type of Frother	Location	Frother Concentrations, ppm	Superficial Gas Rate, cm/s
MIBC	1 and 2	0, 2, 4, 8, 16	0.5, 1.0, 1.5, 2.0, 2.5
AeroFroth® 70	1	0, 2, 4, 8, 16, 32	0.4, 1.2, 2.0
OrePrep® F-507	1	0, 2, 4, 8, 16, 32	0.4, 1.2, 2.0
Flotanol® 9946	1	0, 2, 4, 8, 16, 32	0.4, 1.2, 2.0

Table 2. Flotation tests for testing.

Type of Frother	Location	Frother Concentrations, ppm	Superficial Gas Rate, cm/s
AeroFroth® 70	2	0, 2, 4, 8, 16, 32	0.8, 1.6
		0, 2, 8, 32	0.4, 1.2, 2.0
OrePrep® F-507	2	0, 2, 4, 8, 16, 32	0.8, 1.6
		0, 2, 8, 32	0.4, 1.2, 2.0
Flotanol® 9946	2	0, 2, 4, 8, 16, 32	0.8, 1.6
		0, 2, 8, 32	0.4, 1.2, 2.0

2.2. Semiautomated Image Processing

The BSDs and Sauter mean diameters used as reference (ground truth) were obtained by a semiautomated application based on the Image Processing Toolbox of MATLAB (11.4, The MathWorks Inc., Natick, MA, USA). A field of view of 45 × 35 mm was chosen for image analysis. The images were firstly converted to their binary representation. Bubbles observed as isolated spheres and ellipsoids were identified based on solidity [26]. Objects that presented low solidity were first segmented using Watershed, followed by Hough transforms [13,32]. The previous automated steps were complemented by manual processing: (i) false positives obtained in the automated processing were corrected; (ii) non-identified bubbles (bubbles in clusters and irregular bubbles) were manually estimated. This procedure avoided the biases caused by removing bubbles from the analysis as reported in the literature [13–15]. The size of each identified bubble was estimated as an equivalent ellipsoid diameter. The D_{32} values obtained from the semiautomated algorithm were used as references to evaluate the ability to predict bubble size. We recorded 180 images per experimental condition, from which a subset was randomly chosen to process a minimum of 1500 bubbles per test. However, at least 10 images were processed in all cases. This limit for the number of processed images was especially defined for conditions with a high gas hold-up. All images were analyzed when operating the cell with no frother. For further details on the semiautomated procedure, please refer to Vinnett, Urriola, Orellana, Guajardo and Esteban [29]. Appendix A presents examples of the bubble size distributions obtained by the semiautomated approach.

2.3. Region Properties and Their Association with Bubble Size

The same images that were processed by the semiautomated algorithm were studied in terms of their region properties from their binary representation. The statistical parameters of these properties were analyzed based on their association with the Sauter mean diameter of the BSDs. For example, for the binary image shown in Figure 2a, the object and region properties summarized in Table 3 were calculated. For each experimental condition and all processed images, the statistics of the properties of all objects (e.g., circularity, solidity, aspect ratio, and perimeter) were estimated to obtain the median and some indicators of variability. In addition, the shadow fraction (black region with respect to the region of interest) and the spatial bandwidth were obtained for each image. The spatial bandwidth was an indicator of the average pulse width generated by the black pixels associated with the bubbles (i.e., disturbances of bubbles over the gray line in Figure 2a). This bandwidth was obtained at −20 dB with respect to the peak in the power spectral density, as shown in Figure 2b [26]. The shadow fraction and the spatial bandwidth have been proven to be correlated with D_{32} [25,26]. Most of the object features were directly obtained from the Image

Processing Toolbox of MATLAB (11.4, The MathWorks Inc., Natick, MA, USA). Circularities, aspect ratios, eccentricities, perimeters, solidities, equivalent diameters, and the number of objects were obtained from the *regrionprops* function of this toolbox. The estimations of the shadow fractions and spatial bandwidths have been proven to be straightforward, as reported by Vinnett and Alvarez-Silva [25] and Vinnett, Sovechles, Gomez and Waters [26]. Table 3 also presents the variable symbols for each studied feature. The computations of all region and object properties were limited by the bandwidth estimations, whose processing times were proven to be shorter than those of conventional image analysis [26]. The median (subscript 50) along with variability indexes were correlated with D_{32}. Variability was evaluated by the relative standard deviation (subscript RSD), relative interdecile range (subscript RIDR), relative interquintile range (subscript RIQQR), and relative interquartile range (subscript RIQR). Only one variability indicator was used per feature, which was chosen based on the highest level of association with the D_{32} values.

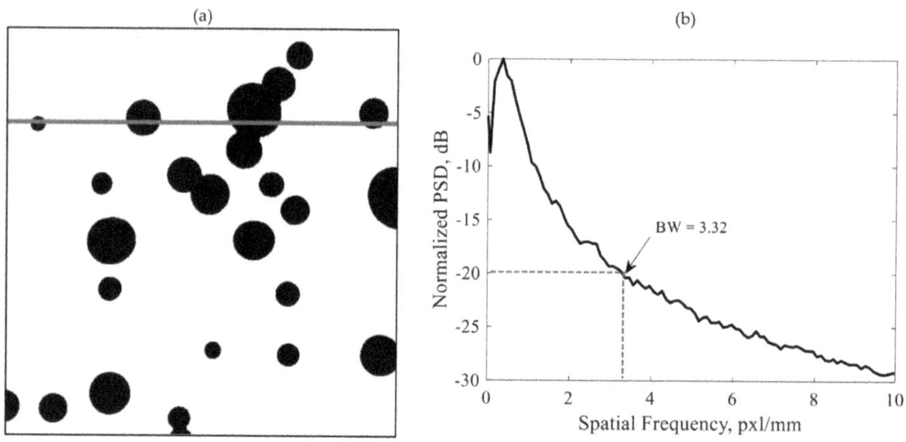

Figure 2. (**a**) Example of binary image, and (**b**) normalized power spectral density and bandwidth estimation.

Table 3. List of studied region and object properties.

Property	Variable Symbol	Statistical Index
Shadow Fraction	*SF*	
Circularity, $4\pi\ area/P^2$	*C*	
Aspect Ratio, *major axis length/minor axis length*	*AR*	Median
Eccentricity	*E*	Relative Standard Deviation
Perimeter, mm	*P*	Relative Interdecile Range
Solidity	*S*	Relative Interquintile Range
Equivalent Diameter, $\sqrt{4\ area/\pi}$, mm	*ED*	Relative Interquartile Range
Number of Objects per mm^2, 1/mm^2	*N*	
Spatial Bandwidth, pxl/mm	*BW*	

All properties were related to the measured D_{32} values using the Pearson coefficient of correlation (R) and the maximal information coefficient [33]. The former measures linear correlation, whereas the latter indicates the level of association between the evaluated variables, not constrained to linear relationships [33]. The maximal information coefficient (MIC) was used to detect variables that were non-linearly related to the Sauter diameter and did not lead to a high coefficient of correlation.

It should be noted that the experimental conditions with no frother and J_G = 0.4, 1.2, and 2.0 cm/s were run six times in different locations of the flotation cell. Manual processing for these three conditions was conducted by 4, 3, and 3 different users, respectively. The relative standard deviations of the estimated D_{32} were 6.2%, 3.7%, and 2.8%, respectively. These variabilities included the experimental and spatial variability, and uncertainties in the manual processing. The latter was considered acceptable for the purpose of this study.

3. Results

All analyzed region and object properties shown in Table 3 were related to the measured Sauter diameters using the Pearson coefficient of correlation and the maximal information coefficient. Only the training datasets were included in this analysis. Table 4 shows the 12 predictors with the most significant (absolute) coefficients of correlation. From these results, the highest absolute coefficients of correlation did not consistently agree with the highest MICs because the latter were able to detect non-linear associations. Figure 3 presents examples of correlations between some region or image properties and the Sauter diameter of the BSDs. Figure 3a,c illustrate the increasing trends between predictors (relative interquintile range for the shadow fraction and perimeter median) and D_{32}. Except for the variability, which is higher at high D_{32}, these trends are compatible with linear dependencies. The coefficients of correlation between these predictors and D_{32} thus resulted in higher values compared with other trends. The maximal information coefficients were also high or moderately high for these trends. Figure 3b,d present non-linear trends between the spatial bandwidth and the number of objects per unit area, and D_{32}, respectively. Although clear relationships were observed, the coefficients of correlation resulted in lower values with respect to Figure 3a,b. The maximal information coefficient was therefore effective in determining non-linear associations between the predictors and the Sauter diameter. It should be noted that D_{32} values greater than 6.0 mm were observed, which were mainly associated with experimental conditions under high superficial gas rates and with MIBC as a frother. These experimental conditions transitioned to churn-turbulent regimes, as exemplified in Appendix B.

Table 4. Twelve properties that led to the highest coefficients of correlation with D_{32}.

Title	P_{50}	SF_{RIQQR}	ED_{50}	N_{RSD}	BW_{50}	EC_{50}	AR_{RSD}	C_{50}	EC_{RIDR}	N_{50}	AR_{50}	BW_{RSD}
R	0.838	0.831	0.829	0.822	−0.820	0.766	0.762	−0.750	−0.730	−0.717	0.700	0.650
MIC	0.942	0.794	0.960	0.859	1.000	0.813	0.969	0.754	0.798	0.953	0.813	0.477

The results from Table 4 and Figure 3 show that different fractions of the D_{32} variability can be explained by the variability of some region and object properties. A multivariable linear model was implemented to obtain the Sauter diameter from all studied predictors. This model was obtained from the training datasets. Predictors that had non-linear trends in relation to D_{32} were transformed to favor linearity, which was applied to BW_{50} (Figure 3b), BW_{SRSD}, and N_{50} (Figure 3d). Thus, $1/BW_{50}$, $1/N_{50}$, and $\ln(BW_{SRSD})$ were employed in the linear regression. The model incorporated a constant term; therefore, 19 parameters were estimated. Robust linear regression was used from the Statistics and Machine Learning toolbox of MATLAB (The MathWorks Inc., Natick, MA, USA). Ordinary least-squares estimation was sensitive to leverage points associated with abnormal gas dispersion conditions, as illustrated in Appendix B. Robust regression performs iteratively reweighted least-squares estimations, assigning a weight to each residual based on its magnitude [34]. This approach was then used to reduce the impact of data points that were far from the main trends. Figure 4 presents the model fitting. A good agreement was observed, except for some tests with $D_{32} > 5.0$ mm. According to the procedure reported by Vinnett et al. [35], experimental conditions with relative standard deviations greater than 0.7 in the shadow fraction were removed from the overall dataset, allowing abnormal gas dispersion conditions to be skipped from the analysis.

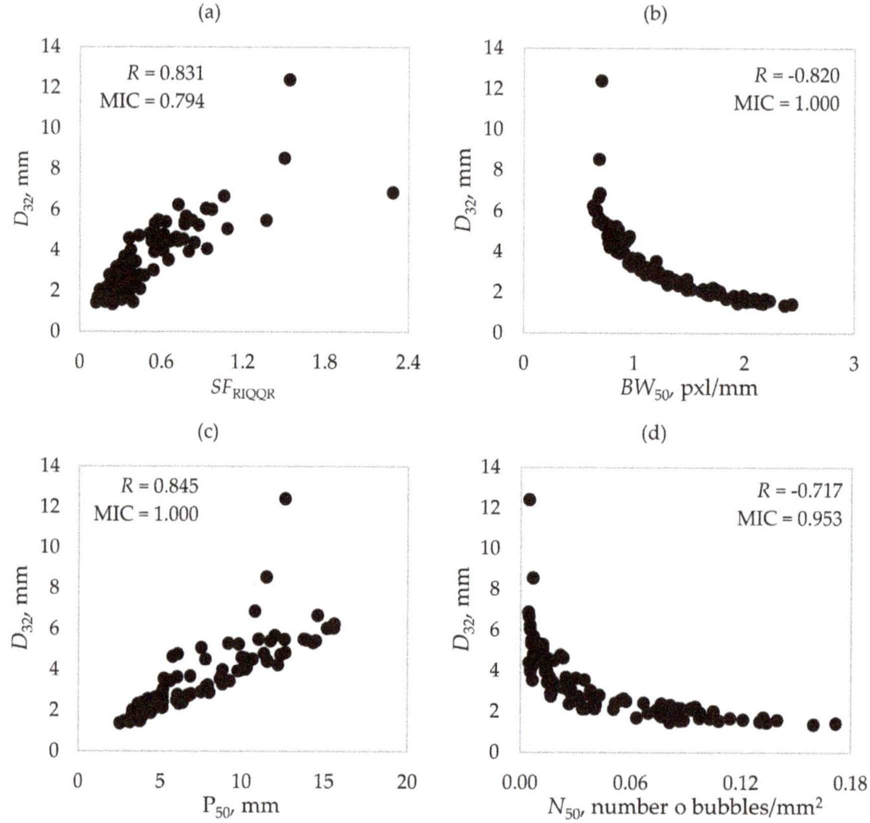

Figure 3. Correlation between different object/image properties and the Sauter diameter: (**a**) D_{32} versus SF_{RIQQR}, (**b**) D_{32} versus BW_{50}, (**c**) D_{32} versus P_{50}, and (**d**) D_{32} versus N_{50}.

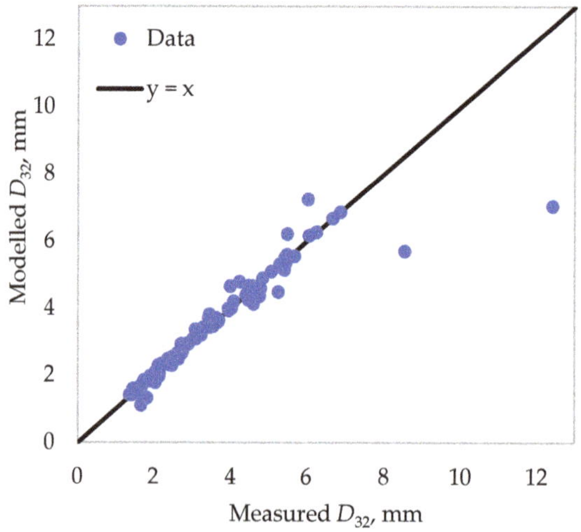

Figure 4. Measured versus modelled Sauter diameters: robust linear regression in the presence of churn-turbulent conditions.

Five data points were removed from the training dataset. Thus, 99 conditions were used for model fitting. To reduce over-parameterization, ordinary linear regression was conducted using the best-subset approach. All combinations of predictors were used in the model fitting to choose the model structure that led to the lowest predicted residual error sum of squares (PRESS). The PRESS is obtained as the sums of squares of the prediction residuals, after removing one data point at a time. Four predictors were then chosen in the model structure: C_{50}, ED_{RIQR}, N_{RSD} and BW_{50} (reciprocal). The root mean squared error was 0.214 with these four predictors, compared with an RMSE of 0.204 when including all predictors in the regression. Figure 5 shows the model fitting for the training datasets along with the comparisons for the testing datasets. An adequate D_{32} description was observed for D_{32} = 1.3–6.7 mm, after removing abnormal gas dispersion conditions. Higher variability was observed for $D_{32} \geq 4.0$ mm, which was caused by the sensitivity of the Sauter diameter to large bubbles. The testing results proved that the model was generalizable to independent datasets. Again, higher dispersion was observed for $D_{32} \geq 4.0$ mm. It should be noted that the modeled D_{32} values were automatically estimated without individualizing every single bubble. Although the D_{32} values observed in the 2D cell corresponded to intermediate- and large-size ranges from industrial databases [18,20,36], poorer bubble size estimations are observed for $D_{32} \geq 2.0$ mm by conventional image analysis, as reported by Vinnett, Yianatos, Arismendi and Waters [20]. Thus, the correlation presented here is suitable for ellipsoidal regimes and in the transition toward turbulent regimes. Conventional image analysis (shape factors and object segmentation) is recommended for $D_{32} < 2.0$ mm, as no significant bias was observed with this method from industrial data.

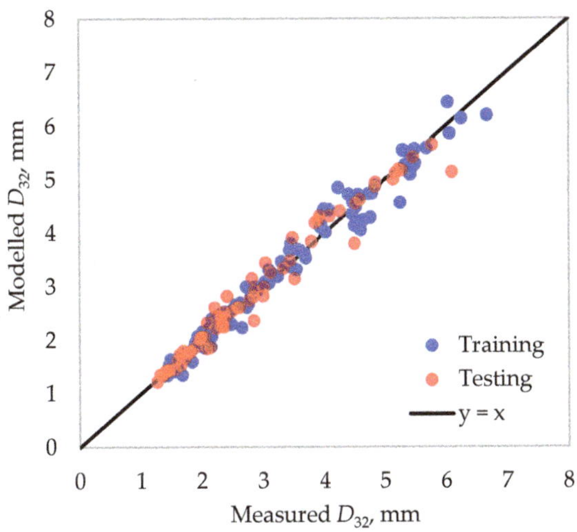

Figure 5. Measured versus modeled Sauter diameters: training and testing datasets, after removing abnormal conditions.

Equation (1) presents the D_{32} model, in which all parameters were significant at the 95% confidence level. All p-values and 95% confidence intervals of the predictors are presented in Appendix C. From Equation (1), a D_{32} increase is related to variability increases in the number of objects per millimeter square (N_{RSD}) and in the equivalent diameter (ED_{RIQR}). The former is caused by the variability in the gas hold-up when transitioning from spherical to ellipsoidal and turbulent conditions [35]. This transition also increased the variability in the equivalent diameter because more irregular bubbles, coexisting with small bubbles, are typically observed at higher D_{32} values. Lower D_{32} values were observed at lower median circularities (C_{50}) as small bubbles are consistently observed as spheres in a bubbly regime. As the spatial bandwidth is an indicator of the horizontal and vertical pulse widths that are caused by bubbles (Figure 2a), wider pulses (and lower BW) lead to higher Sauter diameters [26]. A sensitivity analysis proved that BW_{50} was the most significant predictor. The increases in C_{50}, ED_{RIQR}, N_{RSD}, and BW_{50} from the 25th percentile to the 75th percentile led to D_{32} variations of −0.13, 0.27, 0.20, and 1.90, respectively. The spatial bandwidth was previously tested with industrial data, adequately explaining the D_{32} variability for $D_{32} \approx 2.0$–6.0 mm, under normal gas dispersion conditions. Vinnett, Sovechles, Gomez and Waters [26] proposed $D_{32} = \alpha/BW^{\beta}$ to estimate bubble size, obtaining $\alpha = 3.7$ and $\beta = 1.1$ by non-linear regression. As the linear model proposed here incorporates the bandwidth by its reciprocal ($\beta = 1.0$), any bias in the industrial variability explained by BW will be moderate with respect to the laboratory results in Figure 5 and Equation (1). It should be noted that Equation (1) only allows for the estimation of the Sauter mean diameter; therefore, additional correlations are required to automatically obtain unbiased BSDs.

$$D_{32} = 1.14 - 1.80C_{50} + 0.785ED_{RIQR} + 1.52N_{RSD} + \frac{3.49}{BW_{50}} \quad (1)$$

The results from Table 4 and Figures 3–5 show that some image properties were linearly or non-linearly correlated with the Sauter mean diameter. Except for abnormal gas dispersion conditions, these image properties proved to be applicable as predictors to automatically estimate bubble size, without individualizing all single bubbles or removing irregular objects. Additional predictors can be incorporated into the model structure to improve its predictability, using cross-validation to control over-parameterization. The modeling strategy presented here can also be extended to different machine learning tools, considering the continuous improvement in the training stage after increasing D_{32} and image databases. Thus, gas dispersion data from different flotation machines and scales can be incorporated into the algorithms for model generalizations. Further developments are being made to expand D_{32} estimations using experimental data from different flotation machines, operating conditions, and flotation scales.

4. Conclusions

One hundred and four images and D_{32} datasets were studied, correlating different image and object properties (from binary representations) with bubble size. All properties were automatically determined, whereas the D_{32} values were obtained from a semiautomated approach that did not remove bubbles from the analysis. The main results are summarized as follows:

- Several image and object properties showed moderate or strong correlations, linear and non-linear, with the Sauter diameter.
- The maximal information coefficient was successfully used to detect non-linear associations between image and object properties with bubble size. These associations were not clearly detected with the coefficient of correlation. The strongest associations were observed with the median of the spatial bandwidth, median of the equivalent diameter, relative standard deviation of the aspect ratio, and median of the number of objects per unit area.
- After removing churn-turbulent conditions and linearizing non-linear associations, a multivariable linear model was proposed, which was able to estimate bubble size in the range 1.3–6.7 mm. This model was obtained from four predictors: median of the circularity, relative interquartile range of the equivalent diameter, relative standard deviation of the number of elements per unit area, and median of the spatial bandwidth. These predictors were chosen from the best subset of all possible linear models, minimizing PRESS.
- The linear model was successfully tested on 72 independent datasets, which showed the generalizability of the model structure.

The strategy to indirectly characterize bubble size from image and object properties proved to be applicable at laboratory scale, without individualizing all single bubbles or removing irregular bubbles and clusters. This approach can be continuously improved by including additional predictors and expanding gas dispersion databases from different experimental conditions.

Author Contributions: Methodology, L.V., B.U., C.G. and A.E.; Software, L.V.; Formal Analysis, L.V., B.U., C.G. and A.E.; Writing—Original Draft Preparation, L.V., I.C., J.Y. and C.A.; Writing—Review and Editing, L.V., I.C., J.Y. and C.A; Supervision, L.V.; Project Administration, L.V.; Funding Acquisition, L.V., I.C. and J.Y. All authors have read and agreed to the published version of the manuscript.

Funding: Funding for process modeling and control research was provided by ANID, Project Fondecyt 1201335, and Universidad Técnica Federico Santa María, Project PI_LIR_2021_78.

Conflicts of Interest: The authors declare no conflict of interest.

Appendix A

Figure A1 illustrates five BSDs, which are presented as cumulative distribution functions. These BSDs were associated with the 10, 30, 50, 70, and 90 percentiles of the measured Sauter diameters. Higher D_{32} values were related to higher mean (or median) bubble sizes as well as to longer distribution tails. The latter was caused by the presence of a low percentage of large bubbles in the analyzed populations.

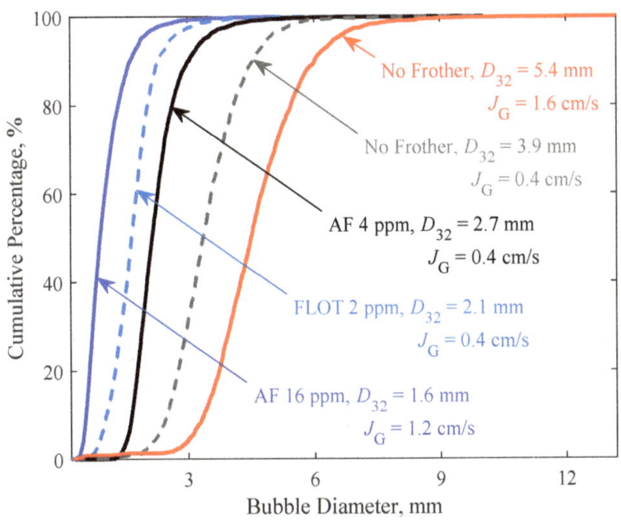

Figure A1. Examples of bubble size distributions in the laboratory tests.

Appendix B

Experimental conditions using MIBC as a frother and J_G = 2.5 cm/s favored the transition to churn-turbulent regimes in the flotation cell. Figure A2 illustrates examples of images from one abnormal flotation test. Although this abnormality can be automatically detected, the incorporation of these datasets into the proposed regression approach distorted the correlations due to leveraging. The high Sauter diameters were influenced by the high sensitivity of this parameter to large bubbles.

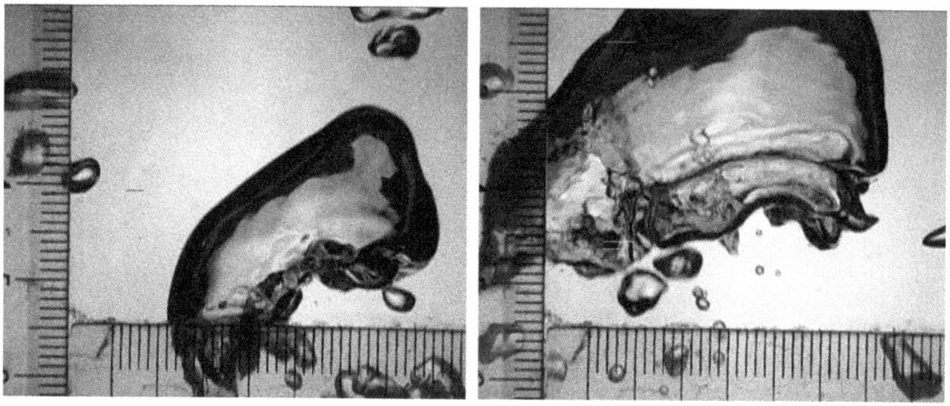

Figure A2. Examples of bubbles under abnormal gas dispersion conditions. Scales in millimeters and centimeters.

Appendix C

Table A1 presents the p-values along with the 95% confidence intervals for the predictors used in Equation (1). All parameters were significant at the chosen confidence level.

Table A1. p-values and 95% confidence intervals for the predictors used in Equation (1).

Title	p-Values	95% Confidence Intervals
Constant	0.0453	(0.0242, 2.25)
C_{50}	0.000420	(−2.78, −0.822)
ED_{IQR}	6.90×10^{-11}	(0.573, 0.997)
N_{RSD}	2.14×10^{-5}	(0.843, 2.19)
BW_{50}	1.11×10^{-43}	(3.22, 3.76)

References

1. Gorain, B.K.; Franzidis, J.P.; Manlapig, E.V. Studies on impeller type, impeller speed and air flow rate in an industrial scale flotation cell. Part 4: Effect of bubble surface area flux on flotation performance. *Miner. Eng.* **1997**, *10*, 367–379. [CrossRef]
2. Gorain, B.K.; Napier-Munn, T.J.; Franzidis, J.-P.; Manlapig, E.V. Studies on impeller type, impeller speed and air flow rate in an industrial scale flotation cell. Part 5: Validation of k-Sb relationship and effect of froth depth. *Miner. Eng.* **1998**, *11*, 615–626. [CrossRef]
3. Finch, J.A.; Dobby, G.S. *Column Flotation*; Pergamon Press: Oxford, UK, 1990.
4. Rojas, I.; Vinnett, L.; Yianatos, J.; Iriarte, V. Froth transport characterization in a two-dimensional flotation cell. *Miner. Eng.* **2014**, *66–68*, 40–46. [CrossRef]
5. Jameson, G.J.; Nam, S.; Young, M.M. Physical factors affecting recovery rates in flotation. *Miner. Sci. Eng.* **1977**, *9*, 103–118.
6. Hernandez-Aguilar, J.; Gomez, C.; Finch, J. A technique for the direct measurement of bubble size distributions in industrial flotation cells. In Proceedings of the 34th Annual Meeting of the Canadian Mineral Processors, Ottawa, ON, Canada, 22–24 January 2002; pp. 389–402.
7. Mesa, D.; Quintanilla, P.; Reyes, F. Bubble Analyser—An open-source software for bubble size measurement using image analysis. *Miner. Eng.* **2022**, *180*, 107497. [CrossRef]
8. Grau, R.A.; Heiskanen, K. Visual technique for measuring bubble size in flotation machines. *Miner. Eng.* **2002**, *15*, 507–513. [CrossRef]
9. Acuña, C.; Vinnett, L.; Kuan, S.H. Improving image analysis of online bubble size measurements with enhanced algorithms. In Proceedings of the 12th International Mineral Processing Conference, Procemin, Santiago, Chile, 26–28 October 2016.
10. Bailey, M.; Gomez, C.O.; Finch, J.A. Development and application of an image analysis method for wide bubble size distributions. *Miner. Eng.* **2005**, *18*, 1214–1221. [CrossRef]
11. Sovechles, J.M.; Waters, K.E. Effect of ionic strength on bubble coalescence in inorganic salt and seawater solutions. *AIChE J.* **2015**, *61*, 2489–2496. [CrossRef]
12. Grau, R.A.; Heiskanen, K. Gas dispersion measurements in a flotation cell. *Miner. Eng.* **2003**, *16*, 1081–1089. [CrossRef]

13. Riquelme, A.; Desbiens, A.; Bouchard, J.; del Villar, R. Parameterization of Bubble Size Distribution in Flotation Columns. *IFAC Proc. Vol.* **2013**, *46*, 128–133. [CrossRef]
14. Karn, A.; Ellis, C.; Arndt, R.; Hong, J. An integrative image measurement technique for dense bubbly flows with a wide size distribution. *Chem. Eng. Sci.* **2015**, *122*, 240–249. [CrossRef]
15. Ma, Y.; Yan, G.; Scheuermann, A.; Bringemeier, D.; Kong, X.-Z.; Li, L. Size distribution measurement for densely binding bubbles via image analysis. *Exp. Fluids* **2014**, *55*, 1860. [CrossRef]
16. Grau, R.A.; Heiskanen, K. Bubble size distribution in laboratory scale flotation cells. *Miner. Eng.* **2005**, *18*, 1164–1172. [CrossRef]
17. Lau, Y.M.; Deen, N.G.; Kuipers, J.A.M. Development of an image measurement technique for size distribution in dense bubbly flows. *Chem. Eng. Sci.* **2013**, *94*, 20–29. [CrossRef]
18. Vinnett, L.; Yianatos, J.; Alvarez-Silva, M. Gas dispersion measurements in industrial flotation equipment. In Proceedings of the 8th Copper International Conference, Copper 2013, Santiago, Chile, 1–4 December 2013.
19. Wang, J.; Forbes, G.; Forbes, E. Frother Characterization Using a Novel Bubble Size Measurement Technique. *Appl. Sci.* **2022**, *12*, 750. [CrossRef]
20. Vinnett, L.; Yianatos, J.; Arismendi, L.; Waters, K.E. Assessment of two automated image processing methods to estimate bubble size in industrial flotation machines. *Miner. Eng.* **2020**, *159*, 106636. [CrossRef]
21. Steinemann, J.; Buchholz, R. Application of an Electrical Conductivity Microprobe for the Characterization of bubble behavior in gas-liquid bubble flow. *Part. Part. Syst. Charact.* **1984**, *1*, 102–107. [CrossRef]
22. Meernik, P.; Yuen, M. An optical method for determining bubble size distributions—Part II: Application to bubble size measurement in a three-phase fluidized bed. *J. Fluids Eng.* **1988**, *110*, 332–338. [CrossRef]
23. Kracht, W.; Emery, X.; Paredes, C. A stochastic approach for measuring bubble size distribution via image analysis. *Int. J. Miner. Processing* **2013**, *121*, 6–11. [CrossRef]
24. Kracht, W.; Moraga, C. Acoustic measurement of the bubble Sauter mean diameter d32. *Miner. Eng.* **2016**, *98*, 122–126. [CrossRef]
25. Vinnett, L.; Alvarez-Silva, M. Indirect estimation of bubble size using visual techniques and superficial gas rate. *Miner. Eng.* **2015**, *81*, 5–9. [CrossRef]
26. Vinnett, L.; Sovechles, J.; Gomez, C.O.; Waters, K.E. An image analysis approach to determine average bubble sizes using one-dimensional Fourier analysis. *Miner. Eng.* **2018**, *126*, 160–166. [CrossRef]
27. Ilonen, J.; Juránek, R.; Eerola, T.; Lensu, L.; Dubská, M.; Zemčík, P.; Kälviäinen, H. Comparison of bubble detectors and size distribution estimators. *Pattern Recognit. Lett.* **2018**, *101*, 60–66. [CrossRef]
28. Bu, X.; Zhou, S.; Sun, M.; Alheshibri, M.; Khan, M.S.; Xie, G.; Chelgani, S.C. Exploring the Relationships between Gas Dispersion Parameters and Differential Pressure Fluctuations in a Column Flotation. *ACS Omega* **2021**, *6*, 21900–21908. [CrossRef]
29. Vinnett, L.; Urriola, B.; Orellana, F.; Guajardo, C.; Esteban, A. Reducing the Presence of Clusters in Bubble Size Measurements for Gas Dispersion Characterizations. *Minerals* **2022**, *12*, 1148. [CrossRef]
30. Saavedra Moreno, Y.; Bournival, G.; Ata, S. Classification of flotation frothers—A statistical approach. *Chem. Eng. Sci.* **2022**, *248*, 117252. [CrossRef]
31. Arends, M.A. *Reactivos de Flotación: Evaluación de Colectores y Espumantes*; Clariant: Muttenz, Switzerland, 2019.
32. Grau, R.A. *An Investigation of the Effect of Physical and Chemical Variables on Bubble Generation and Coalescence in Laboratory Scale Flotation Cells*; Helsinki University of Technology: Helsinki, Finland, 2006.
33. Reshef, D.N.; Reshef, Y.A.; Finucane, H.K.; Grossman, S.R.; McVean, G.; Turnbaugh, P.J.; Lander, E.S.; Mitzenmacher, M.; Sabeti, P.C. Detecting novel associations in large data sets. *Science* **2011**, *334*, 1518–1524. [CrossRef]
34. Holland, P.W.; Welsch, R.E. Robust regression using iteratively reweighted least-squares. *Commun. Stat.-Theory Methods* **1977**, *6*, 813–827. [CrossRef]
35. Vinnett, L.; Yianatos, J.; Acuña, C.; Cornejo, I. A Method to Detect Abnormal Gas Dispersion Conditions in Flotation Machines. *Minerals* **2022**, *12*, 125. [CrossRef]
36. Vinnett, L.; Yianatos, J.; Alvarez, M. Gas dispersion measurements in mechanical flotation cells: Industrial experience in Chilean concentrators. *Miner. Eng.* **2014**, *57*, 12–15. [CrossRef]

minerals

Article

An Improved Python-Based Image Processing Algorithm for Flotation Foam Analysis

Wenkang Zhang [1,2,3], Dan Liu [1,2,3,*], Chunjing Wang [1,2,3], Ruitao Liu [1,2,3], Daqian Wang [1,2,3], Longzhou Yu [4] and Shuming Wen [1,2,3]

1 State Key Laboratory of Clean Utilization of Complex Nonferrous Metal Resources, Kunming 650093, China
2 Yunnan Key Laboratory of Green Separation and Enrichment of Strategic Mineral Resources, Kunming 650093, China
3 Faculty of Land and Resources Engineering, Kunming University of Science and Technology, Kunming 650093, China
4 Yunnan Amade Electrical Engineering Company, Kunming 650033, China
* Correspondence: kgkjxld@kust.edu.cn

Abstract: For industrial flotation foam image processing, accurate bubble size measurement and feature extraction are very important to optimize the flotation process and to improve the recovery of mineral resources. This paper presents an improved algorithm to investigate mineral flotation foam image segmentation for mineral processing. Several libraries implemented for the Python programming language are used for image enhancement and compensation, quantitative analysis of factors influencing the image segmentation accuracy, and suggestions for improvement of the flotation foam image processing. The bubble characteristics-size and morphology-and the influence of the flotation conditions on the flotation foam image are analyzed. A Python implementation of the Retinex image compensation method-region-adaptive and multiscale-is proposed to address known issues of uneven illumination and shadows affecting flotation foam images, thereby improving brightness uniformity. Finally, an improved version of the watershed segmentation algorithm included in the Python Open Source Computer Vision library is used for segmentation analysis. The accuracy of the flotation foam image segmentation is 3.3% higher than for the standard watershed algorithm and the segmentation time is 9.9% shorter.

Keywords: flotation foam; image processing; image segmentation; machine vision; intelligent

check for updates

Citation: Zhang, W.; Liu, D.; Wang, C.; Liu, R.; Wang, D.; Yu, L.; Wen, S. An Improved Python-Based Image Processing Algorithm for Flotation Foam Analysis. *Minerals* **2022**, *12*, 1126. https://doi.org/10.3390/min12091126

Academic Editors: Fardis Nakhaei, Ahmad Hassanzadeh and Luis A. Cisternas

Received: 12 August 2022
Accepted: 30 August 2022
Published: 4 September 2022

1. Introduction

Mineral resources are extremely important for human development. They represent critical materials for the industry, and they ensure security, economic growth and self-sufficiency for a country. China has a long history of mineral resource exploitation. The mineral industry has not only accelerated the Chinese urbanization process, but it has also allowed for the coordinated development of Chinese regions and has been fundamental for employment and social stability [1]. Mineral resources are non-renewable Because of increasing consumption and intensive exploitation of larger, more accessible deposits, new extraction methods focus on smaller deposits of finer, lower-grade ore. Concurrently, extraction efficiency is constantly improved to increase the quantity of available mineral resources. Mineral flotation is the most widely used technology in the mineral processing industry. It was developed in the late 19th Century and gradually applied to industrial production in the early 20th Century. Since then, flotation equipment and technology have been investigated and regularly improved [2].

Currently, more than half of the non-ferrous metal ore is extracted using froth flotation processes. Froth flotation is a method of mineral separation that differentiates minerals by their surface physical and chemical properties. Despite extensive worldwide research on flotation processes, there is currently no mathematical model to characterize such processes accurately or to optimize flotation control [3].

In China, current flotation processing sites rely on human observation to determine the state of the flotation foam and to adjust the dosage and ventilation volume. This method is not appropriate to meet the current demand in mineral separation for social and economic development [4] and to comply with current requirements of energy saving and emission reduction for climate change mitigation. Surface characteristics of flotation foams strongly depend on the flotation conditions and can be used as indicators. Automating the analysis by replacing human observation and combining data acquisition, processing, and display into a single computer operation would greatly simplify the extraction process [5]. Flotation process parameters can be characterized more clearly from the visual features of a computer-generated foam image. These results can then be used to build a prediction and monitoring model of the flotation system, in order to achieve flotation process control and real-time detection, thereby improving flotation efficiency and reducing errors caused by manual observation [6]. Indeed, although experienced operators can infer the flotation system operation status from the foam, more subtle structural changes in the foam cannot be diagnosed manually. Therefore, stability detection and real-time regulation of the flotation system are difficult to achieve [7,8].

Since the late 20th Century, extensive research has been conducted worldwide to replace manual monitoring with machine vision. The European Union—within the European Strategic Programme on Research in Information Technology/Long Term Research framework—launched the "Machine Vision-based Bubble Structure and Color Representation" project, involving several universities and companies in Finland and Sweden [9,10]. Wang et al. [11] proposed using valley edge detection and tracking to segment foam images. High-illumination points were used to define illumination thresholds for foam bubble classification, then valley bottom edges were tracked in four directions [11]. Citir et al. proposed a two-iteration method for bottom edge tracking, first using pixel minima to mark local regions, then applying the results to refine the edges [12]. Sadr-Kazemi and Cilliters [13] applied the morphological watershed method to flotation foam segmentation. Their robust method used histogram equalization and the "marker" concept—the maxima within the foam bright spots become the identifiers [13]. Bonifazi et al. also applied the watershed method to flotation foam segmentation analysis [5]. Forbes postulated that the combination of texture feature analysis and the watershed method could improve segmentation accuracy [14]. Zhang et al. proposed a watershed segmentation algorithm based on optimal labeling for bubble size measurement [15]. Lezoray et al. proposed an unsupervised clustering classification and region-merging method for color image segmentation, which achieved good results [16].

Despite extensive research, the high accuracy of foam image processing is difficult to achieve for all working conditions and in complex environments [17]. Therefore, the complexity and specificities of flotation foam must be studied. Accurate measurements of the flotation foam bubble size are especially important [18]. Recently, the industrial technology for machine vision monitoring and control has become largely automated. To fully exploit limited mineral resources—such as lean ore or refractory mineral dressing—and to improve their recovery and utilization rates, technologies using machine vision must be applied to mineral flotation detection.

This paper presents an in-depth analysis of flotation foam generation mechanisms and of the correlation between bubble size and working conditions. To improve the characterization—size and morphology—of foam bubbles and the measurement accuracy, the paper defines several methods to evaluate flotation foam image properties—clarity, illumination uniformity, and multiscale enhancement. Furthermore, an adaptive foam image segmentation method is proposed for parameter measurement. Finally, segmentation accuracy and efficiency are assessed to improve the accuracy and robustness of the flotation foam segmentation algorithm [19].

2. Error Analysis and Characterization of Flotation Foam Images

The flotation process is the separation of minerals from finely ground ore at a liquid–gas interface. The foam bubble properties—structure, stability, size, shape and number—all affect mineral flotation. For example, even if the ore selection conditions are suitable for flotation, improper adjustment of the foam bubble shape might negatively influence flotation. Flotation foam characteristics are adjusted by modifying the physical and chemical conditions during the flotation process.

2.1. Factors Influencing the Flotation Foam

The characteristics of mineral flotation foam depend mainly on the flotation agent, the pulp concentration, the mineral particle size, the wind pressure, the foam layer thickness and the work environment. For example, the lighting conditions in a milling plant influence the flotation conditions. Furthermore, industrial flotation is a mechanized process that generates noise, with a possible effect on the flotation image bubble extraction accuracy [20]. Therefore, flotation foam image denoising is an important step of the flotation process.

The qualitative relationship between the flotation operation variables and the foam bubble morphology is shown in Table 1.

Table 1. Relationship between flotation operation variables and foam bubble morphology.

Operation Variable Description	Bubble Characteristics
The amount of foaming agent is large	Bubbles are small and bubbles are stable
Small amount of foaming agent	Bubbles are large and bubbles are unstable
The more inhibitor	The bubbles are small and round, and the foam load is small
The less inhibitor	Bubbles are large, elliptic, sticky and slow moving
The pH value of the pulp increases	Bubble increase
The pH value of the pulp decreases	Bubble decreases
High pulp concentration	The bubbles are large, elliptic, slow moving and high bearing rate
Low pulp concentration	The bubbles are small, round and unstable
Air pressure is high	The bubbles are large, elliptic, fast and low mineralization
Air pressure is down	Bubbles are small, round, slow and highly mineralization

2.2. Compensation of Machine Vision Errors during the Flotation Process

2.2.1. Model Definition for Machine Vision Image Recognition

Statistical modeling provides new possibilities for automatic detection, measurement, analysis and identification of flotation foam images with the random accumulation of mineralized bubbles at the surface. Currently, most statistical image models use parameter quantization methods, because they are simple to implement and easy to understand [21,22]. A schematic representation of statistical image modeling by parameter quantization is shown in Figure 1.

During the statistical image modeling process, a specific mathematical image transformation is applied to the original image pixels. The statistical pixel distribution analysis is then performed in the transform domain. The most common image transformation operators for industrial mineral processing are the wavelet transform, the Gabor filter, and high-order Gaussian derivative filter banks. Empirical probability density functions are then used to fit the statistical pixel distribution and characterize the image.

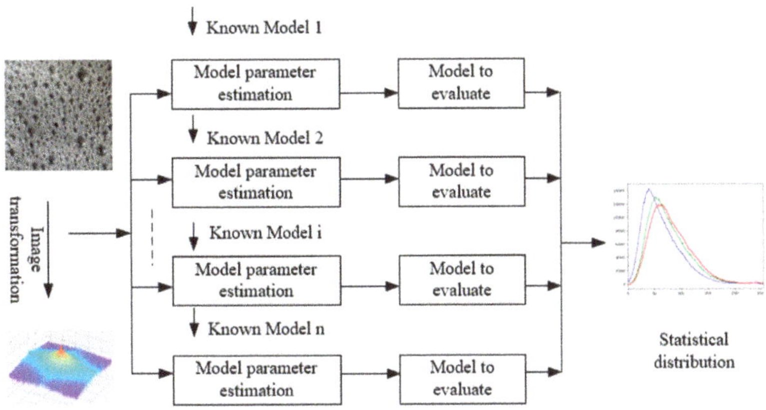

Figure 1. Schematic diagram of statistical image modeling by parameter quantization.

2.2.2. General Principle for Statistical Analysis of Flotation Foam Images

An image, as perceived by the human body, is a form of physical information (light) collected and interpreted by the eye–optical nerve–visual cortex processing system. A digitized (computerized) image, such as the foam image shown in Figure 2, is stored only as a two-dimensional pixel value matrix. Specific information on the image—shapes, granularity—or on the physical process—uneven foam surface, specific flotation conditions—is not included. Therefore, the purpose of machine vision monitoring is to allow the computer to emulate the observation capabilities of the human visual system. If the algorithm is accurate enough to automatically derive surface information from the foam images, flotation control and flotation process operations can be automated.

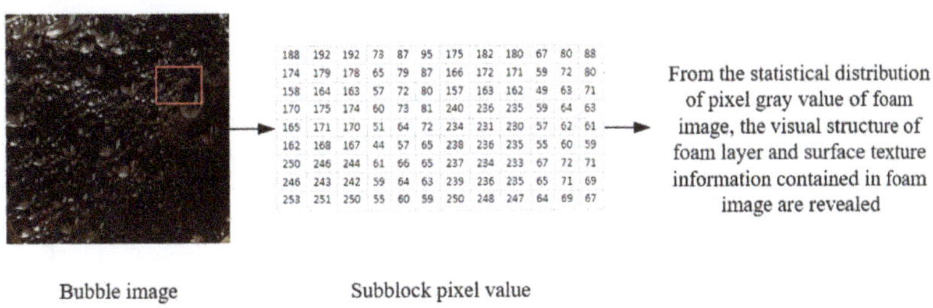

Figure 2. Gray levels and pixel value information for a foam image.

2.2.3. Flotation Foam Image Spectrum Characteristics

For an observation system, the pixels composing a digital image represent discrete point sources. Mathematically, point sources can be represented by the Dirac delta function (or distribution). The two-dimensional delta function $\delta(x, y)$ can be expressed as:

$$\iint_{x,y} \delta(x,y)dxdy = 1 \tag{1}$$

By definition, $\delta(x, y) = \infty$ when $x = 0$ and $y = 0$, and its value is zero everywhere else. The δ function is a generalized function over the real numbers. Therefore, for any function $f(x, y)$ that is continuous at a point (x_0, y_0), the function value at that point is expressed as:

$$\iint_{x,y} f(x,y)\delta(x - x_0, y - y_0)dxdy = f(x_0, y_0) \tag{2}$$

Assuming $f(x, y)$ is an analog image, its amplitude f and coordinates x and y must be discretized to convert it into a digital image for computer image processing. This digital image is the matrix form—obtained by sampling and quantization—of the continuous, analog $f(x, y)$. If $f(x, y)$ is sampled on a uniform grid with sampling intervals Δx and Δy in the x direction and y direction, respectively, the coordinates of the sampling points on the image are $x = m\Delta x$ and $y = n\Delta y$, where m and n are signed integers. The sampling function for one point is defined by a δ function. The discrete image sampling function $s(x, y)$ is thus:

$$s(x,y) = \sum_{m=-\infty}^{\infty} \sum_{n=-\infty}^{\infty} \delta(x - m\Delta x, y - n\Delta y) \tag{3}$$

The sampled image $f_s(x, y)$ is the product of the original simulated image $f(x, y)$ and the sampling function $s(x, y)$:

$$f_s(x,y) = f(x,y)s(x,y) \tag{4}$$

To obtain the image spectrum, the Fourier transform operator is applied to both sides of Equation (4). We define the Fourier transforms of $f(x, y)$ and $s(x, y)$ as $F(u, v)$ and $S(u, v)$, respectively. Applying the convolution theorem, the Fourier transform of $f_S(x, y)$ is then expressed as:

$$F_s(u,v) = F(u,v) * S(u,v) \tag{5}$$

Equation (5) can be written in its specific integral form as:

$$F_s(u,v) = \frac{1}{\Delta x \Delta y} \int_{\omega=-\infty}^{\infty} \int_{\omega=-\infty}^{\infty} F(\omega_1, \omega_2) \sum_{m=-\infty}^{\infty} \sum_{n=-\infty}^{\infty} \delta(u - \omega_1 - m\Delta u, v - \omega_2 - n\Delta v) d\omega_1 d\omega_2 \tag{6}$$

After calculation, Equation (6) becomes:

$$F_s(u,v) = \frac{1}{\Delta x \Delta y} \sum_{m=-\infty}^{\infty} \sum_{n=-\infty}^{\infty} F(u - \frac{m}{\Delta x}, v - \frac{n}{\Delta y}) \tag{7}$$

Equation (7) shows that the spectrum of the sampled image is composed of the original continuous image spectrum and of an infinite number of periodic translation spectra. Therefore, the spectrum derived from the foam image by Fourier transform gives the energy of all frequencies forming the image. The number of frequencies corresponds to the number of pixels in the spatial domain image.

We consider the following continuous image function:

$$f(x,y) = [3\sin(2\pi n_1 x + \frac{\pi}{3}) + 4\cos(2\pi n_2 x)] \cdot [3\sin(2\pi n_1 y + \frac{\pi}{3}) + 4\cos(2\pi n_2 y)] \tag{8}$$

The sampled image and the Fourier spectrum of $f(x, y)$ corresponding to scheelite, coal gangue and lead-zinc ore flotation foams for $n_1 = 3$ and $n_2 = 4$ are shown in Figures 3–5, respectively.

The autocorrelation function of an image characterizes distance and orientation correlations between pairs of image pixels. Thus, the image spectrum is directly related to the autocorrelation function. By examining the spectra in Figures 3–5, we conclude that there are correlations between image pixels for each mineral. Therefore, determining the image autocorrelation function would yield information on the image.

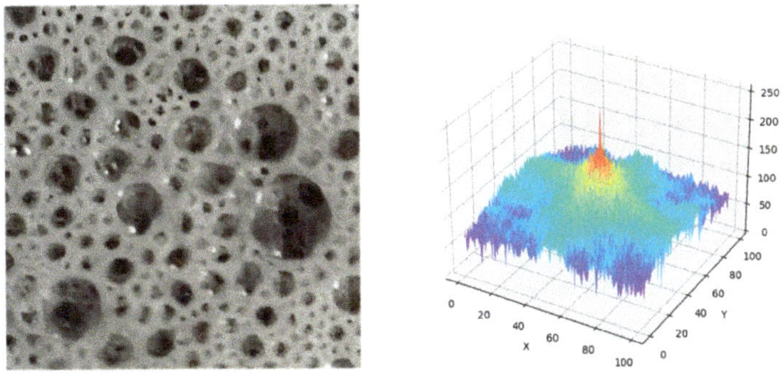

Figure 3. Sampled image of a scheelite flotation foam (**left**) and the corresponding Fourier-transform spectrum ((**right**), with x, y and z in arbitrary units).

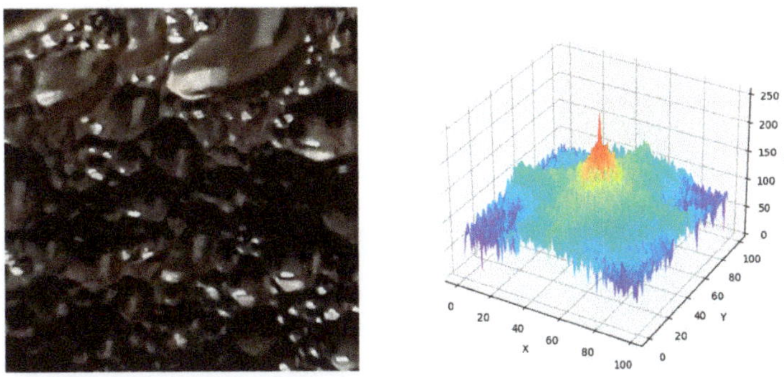

Figure 4. Sampled image of a coal gangue flotation foam (**left**) and the corresponding Fourier-transform spectrum ((**right**), with x, y and z in arbitrary units).

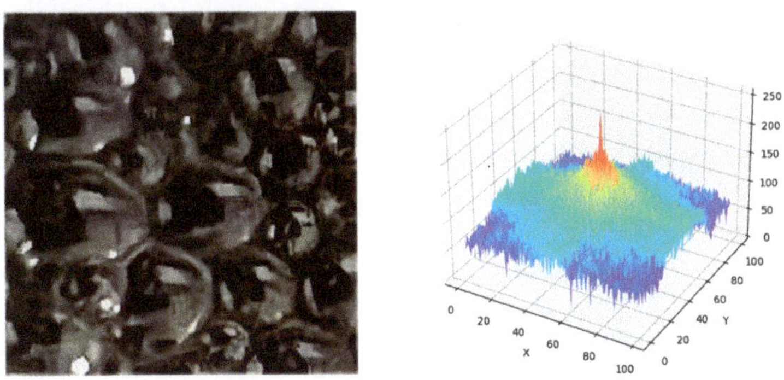

Figure 5. Sampled image of a lead-zinc ore flotation foam (**left**) and the corresponding Fourier-transform spectrum ((**right**), with x, y and z in arbitrary units).

We define a foam image $I(x, y)$ and its Fourier transform $\hat{I}(\xi, \eta)$. The frequency energy spectrum $A(f)$ is equal to $|\hat{I}^2(\xi, \eta)|$, where $f = \sqrt{\xi^2 + \eta^2}$ is the direction of average energy in the frequency domain. The spectrum A and the spatial frequency f are related by:

$$A = b/f^a \tag{9}$$

This is equivalent, in logarithmic coordinates, to:

$$\log A = \log b - a \log f \tag{10}$$

Because the image spectrum shows the same energy in each frequency band, the image is scale-invariant.

$$E_{f-2f} = \iint_{f^2 \le \xi^2 + \eta^2 \le (2f)^2} \left| \hat{I}^2(\xi, \eta) \right| d\xi d\eta = \int_f^{2f} 2 * \pi * x * \frac{const}{x^{a/2}} dx \tag{11}$$

Substituting the approximate value of constant a (a = 2) into Equation (11), we note that $E_{f\text{-}2f}$ depends only on frequency. To investigate the frequency spectrum of flotation foam images, an image database containing images acquired under different flotation production conditions is first constructed. The size of each image is 300 × 300 pixels with 256 levels of gray (8-bit coding).

2.2.4. Retinex Image Compensation

The Retinex—composite of retina and cortex —theory was developed by E. H. Land in 1963 [23]. It is used for image enhancement under varying illumination conditions [24].

Jobson et al. later defined a multiscale version of the Retinex algorithm [25]. By adjusting a scale parameter, the algorithm produces de-illuminated images at different scales. The final output is expressed as:

$$R(x,y) = \sum_{k=1}^{N} W_k \{\log S(x,y) - \log[S(x,y) * G_k(x,y)]\} \tag{12}$$

where N represents the number of scales and W_k is the weight corresponding to scale k. For N = 1—when W_1 = 1—the algorithm reverts to the standard single-scale Retinex algorithm.

When performing operations in the logarithmic domain, pixel values can become negative or fall outside of the display range. To ensure that the range of values is suitable for display, a compensation operation—consisting of a translation and a compression—is often applied to the initial range. For the output image R_{Mi} of the multiscale Retinex algorithm at scale M_i (for color spectrum component i), the "compensated" image is:

$$R'_{M_i}(x,y) = GR_{M_i}(x,y) + b \tag{13}$$

where G is the gain (compression factor) and b is the compensation (translation value). A common compensation method is the automatic gain compensation method, for which a linear stretching is applied to the range of gray values to map values outside of the range onto a new range suitable for display. The mapping operation is defined as:

$$R'_{M_i}(x,y) = \frac{R_{M_i}(x,y) - R_{\min}}{R_{\max} - R_{\min}} \times d_{\max} \tag{14}$$

where R'_{Mi} is the image obtained after linear gray stretching, R_{min} and R_{max} are the minimum and maximum gray values of the input image, respectively, and d_{max} is the dynamic gray value range of the output display device. For 8-bit devices, the maximum number of gray values is 255 and the gain and compensation are constant:

$$G = \frac{d_{\max}}{R_{\max} - R_{\min}}, b = -\frac{R_{\min}}{R_{\max} - R_{\min}} \tag{15}$$

To improve the image appearance on the display, it is generally necessary to truncate the gray value range of the original image by excluding the minimum and maximum gray values of each color spectrum component, then to stretch the truncated range onto the dynamic range of the display device:

$$R' = \frac{R_{ci} - R_{low}}{R_{up} - R_{low}} \times d_{\max} \tag{16}$$

where R_{ci} is the output gray value corresponding to R_i after truncation, and R_{low} and R_{up} represent the minimum and maximum truncated gray values, respectively. The final gray value range—after truncation and stretching—is $[R_{low}, R_{up}]$ and the gain and compensation are constant:

$$G = \frac{d_{\max}}{R_{up} - R_{low}}, b = -\frac{R_{low}}{R_{up} - R_{low}} \tag{17}$$

2.2.5. The LoG Edge Detection Operator

The Laplacian of Gaussian (LoG) operator is often used in edge extraction and binarization of digital images. A target image is first smoothed using a Gaussian filter, then the image Laplacian—second-order derivative—is calculated to enhance the image features. Finally, the edges are determined by detecting zero crossings—where the values of neighboring pixels change sign—in the filtered image. The zero-crossing points of the Laplacian (second-order derivative) correspond to the local maxima or minima of the first-order derivative. If necessary, the detected edge positions can be refined at the subpixel resolution level using a linear interpolation method.

The edge detection algorithm is detailed hereafter:

(1) Image Smoothing

The image $f(x, y)$ is smoothed with a two-dimensional Gaussian filter that approximates the characteristics of the human visual system:

$$G(x,y) = \frac{1}{2\pi\sigma} \exp(-\frac{1}{2\pi\sigma^2}(x^2 + y^2)) \tag{18}$$

$G(x, y)$ is a circular symmetric function. The strength of the smoothing is controlled by the scale parameter—standard deviation—σ. A smoothed image is obtained by convolving the initial image $f(x, y)$ with the Gaussian filter:

$$g(x,y) = f(x,y) \times G(x,y) \tag{19}$$

(2) Image Enhancement

After Gaussian smoothing, the Laplacian operator is applied to the smoothed image $g(x, y)$ to calculate the second-order image derivative:

$$h(x,y) = \nabla^2(f(x,y) \times G(x,y)) \tag{20}$$

(3) Edge Detection

The edge detection criterion is a change of sign of the second-order derivative, i.e., between adjacent pixels—the zero-crossing points—in the filtered image $h(x, y)$, corresponding to local maxima or minima of the first-order derivative. Practically, only zero-crossing points with a first-order derivative greater than a fixed threshold are selected as edge points to avoid the detection of non-significant edges. Applying the Laplacian operator to the smoothed image $g(x, y)$ is equivalent to applying the Laplacian to the Gaussian operator and convolving the result with the initial image $f(x, y)$. Therefore, Equation (20) is equivalent to:

$$h(x,y) = f(x,y) \times \nabla^2 G(x,y) \tag{21}$$

$\nabla^2 G(x, y)$ is the LoG filter, defined as:

$$\nabla^2 G(x,y) = \frac{\partial^2 G}{\partial x^2} + \frac{\partial^2 G}{\partial y^2} = \frac{1}{\pi \delta^4}(\frac{x^2+y^2}{2\delta^2} - 1)\exp(-\frac{1}{2\delta^2}(x^2+y^2)) \quad (22)$$

Second-order derivative filters are very sensitive to noise. By applying a Gaussian smoothing filter before the Laplacian sharpening filter, the image noise is smoothed out before edge detection, resulting in better detection performance. Results of the edge detection algorithm for flotation foam images of scheelite, coal gangue and lead-zinc ore are shown in Figures 6–8, respectively.

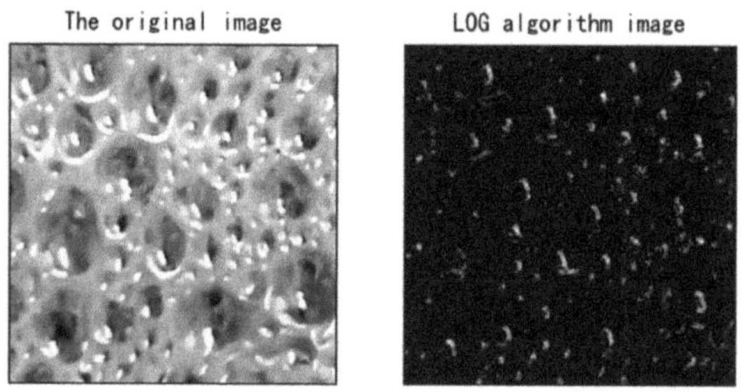

Figure 6. Original image of a scheelite flotation foam (**left**) and result of the LoG detection algorithm (**right**).

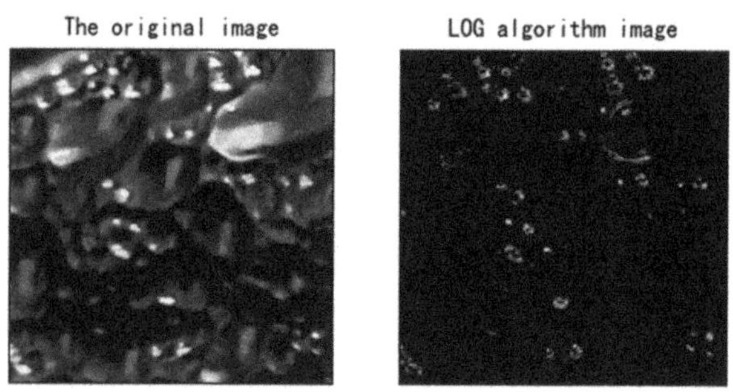

Figure 7. Original image of a coal gangue flotation foam (**left**) and result of the LoG detection algorithm (**right**).

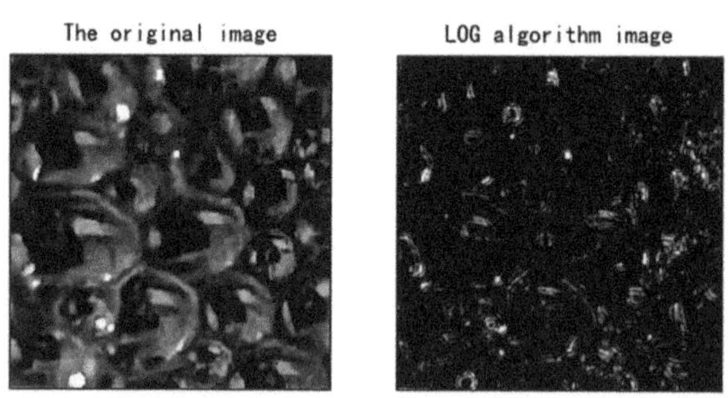

Figure 8. Original image of a lead–zinc ore flotation foam (**left**) and result of the LoG detection algorithm (**right**).

2.3. *Screening and Analysis of Factors Influencing the Flotation Process*

2.3.1. Image Preprocessing

(1) Grayscale Transformation

A grayscale transformation is a point-by-point image transformation that corrects the pixel gray values of the original input image to determine a corresponding output gray value for each pixel. The general expression is:

$$s = T(r) \tag{23}$$

where T is the grayscale transformation function, R is the matrix of gray level values before transformation, and S is the output matrix containing the transformed gray values.

If T is determined, then the output gray values are also known. The properties of function T determine the effect of the grayscale transformation.

(2) Gray-Level Histogram

The histogram of a grayscale image counts the number of image pixels for each gray level value. When divided by the total number n of image pixels, the resulting normalized histogram also represents the probability density function $p(r_k)$—in the interval [0, 1]—for each gray value r_k and directly reflects the occurrence probability of each gray value. For an image with n = 256, if n_k is the number of pixels with a gray value r_k, then the occurrence probability of r_k is:

$$p(r_k) = \frac{n_k}{n}, (k = 0, 1, \cdots, L-1, n_k \geq 0) \tag{24}$$

where L represents the odd gray level in the image.

The histogram is easily calculated by counting the number of pixels for each gray level and sorting them in ascending order of gray value. The Open Source Computer Vision (OpenCV) library is a software package dedicated to machine vision and developed for the Python open-source programming language. OpenCV provides a histogram calculation function (calcHist) that calculates gray-level histograms for multiple images, multiple channels and different ranges of gray values simultaneously. The histograms calculated with calcHist for flotation foam images of scheelite, coal gangue and lead-zinc ore are shown in Figures 9–11, respectively.

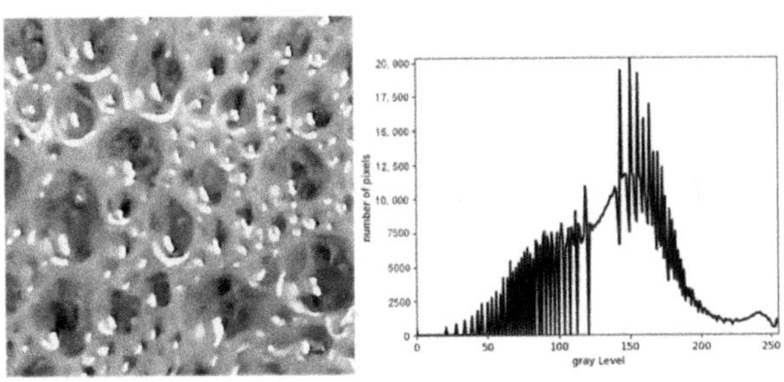

Figure 9. Original image of a scheelite flotation foam (**left**) and the corresponding histogram (**right**).

Histograms form the basis of image recognition. Figures 9–11 clearly show marked histogram differences between the gray-level distributions of the three flotation foams. Such information is fundamental for parameter adjustment in the next image processing step, thereby strongly improving the efficiency and accuracy of the image recognition algorithm.

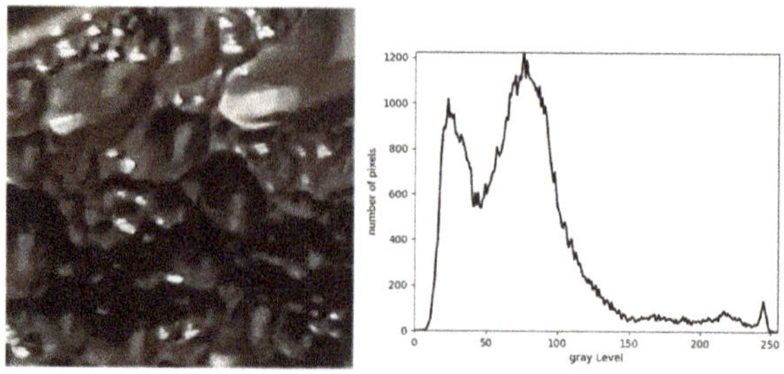

Figure 10. Original image of a coal gangue flotation foam (**left**) and the corresponding histogram (**right**).

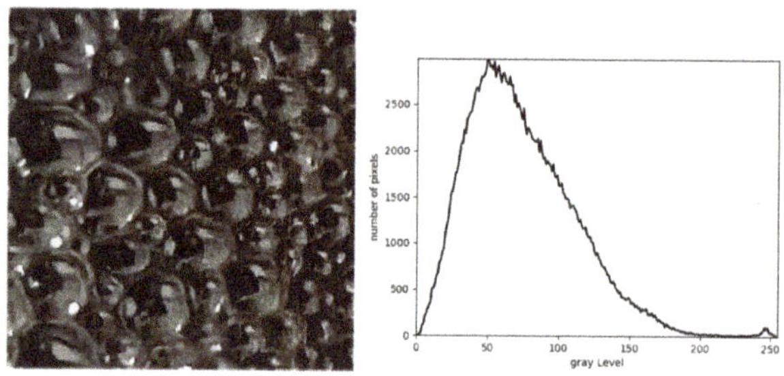

Figure 11. Original image of a lead-zinc ore flotation foam (**left**) and the corresponding histogram (**right**).

2.3.2. Linear Regression Model for Factors Influencing the Flotation Process

NumPy is the most widely used numerical analysis library for scientific computing in Python. NumPy provides high-level mathematical functions and algorithms and is very useful to build linear regression models. Furthermore, to eliminate model errors caused by algorithm randomness, the scikit-learn package—a Python package for machine learning—was used to simulate observations, train and evaluate the model independently, and assess the possibilities for further model optimization [26,27]. Unary linear regression models for light and noise intensity were built and trained on the dataset. Output models derived for light and noise intensity from 1000 simulation experiments are shown in Figures 12 and 13, respectively.

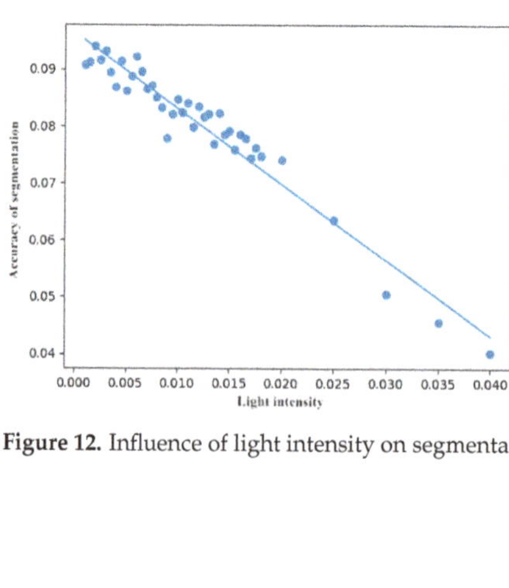

Figure 12. Influence of light intensity on segmentation accuracy and the linear regression model.

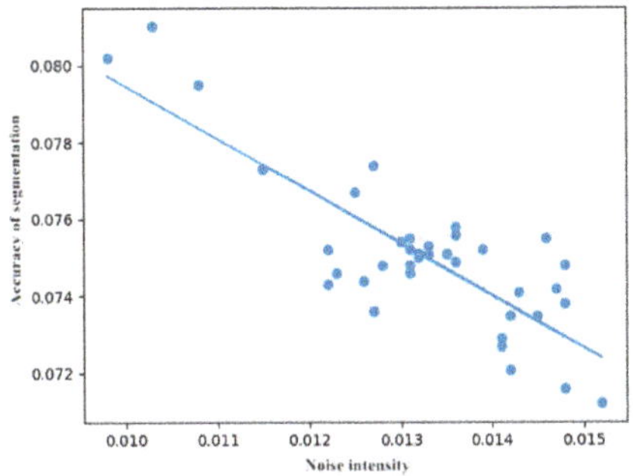

Figure 13. Influence of noise intensity on segmentation accuracy and the linear regression model.

There are discrepancies between the regression models and the data, notably for the noise intensity. This is expected because light and noise intensity are only two of the main factors affecting image segmentation. Without accounting for other factors such as dosage, pulp density, wind pressure, and foam layer thickness, the fitted values necessarily differ from the original data. To evaluate the quality of the regression models and the correlation between influencing factors and segmentation accuracy, the goodness-of-fit is estimated from the coefficient of determination and the linear correlation is evaluated from the correlation coefficient. Results for the light and noise intensity models are shown in Table 2.

Table 2. Coefficients of determination and correlation coefficients for both regression models.

Influence Factor	Coefficient of Determination	Coefficient of Association
Noise intensity	0.6764	−0.7639
Light intensity	0.9431	−0.9169

Table 2 shows that the coefficient of determination for light intensity is higher than 0.9, indicating high goodness-of-fit, whereas the coefficient of determination for noise intensity is lower than 0.7, indicating low goodness-of-fit. The absolute value of the correlation coefficient for light intensity is also higher than 0.9, indicating a strong linear correlation. The corresponding value for noise intensity is higher than 0.75, also showing a good linear correlation. However, because of the low goodness-of-fit for the noise intensity regression model (67.64%), only light intensity is considered hereafter.

3. Improvement of the Watershed Segmentation Algorithm Using the OpenCV Library

Image segmentation is an important part of image analysis. To retrieve and measure all features of interest from the full image, the original must be separated into meaningful subsets or regions.

3.1. The OpenCV Watershed Segmentation Algorithm

The purpose of image thresholding is to separate the pixel set into subsets defined by their gray level value, with each subset or region corresponding to the real scene. Each region is described by consistent internal attributes that are different from those of adjacent regions. Subsetting can be achieved by selecting appropriate gray-level thresholds [28].

The watershed algorithm in OpenCV improves on the original algorithm by adding a preprocessing step: converging and non-converging valleys are selected before segmentation. To avoid a known oversegmentation issue affecting the original watershed algorithm, OpenCV uses a series of predefined markers to guide the image segmentation definition. As input, the OpenCV algorithm requires a labeled image for which pixel gray values are 32-bit signed positive numbers and each non-zero pixel represents a label.

During the segmentation process, the top regions of the bubbles are identified first and optimal structural elements are selected using an adaptive segmentation parameter selection method. Then, the output of the watershed algorithm is obtained by morphological processing of the foam image, which becomes coarsely segmented.

Uneven bubble mixing produces under- and oversegmented regions. To minimize this effect, we use a fuzzy texture spectrum algorithm—with good real-time texture feature extraction and area segmentation (fuzzy c-means clustering algorithm)—and a support vector machine for regional identification. After identifying undersegmented and oversegmented areas, finer segmentation is applied to the undersegmented areas. This method can effectively be used to solve the problem of inappropriate segmentation [29].

3.2. Flow Improvement for the OpenCV Watershed Segmentation Algorithm

As mentioned in Section 3.1, the foam image is first clustered with the Fuzzy C-means clustering algorithm—implemented in the "fuzzy-c-means" Python module [30]—and the top regions of the bubbles are identified. An adaptive segmentation parameter selection method is used to select the optimal structural elements, morphological processing is applied to the foam image to derive the watershed algorithm output image, then the foam image becomes coarsely segmented. The algorithm flow is shown in Figure 14.

3.3. Comparison of Experimental Results from Segmentation Simulations

For this analysis, a library of 80 flotation foam images acquired in December 2021 was selected from the image database of a lead-zinc mine in Yunnan Province. The selected images were extracted from video footage recorded during two work shifts in different working conditions. Because the images show a flotation foam without background nonuniform adhesion, segmentation must meet the real-time performance requirements of the flotation process. To evaluate the segmentation performance, this manuscript compares the original and improved versions of the OpenCV watershed segmentation algorithm with the result of expert manual segmentation.

Full evaluation of the algorithm performance includes a subjective evaluation and an objective assessment. The subjective evaluation is conducted by skilled personnel who visually evaluate the segmentation results. The objective evaluation is a statistical analysis of the automatic segmentation results using expert manual segmentation as a reference. A segmentation evaluation index k is defined to compare the automatic segmentation algorithms and the manual segmentation reference:

$$k = 2 \times \frac{N(M \cap S)}{N(M) + N(S)} \times 100\% \tag{25}$$

Here, M and S are the results from expert manual segmentation and from the automatic segmentation algorithms, respectively; the intersection $M \cap S$ represents the segmentation regions common to the manual and automatic results; and N represents the number of regions identified during the manual—$N(M)$—and automatic—$N(S)$—segmentation experiments or common to both—$N(M \cap S)$. Operation time is also compared.

An image acquired in typical working conditions is randomly selected from the 80-image library for a segmentation simulation. The original gray image is shown in Figure 15a and the enhanced image after Retinex compensation is shown in Figure 15b. Segmentation results for the standard watershed segmentation algorithm are shown in Figure 15c. Results for the improved watershed segmentation algorithm proposed in this paper, after enhancement with the Retinex image compensation method, are shown in Figure 15d. For the selected image, representative of typical working conditions, the standard watershed algorithm segments the foam image into 55 bubbles, but the improved watershed algorithm yields 66 bubbles.

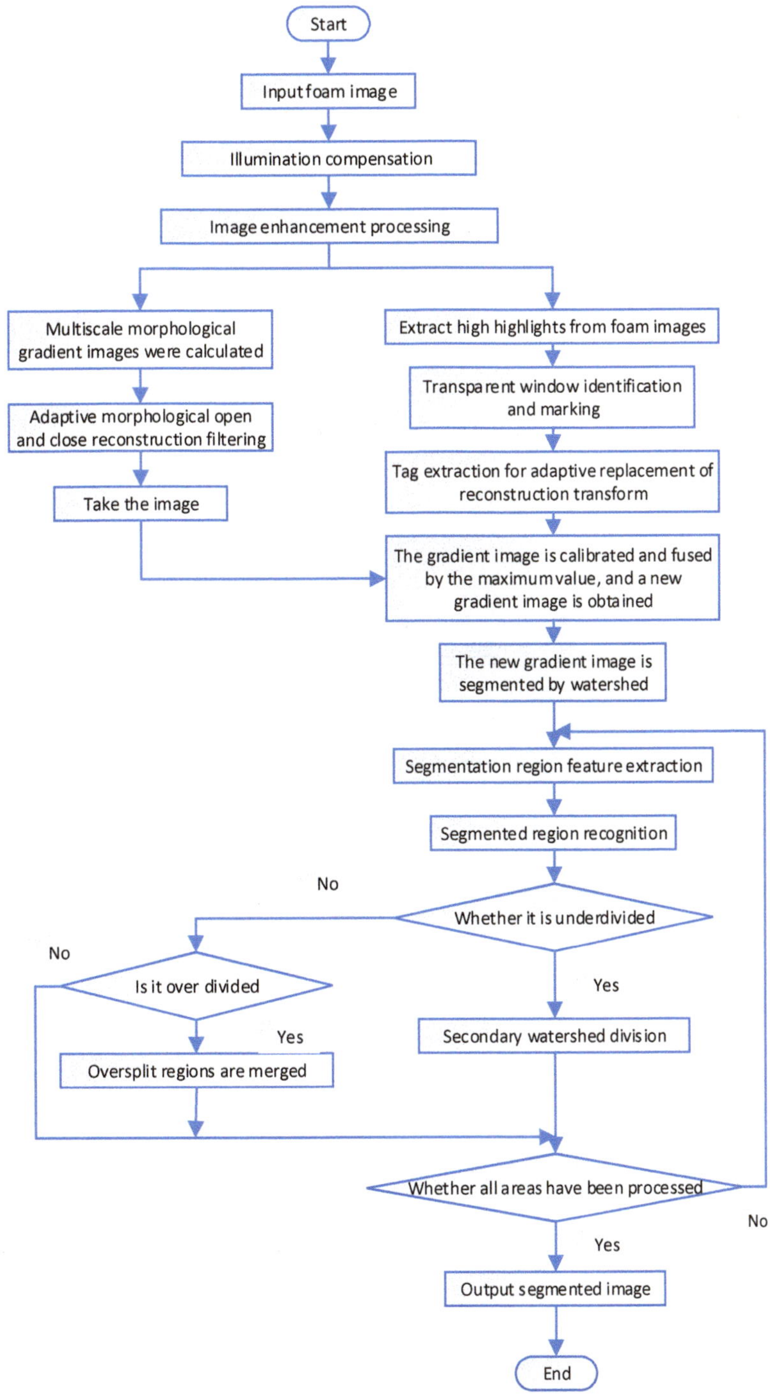

Figure 14. Flowchart of the improved OpenCV watershed segmentation algorithm.

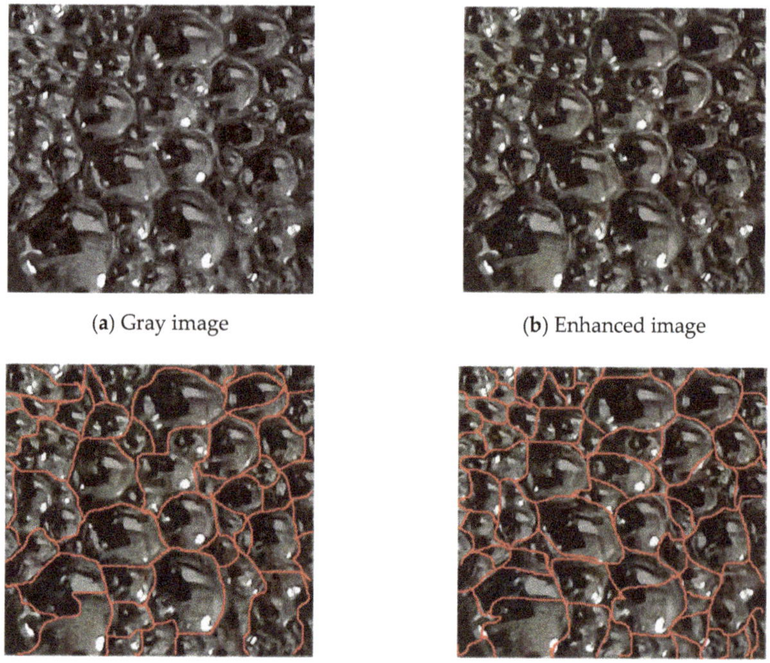

(**a**) Gray image (**b**) Enhanced image

(**c**) Watershed segmentation algorithm (**d**) The improved watershed algorithm

Figure 15. Flotation foam image segmentation simulation. (**a**) Gray image; (**b**) Enhanced image; (**c**) Watershed segmentation algorithm; (**d**) The improved watershed algorithm.

The difference between the standard and improved versions of the watershed segmentation algorithm is apparent in Figure 15c,d. Both versions yield good results, with generally accurate segmentation of the foam image into conventional regions. In good working conditions, segmentation results are similar for both algorithms. However, the selected image (Figure 15a) includes several bright spot areas near the edges of the frame. In this case, operating assumptions for the watershed algorithm are not met. Thus, uneven illumination causes improper segmentation for both algorithm versions relative to the manual segmentation reference, but with comparatively better performance of the improved watershed algorithm. The improved algorithm developed in this work is markedly more robust and accurate, and yields results closer to manual segmentation. Therefore, it is more suitable for foam image segmentation under less favorable working conditions.

3.4. Statistical Analysis of the Segmentation Simulation Results

Simulation results for the standard and improved versions of the watershed segmentation algorithm are given in Table 3.

Table 3. Evaluation of the segmentation results for the standard watershed algorithm and for the improved version defined in this work.

Algorithm	Mean Split Time (s)	Average Segmentation Accuracy (%)
Watershed algorithm	5.863	88.73
Improved watershed algorithm	5.285	92.06

To compare the real-time performance of both algorithms, 50 flotation foam images were randomly selected for segmentation from the image library. To reduce the manual segmentation error, the results of five manual segmentation operations were averaged to produce a reference segmented image. Segmentation accuracy relative to the reference and operation time is calculated for both algorithms. As seen in Table 3, the accuracy of the improved watershed algorithm is 3.3% higher than that of the standard algorithm and the operation time is reduced by 9.9%. Segmentation results from both algorithms are comparable when the image is segmented. However, the proposed algorithm is faster and yields markedly higher segmentation accuracy for non-uniform flotation foam images under fluctuating working conditions.

4. Conclusions

During the industrial mineral flotation process, bubble size and morphology reflect real-time changes in the flotation conditions. Accurate foam image segmentation algorithms must be able to adapt to such fluctuating conditions. In this paper, we proposed an improved version of the watershed segmentation algorithm implemented in the OpenCV library of the versatile, open-source Python programming language. The improved algorithm was applied to flotation foam images and segmentation results were compared with results from the OpenCV standard watershed segmentation algorithm and from manual segmentation.

(1) First, the flotation foam images were enhanced with the Retinex image compensation method—also implemented in a Python module. Strong-contrast area recognition and illumination compensation of the flotation foam image were improved, with a better visual result and more possibilities to extract useful details from the foam image.
(2) A linear regression model was developed, also in Python, to analyze major factors influencing the segmentation accuracy. Application of the model to light and noise intensity showed that both factors had an influence on flotation foam segmentation, larger for light intensity.
(3) The improved version of the OpenCV watershed segmentation algorithm proposed in this paper, also written in Python, yielded better results than the standard version. Segmentation time was 3.3% shorter and segmentation accuracy increased by 9.9%. Comparison with results from the standard watershed algorithm and from manual segmentation showed that the proposed algorithm is accurate and robust.

Author Contributions: D.L. conceived of and designed the experiments. W.Z. and C.W. performed the experiments and analyzed the data. R.L. and D.W. contributed materials. W.Z. wrote the paper. L.Y. and S.W. modified the paper. All authors have read and agreed to the published version of the manuscript.

Funding: This research received no external funding.

Acknowledgments: We thank Eric Dupuy for editing the English text of a draft of this manuscript.

Conflicts of Interest: The authors declare no conflict of interest.

References

1. Beneventi, D.; Benesse, M.; Carré, B.; Saint Amand, F.J.; Salgueiro, L. Modelling Deinking Selectivity in Multistage Flotation Systems. *Sep. Purif. Technol.* **2007**, *54*, 77–87. [CrossRef]
2. Bergh, L.; Yianatos, J. The Long Way Toward Multivariate Predictive Control of Flotation Processes. *J. Process Control* **2011**, *21*, 226–234. [CrossRef]
3. Aldrich, C.; Marais, C.; Shean, B.; Cilliers, J. Online Monitoring and Control of Froth Flotation Systems with Machine Vision: A Review. *Int. J. Miner. Process.* **2010**, *96*, 1–13. [CrossRef]
4. Moolman, D.W.; Eksteen, J.J.; Aldrich, C.; Van Deventer, J.S.J. The Significance of Flotation Froth Appearance for Machine Vision control. *Int. J. Miner. Process.* **1996**, *48*, 135–158. [CrossRef]
5. Bonifazi, G.; Serranti, S.; Volpe, F.; Zuco, R. Characterisation of flotation froth colour and structure by machine vision. *Comput. Geosci.* **2001**, *27*, 1111–1117. [CrossRef]
6. Shean, B.; Cilliers, J. A Review of Froth Flotation Control. *Int. J. Miner. Process.* **2011**, *100*, 57–71. [CrossRef]

7. Kaartinen, J.; Hätönen, J.; Hyötyniemi, H.; Miettunen, J. Machine-Vision-Based Control of Zinc Flotation—A Case Study. *Control Eng. Pract.* **2006**, *14*, 1455–1466. [CrossRef]
8. Van Brown, O.; Bourke, R.J.A. Improving Flotation Plant Performance at Cadia by Controlling and Optimising the Rate of Froth Recovery Using Outokumpu Frothmaster. *Australas. Inst. Min.* **2000**, *6*, 127–135.
9. Bonifazi, G.; Giancontieri, V.; Meloni, A.; Serranti, S.; Volpe, F.; Zuco, R.; Koivo, H.; Hätönen, J.; Hyötyniemi, H.; Niemi, A.; et al. Characterization of the Flotation Froth Structure and Color by Machine Vision (ChaCo). *Dev. Miner. Process.* **2000**, *13*, C8a–39.
10. Wang, W.X.; Stephansson, O.; Wan, S.C. On-Line System Setup in a Cellar of a Flotation Plant. In Proceedings of the 15th International Conference on Pattern Recognition. ICPR-2000, Barcelona, Spain, 3–7 September 2000; Volume 4, pp. 791–794.
11. Wang, W.; Stephansson, O.A. Robust Bubble Delineation Algorithm for Froth Images. In Proceedings of the Second International Conference on Intelligent Processing and Manufacturing of Materials. IPMM'99 (Cat. No.99EX296), Honolulu, HI, USA, 10–15 July 1999; Volume 1, pp. 471–476.
12. Citir, C.; Aktas, Z.; Berber, R. Off-Line Image Analysis for Froth Flotation of Coal. *Comput. Chem. Eng.* **2004**, *28*, 625–632. [CrossRef]
13. Sadr-Kazemi, N.; Cilliers, J. An Image Processing Algorithm for Measurement of Flotation Froth Bubble Size and Shape Distributions. *Miner. Eng.* **1997**, *10*, 1075–1083. [CrossRef]
14. Forbes, G.; De Jager, G. Texture Measures for Improved Watershed Segmentation of Froth Images. In *Fifteenth Annual Symposium of the Pattern Recognition Association of South Africa*; OpenUCT: Cape Town, South Africa, 2004; pp. 1–6.
15. Zhang, H.; Tang, Z.; Xie, Y.; Gao, X.; Chen, Q. A Watershed Segmentation Algorithm Based on an Optimal Marker for Bubble Size Measurement. *Measurement* **2019**, *138*, 182–193. [CrossRef]
16. Lézoray, O.; Charrier, C. Color Image Segmentation Using Morphological Clustering and Fusion with Automatic Scale Selection. *Pattern Recognit. Lett.* **2009**, *30*, 397–406. [CrossRef]
17. Marais, C.; Aldrich, C. Estimation of Platinum Flotation Grades from Froth Image Data. *Miner. Eng.* **2011**, *24*, 433–441. [CrossRef]
18. Henry, T.; Ngan, Y.; Grantham, H.; Pang, K.; Nelson, C. Computing, Automated fabric defect detection—A Review. *Image Vis. Comput.* **2011**, *29*, 442–458.
19. Bask, M.; Johansson, A. Robust Time-Varying Thresholds for Supervision of Valves in a Flotation Process. In Proceedings of the 2004 43rd IEEE Conference on Decision and Control, Nassau, Bahamas, 14–17 December 2004.
20. Liu, J.J.; MacGregor, J.F. Froth-Based Modeling and Control of Flotation Processes. *Miner. Eng.* **2008**, *21*, 642–651. [CrossRef]
21. Kim, J.; Fisher, J.; Yezzi, A.; Cetin, M.; Willsky, A. A Nonparametric Statistical Method for Image Segmentation Using Information Theory and Curve Evolution. *IEEE Trans. Image Process.* **2005**, *14*, 1486–1502. [CrossRef]
22. Torralba, A.; Oliva, A. Statistics of Natural Image Categories. *Netw. Comput. Neural Syst.* **2003**, *14*, 391. [CrossRef]
23. Land, E.H. An Alternative Technique for the Computation of the Designator in the Retinex Theory of Color Vision. *Proc. Natl. Acad. Sci. USA* **1986**, *83*, 3078–3080. [CrossRef]
24. Kimmel, R.; Elad, M.; Shaked, D.; Keshet, R.; Sobel, I. A Variational Framework for Retinex. *Int. J. Comput. Vis.* **2003**, *52*, 7–23. [CrossRef]
25. Hao, W.; He, M.; Ge, H.; Wang, C.-J.; Gao, Q.-W. Retinex-Like Method for Image Enhancement in Poor Visibility Conditions. *Procedia Eng.* **2011**, *15*, 2798–2803. [CrossRef]
26. Swami, A.; Jain, R.J. Scikit-Learn: Machine Learning in Python. *J. Mach. Learn. Res.* **2013**, *12*, 2825–2830.
27. Güney, Y.; Bozdogan, H.; Arslan, O. Robust Model Selection in Linear Regression Models Using Information Complexity. *J. Comput. Appl. Math.* **2021**, *398*, 113679. [CrossRef]
28. A Full Color Digital Imaging Based Approach to Characterize Flotation Froth: An Experience in Pyhasalmi (SF) and Garpenberg (S)Plants. In In Proceedings of the Beijing International Conference on Imaging Technology and Applications in the 21st Century, Beijing, China, 23–26 May 2005.
29. Luo, J.; Tang, Z.; Zhang, H.; Fan, Y.; Xie, Y. LTGH: A Dynamic Texture Feature for Working Condition Recognition in the Froth Flotation. *IEEE Trans. Instrum. Meas.* **2021**, *70*, 1–10. [CrossRef]
30. Kuo, B.-C.; Huang, W.-C.; Liu, H.-C.; Tseng, S.-C. A Novel Fuzzy C-Means Method for Hyperspectral Image Classification. *IEEE Int. Geosci. Remote Sens. Symp.* **2008**, *2*, II-1002–II-1005. [CrossRef]

MDPI
St. Alban-Anlage 66
4052 Basel
Switzerland
www.mdpi.com

Minerals Editorial Office
E-mail: minerals@mdpi.com
www.mdpi.com/journal/minerals

www.ingramcontent.com/pod-product-compliance
Lightning Source LLC
Chambersburg PA
CBHW040803250926
51850CB00007BA/1065
* 9 7 8 3 7 2 5 8 1 3 9 6 4 *